Construction
site studies

G. Forster
FCIOB Chartered Builder

Construction site studies
production, administration and personnel

An imprint of **Pearson Education**

Harlow, England · London · New York · Reading, Massachusetts · San Francisco · Toronto · Don Mills, Ontario · Sydney
Tokyo · Singapore · Hong Kong · Seoul · Taipei · Cape Town · Madrid · Mexico City · Amsterdam · Munich · Paris · Milan

Pearson Education Limited
Edinburgh Gate
Harlow
Essex CM20 2JE
England

and Associated Companies throughout the world

Visit us on the World Wide Web at:
www.pearsoned.co.uk

First published 1981
Second edition 1989

British Library Cataloguing in Publication Data

Forster, G. (George), *1938–*
    construction site studies.—2nd ed.
    1. Construction. Sites. Management — Manuals
    I. Title
    624'.068

ISBN 0-582-01971-0

18   17   16   15
07   06   05   04

Set in Garamond 9/10½

Printed in Malaysia, PPSB

Contents

Acknowledgements

We are indebted to the following for permission to reproduce copyright material:

British Standards Institution for our Figs 3.3.2/3 from BS 1192 reproduced by permission of the British Standards Institution; Her Majesty's Stationery Office for our Fig. 3.4.1 reproduced by permission of the Controller of Her Majesty's Stationery Office; National Joint Council for the Building Industry for extracts from the *Working Rule Agreement*; Riba Publications Ltd for our Fig. 8.4.1 from the *C.O.W. Manual* by permission of Riba Publications Ltd.

The author would also like to thank his wife for typing the manuscript and for her continued assistance and support.

Chapter 1
Supervisor's role regarding policies of the firm and other procedures

1.1   Areas in which policies apply

Whether it is a sole trading concern, partnership or company a business must have an aim or goal to achieve, and to this end there must be objectives to which each individual plays a part in achieving. These objectives may be in either written or implied terms.

The main objectives must be determined by the owners or directors of a business. Assuming that a company is involved, objectives are determined by the board of directors, guided by the chairman, and then it would be their responsibility to have a policy document prepared outlining these objectives and how they should be achieved.

The preparation of a general comprehensive policy for any business would be a long and difficult exercise, and even if one was prepared it would, to cover every aspect of how a business should operate, be virtually impossible to implement. A too comprehensive policy means a too centralised straitjacket from which everyone would have to work; it would leave little scope for each executive manager, supervisor and employee to improvise and operate within.

The more usual type of policy would be of a decentralised nature, which allows the optimum of discretion for each superior in organising and controlling his immediate subordinates.

The Memorandum of Association and Articles of Association are documents which are submitted to the Registrar of Joint Stock Companies prior to the commencement of trading by a company. These documents govern the lines by which the business should operate and by which the directors and shareholders must conform. Bearing these important documents in mind, the board of directors would prepare policies, the parameters of which the managing director would in turn interpret and ensure that his/her immediate subordinates (executives and managers) work within to achieve the objectives laid down.

Firms seldom write down their policies but, nevertheless, policies exist and new ones are introduced at each level of responsibility from the managing director and executives, down to the shop floor level.

Each time a managing director makes a decision he/she introduces new or modifies existing policies.

A managing director's subordinates would then be in a position to know what the outcome should be to similar future situations and would therefore not deem it necessary to approach the managing director for subsequent decisions. They follow the precedents set, i.e. the policies.

Each new management problem solved sees the introduction of a policy for dealing with a similar recurrence in the future. Actions or inactions of a superior determines policies by which his/her subordinates would abide. Take, for instance, a site manager who disciplines operatives each time they are observed arriving late for work. He/she has by his/her actions set a policy for subordinates to observe — that is, lateness will not be tolerated. Another site manager with a similar problem may be more tolerant with the late arrivers. The policy adopted by his/her inaction would be interpreted by the site work-force as normal company acceptability, leading inevitably to a breakdown in discipline later. Policies, therefore, are created at all levels depending on the degree of decentralisation allowed.

The general policy of a company is formulated usually by the unanimous agreement of the board of directors. It is beneficial to those in authority that such a policy is incorporated into a policy document, the contents of which should be available for observance and implementation. Where a policy document exists it should be added to or modified when required in the light of new technology, experiences, developments and problems encountered.

The general policy embraces the total organisation and its activities, and details the strategy by which a company should achieve its objectives. The principal areas which should be covered by a policy are:

1.   The general purposes and objectives of the company.
2.   Executives' responsibilities.
3.   Marketing and sales.
4.   Production/construction.
5.   Finance.
6.   Personnel.

It must be remembered that each area, as shown above, cannot be dealt with in isolation within a company; they must be interrelated. The objectives in each area must be geared to the achievement of the general objectives of the company. As an example, enlarged markets should not be the aim if production capacity and finance is, and will remain, limited.

The areas within each of the aforementioned should deal with some or all of the points as listed below.

The general purposes and objectives of the company
(a) The recapitulation of the main details so contained in the Memorandum and Articles of Association.
(b) The maintenance of competitiveness with rival companies.
(c) The improvement of facilities and employment for the area and integration of the firm into the community.
(d) The achievement of suitable dividends for the shareholders.
(e) The maintenance of suitable salary and wage levels, and to help in the economic needs of the community.
(f) The observance of legal requirements.
(g) The expected expansion in the future.

Executive responsibilities
(a) The role of each executive position.
(b) The parameters of responsibility, accountability and power of each executive position.
(c) The methods whereby changes can be made to the main policy – usually by discussion.
(d) The actions expected by executives regarding outside interests, e.g. not to have interests in rival companies.
(e) The prevention of conduct which may be unlawful.
(f) The service conditions of each executive and methods whereby grievancies, disputes or problems may be resolved.
(g) The compulsory membership of committees and attendance at meetings.
(h) The areas in which decisions can be taken, and those areas which must have the approval of the managing director or board.
(i) The supervision of work and assessment of performance of subordinates.
(j) The disciplining of employees and their positions regarding labour relations.

Marketing and sales
(a) The methods of marketing and percentage expenditure thereon.
(b) The local, regional, national and international requirements for obtaining work.
(c) The type of activity/commodity to be offered, and commodity mix.
(d) The type of client to aim for.

(e) The method of advertising (special offers, promotions, etc.) and percentage of turnover expended thereon.
(f) The types of public relations exercises.
(g) The salesmanship techniques, and whether agents are to be used or not.
(h) The insistence of keeping statistics for observation on trends, degree of competition, etc., in various fields of work.

Production/construction
(a) The levels of expenditure on research and development.
(b) The purchase of materials by centralised or decentralised methods.
(c) The stocks of materials: optimum levels.
(d) The use of plant either by hiring or buying, or both, and other fixed assets.
(e) The percentage use of subcontractors and labour-only subcontractors.
(f) The quality assurance expected.
(g) The use of work study.
(h) The use of incentives and bonus systems.
(i) The methods of distribution of the commodity.
(j) The security measures to be taken regarding the works from outsiders and employees.
(k) The production mix (public or private sector).

Finance
(a) Budgeting.
(b) Cost planning.
(c) The use of resources.
(d) The methods of attracting investors, and financing projects.
(e) The tax situation.
(f) The spreading of the risk.
(g) Methods of investment.
(h) Credit control (debtors and creditors).
(i) The maintenance of statistics on the economic climate.
(j) Depreciation (sinking funds/inflation).
(k) Capital and revenue expenditure.
(l) Dividends to shareholders.
(m) Overseas financing of work (Export Credit Guarantee Department).

Personnel
(a) The percentage of revenue to be allocated to this area.
(b) Joint consultation/industrial relations.
(c) Recruitment/forecasting needs.
(d) Promotion – from within or outside the firm.
(e) Dismissals.
(f) Remunerations.
(g) Motivation/morale.
(h) Extra payments (incentives, overtime, etc.).
(i) Dismissals and appeal system.
(j) Training/education.

(*k*) Safety-representatives.
(*l*) Health and welfare.
(*m*) Holidays, etc.

Policies on-site

The site manager implements the company's general policy in numerous areas – personnel, production, health and welfare, safety, industrial relation, security, public relations, quality control, communication, conditions of contract, purchasing, etc. The amount of knowledge and experience a site manager must have to conduct the affairs of the company successfully at site level is vast, his/her subordinates, the supervisors, have delegated responsibilities in order to assist him/her.

At each stage of the work on-site policies are being formulated by decisions and actions taken by the site manager or supervisors. At all times, therefore, it is essential that suitable and sensible decisions are made because they set precedents by which everyone, the site staff and operatives, would conform to during their period on the site. The precedents set for a particular site become the policies for that site.

1.2 Effective operating by individuals in formal organisations

Businesses operate with a view to making profits and to these ends it is essential, if success is to be achieved and maintained, that they are managed on formal lines. This is to say that they should be operated according to recognised rules whereby everyone is left in no doubt to whom and for what they are responsible. It is also usual for a firm to either give a contract to each new employee to show where he/she fits within an organisation, or produce handbooks outlining each and everyone's place within the firm.

There are, of course, relationships which exist within a formal organisation which are:

1. Line or direct relationships

This implies that certain individuals within a formal organisation have the authority to take action and make decisions. It means generally that there is an organisation structure which starts with the most powerful individual within the organisation at the top as the superior, and stretches down to the least important subordinate at the lowest end. Whether there is an organisation structure diagram drawn up for everyone to observe or not, superiors should not normally be expected to direct or deal with more than eight subordinates if he/she is to be able to control, coordinate, and execute his/her duties successfully; this is referred to as 'span of control'. (See Fig. 1.2.1 (A) and (B).)

2. Lateral or equal level relationships

These relationships exist between persons on a similar level within an organisation, e.g. one site manager with another site manager – those with equal status, responsibility and accountability, etc. At site level foremen have a lateral relationship with each other. (See Fig. 1.2.1 (A) and (B).)

3. Functional relationships

Generally there are individuals within a formal organisation who may be responsible for a section or a department (acting as superior to the others within the department who are subordinates), but who act in an advisory capacity to other individuals within the organisation who are at a more senior level. The marketing manager, personnel manager, security manager, safety officer, etc., each can advise top management of what actions to take in their respective areas. Technical, scientific and financial experts have functional relationships with their superiors. On-site individuals such as programming officers (planners) and bonus surveyors are each involved in an advisory capacity. (See Fig. 1.2.1 (A) and (B).)

4. Staff relationships

This relationship exists between those who act as assistants to a superior and the superior. An individual would have no authority of his/her own but would operate on the authority of the superior to whom he/she is assigned. Neither would the person have any subordinates. One can take as an example an 'assistant to' the site manager – but not an 'assistant' site manager. The former would normally be a trainee site manager who may have very little experience and in no way would he/she be expected to deputise in the site manager's absence. The assistant site manager, on the other hand, would be greatly experienced and in his/her own right would act as the deputy for the site manager (his/her superior) and would be superior to all others on-site.

It could also be said that a staff relationship would exist between an individual and his/her personal secretary or personal assistant. (See Fig. 1.2.1 (A) and (B).)

Span of control

Organisational structures exist in all firms, the shape of which normally depends on the type and size. Each person generally has a clear understanding of theirs and others responsibilities in a small firm, but in a large firm employees tend to be unsure of exactly where they fit in, and do not readily appreciate other individual's positions regarding line, lateral, functional or staff relationships.

Although it is firmly believed by many organisations that organisation structure diagrams should be drawn up and displayed for everyone to observe, many disagree and prefer to withhold the display of such diagrams because they feel it is too clinical and places personnel into tight compartments from which they are unwilling to venture – the attitude being that they then know their exact positions and duties and prefer not to accept other work or responsibilities beyond their defined ones, which generates too rigid compartmental attitudes. It is believed that a more flexible attitude would result, leading to a more

4

(A) Head office level

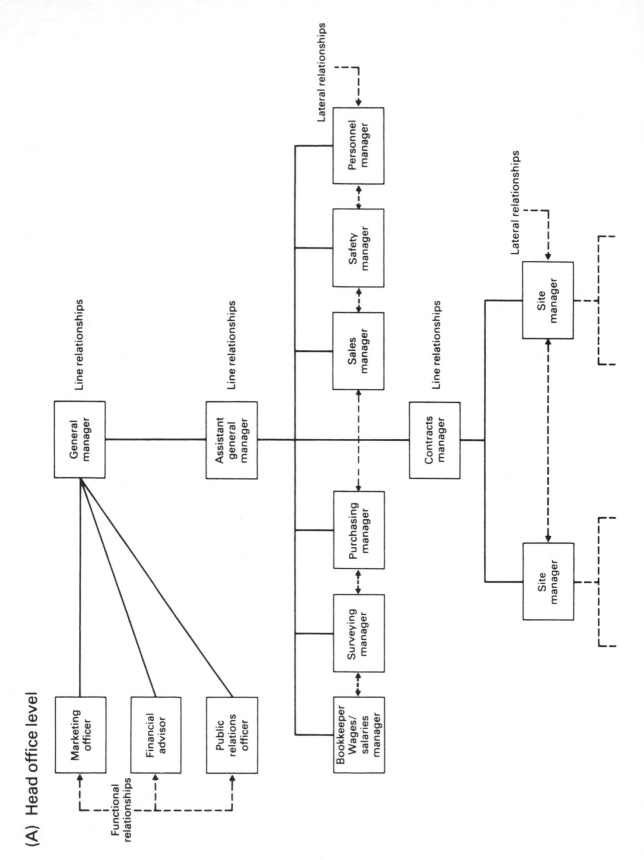

(B) Site level

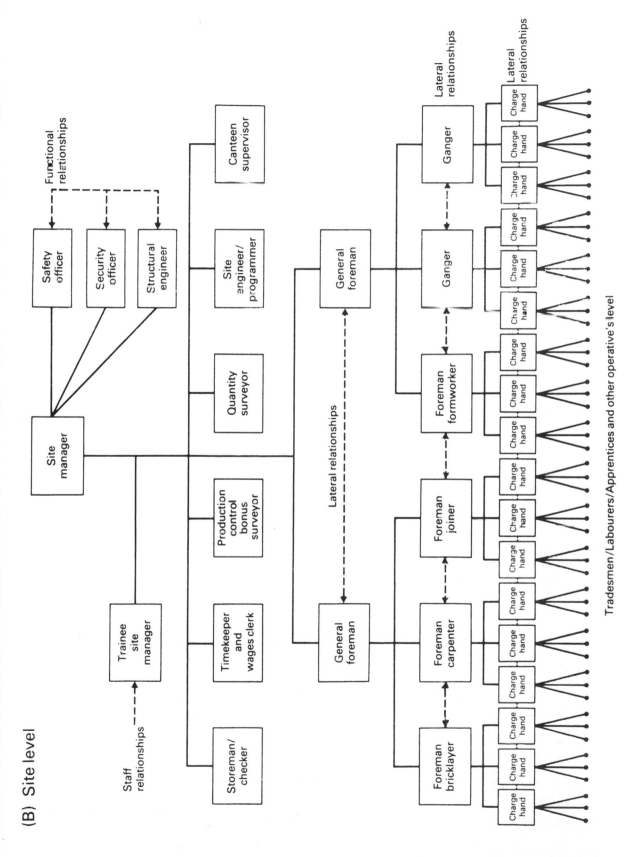

Fig. 1.2.1 Formal organisation structures

Tradesmen/Labourers/Apprentices and other operative's level

effective team spirit if everyone was not given strict demarcation points. In either case whether the structure diagram is displayed or not the span of control within an organisation may be one of two types:

1. The shallow line structure.
2. The deep military line structure (see Fig. 1.2.2).

The shallow line structure

This structure tends to be operated within a very small organisation where perhaps the entrepreneur (risk taker/owner) directs the work and operations of all the employees. The disadvantage of such a system is that the superior would find it burdensome where the structure is not only shallow but very wide – the superior being directly responsible for too many subordinates.

It is believed that superiors should not be expected to control more than eight subordinates otherwise he/she would be ineffective in his/her attempts to:

- coordinate;
- motivate;
- discipline;
- communicate;
- command;
- set an example.

Figure 1.2.2 (A) shows a shallow but wide structure diagram which exists in many small businesses.

Deep military line structure

As in a regiment where the colonel is superior to the majors and the majors in turn are superior to the captains and so on down to the private soldier, so a similar situation or structure exists in the larger companies. The managing director in a company is the all powerful superior, with managers, supervisors and chargehands, etc., being in turn in the line of command. The more that responsibilities are delegated by line managers to subordinates the more the managers can direct their energies to more important and pressing problems. Those with delegated responsibilities, however, would still be answerable to the delegator.

Where delegated responsibilities exist or where an organisation structure operates, the more it opens up a promotion ladder which subordinates would generally be motivated to climb. Individuals who are motivated generate new ideas into an organisation which helps it to compete more efficiently with competitors (see Fig. 1.2.2 (B)).

Responsibility, authority and accountability

A manager or any other supervisor is entrusted with a charge for which he/she is accountable. The success or failure of the charge/task will be entirely his/her responsibility. It is necessary, therefore, for each to know clearly and precisely what his/her duties are. Normally he/she must bear in mind the policies as laid down by the 'Memorandum of Association' and by the company as a whole regarding, sales, plant, organisation, personnel and many others (see section 1.1).

Some of the responsibility given to an individual may be delegated to subordinates who become answerable to that individual. This will allow the superior to concentrate on the more critical and demanding matters.

The total responsibility still remains with the individual/supervisor even for the duties delegated.

Authority must be sufficient to allow a supervisor, etc. to carry out his duties properly, otherwise it will lead to a build up of frustration. One must be given the power to act, and the right to enforce obedience. Responsibility without authority leads to a confused situation and a breakdown in discipline.

Too much authority given to a site supervisor can, however, sometimes lead to embarrassing situations which could also be costly to correct. One notable example can be related to where a supervisor employed a gang of bricklayers at the same time as head office transferred another gang to his site from a site where their services were no longer required. Immediately the head office discovered what the site supervisor had done, they cancelled the engagement of the site supervisor's choice of bricklayers within the legally accepted time limits. The dismissed gang went to the industrial tribunal to prove unfair dismissal, claiming they were sacked because they belonged to a union, and stated they were being victimised under the then Trade Union and Labour Relations Act and the Employment Protection Act 1975. (The gang had been in dispute with a previous employer.) The company had great difficulty proving they were not victimising the gang because of their past record.

Accountability can never be delegated, and therefore accountability to someone in higher authority for decisions and actions of subordinates is a charge to which managers and supervisors are subjected.

Although organisation charts are to illustrate each individual's standing in an organisation, details of each officer's responsibilities, immediate superiors, subordinates, special duties, limitations, relevant contacts and compulsory memberships of committees must be drawn up. These statements and charts will then be issued or displayed to each executive or supervisor.

1.3 Problems, incidents and emergencies

There are a thousand and one problems, incidents and emergencies which could occur on-site and for each the site supervisor should be prepared to deal with them in a cool and calculated way. This can only normally be possible if the supervisor has been forewarned, trained or educated on how to deal with the unexpected. Some supervisors/managers intuitively know how to deal with problems/emergencies and respond well in dealing with or solving them through their natural abilities and good common sense. But as a precaution there should always be a handbook available to which supervisors can refer when problems are encountered on-site.

The problems and incidents can generally be divided into the following areas:

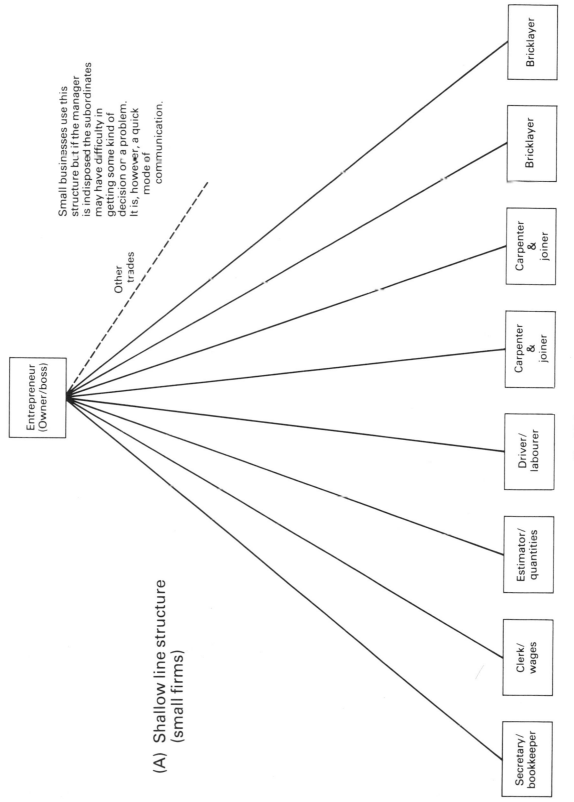

Small businesses use this structure but if the manager is indisposed the subordinates may have difficulty in getting some kind of decision or a problem. It is, however, a quick mode of communication.

(A) Shallow line structure (small firms)

Fig. 1.2.2 Span of control

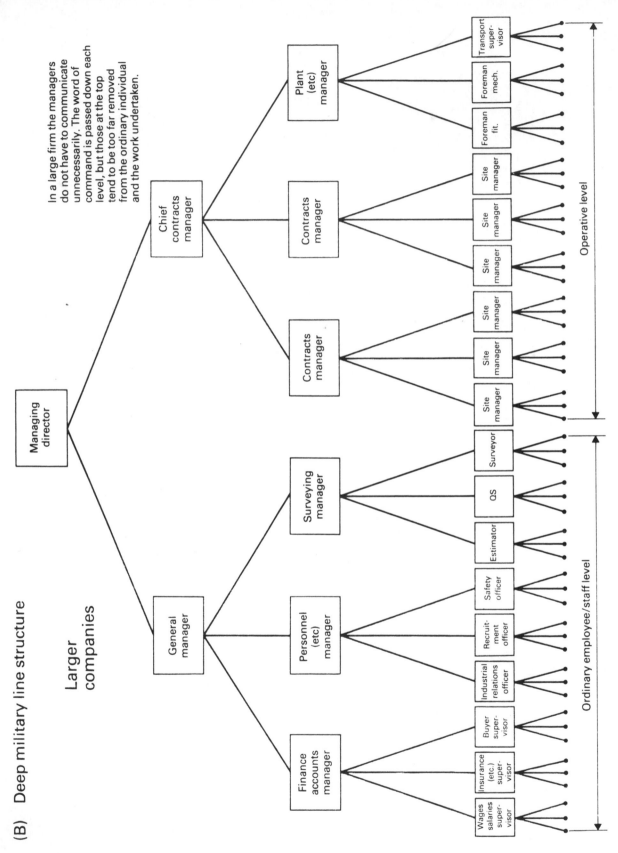

8

(B) Deep military line structure

Larger companies

In a large firm the managers do not have to communicate unnecessarily. The word of command is passed down each level, but those at the top tend to be too far removed from the ordinary individual and the work undertaken.

Managing director

Chief contracts manager

General manager

Plant (etc) manager

Contracts manager

Contracts manager

Surveying manager

Personnel (etc) manager

Finance accounts manager

Transport supervisor

Foreman mech.

Foreman fit.

Site manager

Site manager

Site manager

Site manager

Site manager

Surveyor

QS

Estimator

Safety officer

Recruitment officer

Industrial relations officer

Buyer supervisor

Insurance (etc.) supervisor

Wages salaries supervisor

Operative level

Ordinary employee/staff level

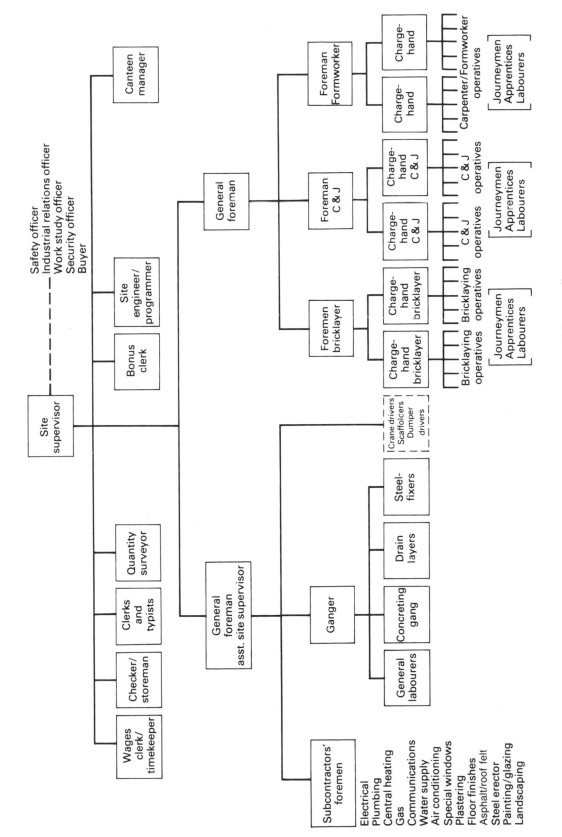

Fig. 1.2.3 Structure diagram (showing span of site control)

1. Industrial relations.
2. Misconduct and breaches of discipline.
3. Personality problems.
4. Emergencies.

Industrial relations

Most companies have policies which lay down procedures for site managers to observe regarding industrial relations problems. It is usual that where there is a labour stoppage of work or it is believed that a stoppage is about to take place that a site manager must contact the personnel officer/industrial relations officer immediately. These officers are generally trained to deal with such problems and they have the authority to act on behalf of the company on such matters (see section 5.6).

Misconduct and breaches of discipline

(*a*) *Insubordination (refusing to comply with reasonable instructions):* Site supervisors should be firm but fair in their approach to insubordination. This kind of behaviour is rare, but when it does occur it is a serious breach of discipline and the normal rule is that the site supervisor has the power to instantly dismiss an individual. Generally, however, an individual who is insubordinate would be warned about his/her actions, and it could be the practice, especially where unions dominate, that the individual would be suspended pending an enquiry, which although it is time-consuming, ensures that any subsequent action or decision is not taken lightly, vexatiously or unfairly.

(*b*) *Threatening behaviour or physical assault to supervisors:* This behaviour cannot be tolerated and the site supervisor is generally within his/her power to instantly dismiss an offending individual, otherwise this behaviour, if it goes unchecked, could seriously undermine the supervisor's authority and that of the company's.

(*c*) *Alcoholism:* This is both dangerous to the person while he/she is on-site and to those with whom the individual has to work. The site supervisor should order the subordinate who is affected by alcohol off the site, or where possible, arrange to have the person taken home. Later, the individual should be verbally warned that the next time he/she may be given a written warning which could subsequently lead to dismissal.

(*d*) *Under the influence of drugs:* This may be due to an accidental overdose by a diabetic. Other ailments could lead to the careless use of drugs as a medicine. One should be cautious in dealing with affected individuals and medical treatment may be the answer. Where a subordinate acts strangely and has been known to act strangely on a previous occasion, and it is suspected he/she is affected by drugs, similar steps should be taken as with a person who is affected by alcohol.

(*e*) *Fighting:* This could lead to a very serious injury. It affects morale and production on-site. It is sometimes encouraged by other workers while others prefer not to get involved. A firm command by a supervisor for the participants in a fight to stop is normally necessary, which should be followed by a statement as to how their actions could affect their positions with the company. An enquiry should be carried out to ascertain the cause, with a warning being given to the one who was responsible for the affray.

(*f*) *Obscenities:* Apart from an individual disgusting other site employees his/her actions may lead to complaints being made by members of the public, especially those neighbouring on to the site. This action brings the company into disrepute and causes embarrassment to all concerned. The supervisor should warn any person who acts obscenely that any future similar actions would be a breach of discipline.

(*g*) *Absenteeism:* If an individual absents himself/herself from the site during working hours without permission from the site supervisor a verbal warning should be given and time should be deducted from his/her total attendance for the day. Where similar action continues a written warning would be necessary. Unless suitable reasons can be given for persistant absenteeism (non-attendance on certain days) the appropriate levels of warnings should be given followed by subsequent dismissal. (See Figs 6.2.2 and 6.2.3 for typical warning and dismissal letters.)

(*h*) *Wilful and malicious damages to company property:* Proof and witnesses may be required. Individuals responsible for such damages are subject to instant dismissal.

(*i*) *Theft of company property:* It would obviously be a requirement that a suspected person should be caught in the act of stealing. Once again action of this nature is subject to instant dismissal, and if serious losses had been sustained by the company it may be necessary to involve the police who generally carry out searches of a suspect's home. Colour coding of equipment frustrates and deters the thief, and if all serial numbers are recorded of equipment, plant, etc., at site or plant department levels, when recovered the stolen items could easily be identified and claimed.

(*j*) *Theft of employees' property:* This is a particularly mean action by anyone. It brings suspicion on to everyone and has a damaging effect on site morale. Employees identified as thieves of this kind should instantly be dismissed.

(*k*) *Serious breaches of the Safety Regulations:* This action cannot be tolerated because of the danger to life and property, which could lead to a high financial loss to the company. Where the breach of the regulations is deliberate instant dismissal is the normal procedure in some companies.

(*l*) *Falsification of records, etc., for personal gain:* This is a calculated and very deliberate misdemeanour, and once it is discovered and a case is proven instant dismissal should result.

(*m*) *Removal of company property without written approval:* The usual excuse by employees discovered in the possession of company property (plant, equipment or material) is that they were only borrowing it. So that no one is in any doubt about the company's policy regarding such matters, employees' handbooks outlining the company's rules should be issued periodically (yearly) to existing employees. The need for written approval prevents the abuse of company property in this way.

Individuals found in possession of company property obviously intended for their own use without written approval should be instantly dismissed.

Note: Each problem outlined under the heading of 'misconduct and breaches of discipline' should be dealt with on its own merits, and the company may have a policy for all events that an employee should be suspended first pending an enquiry. This then allows everyone a fair hearing as in many cases events are not always what they appear.

Personality and personal problems

(*a*) *Victimisation:* A subordinate may claim he/she is receiving treatment of this kind. It is foolish for site supervisors to adopt this principle with anyone. Not only is it unfair, but if it is noticed by other subordinates a site supervisor's reputation is diminished. A union representative could take the case up and unnecessary aggravation can have unpleasant consequences.

Site managers should be in a position to ensure that his/her immediate subordinates (foremen, gangers, etc.) do not cause any aggravation by victimising other employees/operatives.

(*b*) *Bullying:* This is a personal problem of individuals but, obviously, where such action affects their productivity, safety, health or general performance, the site supervisor should make a discreet approach to the guilty party, advising them to relent.

(*c*) *Weak persons:* If the physique of a individual is not adequate for the work for which he/she was employed alternative work should be provided. This may be the personnel officer's responsibility if the individual was taken on at head office. The only other solution would be to go through the normal warning procedure that the individual's work is unacceptable and must show signs of improvement to prevent his/her termination of employment.

(*d*) *Accident proneness:* Alternative employment which is less dangerous than on-site may be offered, and the personnel officer should be approached by the site manager if an individual is accident prone.

(*e*) *Health problems:* If it is suspected that an employee has a health problem arrangements can be made for him/her to attend a medical examination which he/she should not unreasonably refuse to have – under the Health and Safety at Work, etc., Act 1974.

If the person is unfit for the types of work he/she normally does, he/she cannot reasonably object to suitable alternative work being found within the firm. If there is a refusal a warning may be given which can lead to subsequent dismissal. Care should be taken here, and reference should be made to the Employment Protection Acts.

(*f*) *Illnesses:* Doctors' certificates are normally required from individuals who absent themselves from work for more than seven days. Upon their return they should also present doctors' certificates to prove their fitness for work. Employees who cannot prove their fitness for work should not be allowed to work until they can.

Emergencies

It is the events which should follow serious accidents that will be considered here, and not those emergencies which, if not solved, will seriously affect production or slow down a particular process, e.g. shortages of plant, materials or labour.

This section is by no way an attempt to instruct individuals in the art of first aid – only attendance at a recognised course can do this – but it is an attempt to make site supervisors more aware of the difficulties accident emergencies present on site. It is an attempt to encourage supervisors to be more positive in their approaches to problems of this kind.

First-aiders (in many cases the site supervisor) must have obtained a recognised first-aid certificate within the last three years through attendance on special courses approved by the Health and Safety Executive where certificates of first aid are awarded. Boy Scout and Girl Guide certificates in first-aid are not recognised as suitable certificates in the work place.

Site supervisors must have some knowledge of how to deal with an emergency where the health and safety of individuals are concerned, and he/she must be prepared to take charge in such events. The prompt administration of first-aid to accident victims and to those who become ill until medical aid can be given by a doctor is paramount if the life of a casualty is to be sustained and recovery is to be promoted.

It is generally thought imperative that all site managers should have attended a course in first-aid, and have some knowledge of what to do in an emergency situation.

Action to be taken
The site supervisor or the first-aider in the event of an accident should do the following:

(*a*) Remain calm and assess what has happened.
(*b*) Talk to and listen to the casualty, and apart from instilling confidence and reassurance, try to diagnose what injuries have been sustained.
(*c*) **Give immediate treatment after making sure there are no further dangers (collapsing wall, fire, etc.). If**

there are immediate dangers deal with them if possible, or remove the casualty to the safest minimum distance before given treatment. Ensure the casualty is breathing properly, and then place him/her in what is known as the recovery position, if he/she is unconscious or likely to pass out. (See Fig. 1.3.1 (A).)

(d)   The ambulance should be sent for immediately if there are other persons on hand, but if working as an individual without help, ensure the casualty's breathing is unrestricted, the heart beat is satisfactory, serious bleeding is arrested and that he/she is comfortable and covered for warmth before calling the ambulance or seeking medical aid.

Calling the ambulance

State the following when calling an ambulance – the number to call should be situated on or near to the first aid box on site:

(a)   The exact place of the accident and suitable directions to get there, if necessary. (The Ambulance Authority will have particulars of the site if the work is to last for more than six weeks. Notification should have been sent at the commencement of site work under the Factories Act 1961.)

(b)   The number of casualties.

(c)   The type and seriousness of the accident.

Note: If a first-aider is on hand and there are sufficient first-aid facilities, before any casualty is moved who has suspected or obvious broken limbs immobilisation of the affected limbs should be made to help prevent further pain during transit. This may be done while waiting for the ambulance.

Particulars of the casualty

Obtain particulars of the casualty – name, age, address, marital status and details of the injuries and how the accident was caused, and any other relevant information which will assist those at the hospital, i.e. pulse rates obtained at reasonable and regular intervals. These particulars should be written on a piece of paper which should be fastened by a safety pin, etc., to the casualty's clothing.

After the accident

(a)   do not move or touch anything and notify head office and the casualty's next of kin;

(b)   obtain statements from witnesses;

(c)   photograph and sketch diagrams, where necessary, of existing conditions;

(d)   fill in a page of the Accident Book (B1 510);

(e)   complete the appropriate section of the General Register under the Factories Act 1961, if the accident is serious;

(f)   send the Accident Report, Form 43B, to the local factory inspector duly completed with all the accident particulars;

(g)   notify the police if foul play is suspected.

Claims

The next of kin of an operative who has been killed or seriously injured can claim for benefits through:

(a)   The Death Benefit Scheme – outlined in the Working Rules Agreement and paid for through the additional payment in the holiday stamp scheme.

(b)   The employer's insurance – laid down in the Employers Liability (Compulsory Insurance) Act 1969.

(c)   Their own personal National Insurance, as laid down in the Social Security Act 1975.

(d)   A company's own special injuries insurance scheme, if there is one.

(e)   The operatives union, if it operates a compensation for injury scheme.

Further points regarding first aid

The following details relate to the immediate needs of those unfortunate individuals who suffer some accident, etc., which leads to injuries, or breathing, heart beat or circulation failures.

Emergency resuscitation

Each and every one of us should have knowledge of how to revive casualties and the only sure way is to attend a recognised first-aid course, but above all, one should bear in mind that the vital needs of a casualty is to have his/her airway maintained (clear of obstructions such as broken or false teeth which have moved and other debris), breathing is reasonably constant (artifically resuscitate if necessary), and that the blood circulation is adequate.

It is essential for the brain to receive a continuous supply of oxygenated blood, and when giving artificial respiration continue at a careful speed, waiting for the chest to rise with each blow into the lungs, and to deflate after removing mouth (using the mouth to mouth or mouth to nose method of artificial respiration). (See Fig. 1.3.1 (C).)

If the heart stops (and only when it is certain it has stopped) strike the chest of the casualty smartly with the edge of the hand. This may start the heart beating again, but if it does not then give external heart compressions (see Fig. 1.3.1 (D)), using two hands at the rate of 60 compressions per minute for adults, one hand at 80 compressions for children, and about 100 compressions with two fingers for infants – firm pressure should be applied in each case. If working alone, where the casualty has stopped breathing and the heart has stopped beating, one should give about 15 heart compressions to two lung inflations until the casualty recovers or the ambulance or medical aid arrives. Where there are two persons undertaking first aid in such a situation, one can apply artificial respiration while the other applies external heart compressions, if necessary. It must be remembered that if breathing is failing it is safe to apply mouth to mouth, etc., but it must be obvious that the heart has stopped functioning before applying heart compressions.

Electrocutions

Where a casualty has been electrocuted by a large voltage (up to 400 000 V), stand clear, and render first aid after the current has been switched off. No one should attempt to render first aid until the casualty is free from contact.

For domestic or low voltage contact by casualties one should switch off the supply; pull out the cord/cable from the socket or pull it away from the casualty; or attempt to break contact with some dry insulation material, e.g. piece of wood, rope or folded newspaper, rubber, etc., always remembering to stand on some dry insulating material.

One should not touch the casualty while he/she is in contact with electricity.

The immediate treatment is to ensure that the casualty is breathing and that the heart is beating properly. Artificial respiration and/or heart compressions may be required.

Treat for burns where the electricity passes into and out of the body (two possible places). Alleviate the heat by submerging or swilling the burnt area for 10 minutes with cool, clean water. Do not apply ointment. Ensure the casualty is warm and comfortable until the ambulance arrives.

Struck by lightning

Treat as for electrocutions, applying artificial respiration and/or heart compressions where necessary, and afterwards cooling the burnt areas with cool water (and covering with clean sterilised lint where possible).

Burns or scalds

Alleviate the heat by submerging or swilling the affected areas for 10 minutes with clean cool water. Do not apply ointments. It is within the first 10 minutes that the damage, if unattended, is done. Remove rings, etc., and dirty clothing from the affected parts. Lay the casualty down and cover the injured parts with clean sterilised lint or similar, and give the casualty, if conscious, small quantities of cold drink at frequent intervals.

Remove to hospital as soon as possible.

Chemical, etc., burns

Flood the area with clean water to dilute the chemical and to cool. Treat as in burns.

Diabetic emergencies

Deal with the emergency need of the diabetic (breathing, etc.) if unconscious, and place in the recovery position and arrange for admittance into hospital.

If the diabetic is conscious give him/her a sweetened drink (two tablespoonsful of sugar or give sugar lumps; if necessary). Whether the diabetic has given himself/herself too much insulin or there is a lack of insulin it is harmless to give sugar.

Unconsciousness of an individual could be due to many reasons, so take care with the immediate needs and ensure that there is ample fresh air; there is a clear airway; loosen clothing; the casualty is comfortable and warm; and call the ambulance.

Bleeding and circulation failure

Severe bleeding: The aim is to stop the bleeding immediately and obtain medical aid urgently. Therefore:

1. Press a dressing over the bleeding part and hold until a bandage can be positioned ensuring that there is no glass or debris in the wound. A large wound may require both sides of it being pressed together until a dressing is applied.
2. Lay casualty down and lower the head.
3. Raise and immobilise the injured part.
4. Remove to hospital

Severe cuts to arteries are dangerous and at all costs the blood flow should be stemmed until a substantial dressing can be applied. The important arteries to stem the flow of blood to the limbs are situated on the inside edges of the arms and in the leg groins. These should be pressed against the bone on the heart edge of the wound (maximum 15 minutes) until a dressing can be applied. (See Fig. 1.3.1 (E).)

Bleeding from the nose

The casualty should be placed in the sitting position and should breathe through the mouth. Pinch the nose at the softest lowest part for 10 minutes. Loosen clothing and warn the casualty not to blow his/her nose for a few hours after the bleeding stops.

If bleeding does not stop after a short time get medical help.

Heart attacks

Place the individual comfortably either in a semi-recumbent position (see Fig. 1.3.1(B)) or supported in a sitting position depending on which allows the easiest breathing. Send for the ambulance or medical aid.

Clothing should be loosened around neck, chest and waist.

If breathing is failing give artificial respiration immediately, and if the heart has stopped give heart compressions (see Figs 1.3.1 (C) and 1.3.1 (D)) until breathing and heart begins to function properly.

Remember – if breathing is failing it is safe to give mouth to mouth respiration, but under no circumstances should heart compressions be given until one is sure that the heart has stopped. Keep the casualty warm until the ambulance arrives.

Fainting

Lay the individual down and raise his/her legs slightly. Ensure there is a plentiful supply of fresh air and the individual is in the shade. Loosen clothing, and if breathing is difficult place in the recovery position until consciousness returns. Gradually sit individual up and then give sips of water.

There are many health problems which individuals can experience while at work (crushed and broken limbs, etc.) and it is foolhardy for employers to allow numerous individuals to be employed in a work environment without someone being trained or having knowledge of first-aid procedures. It is imperative that employers think again and ensure that either the site manager or a subordinate is sent on a first-aid course to meet the requirement of The Health and Safety (first aid) Regulations 1981.

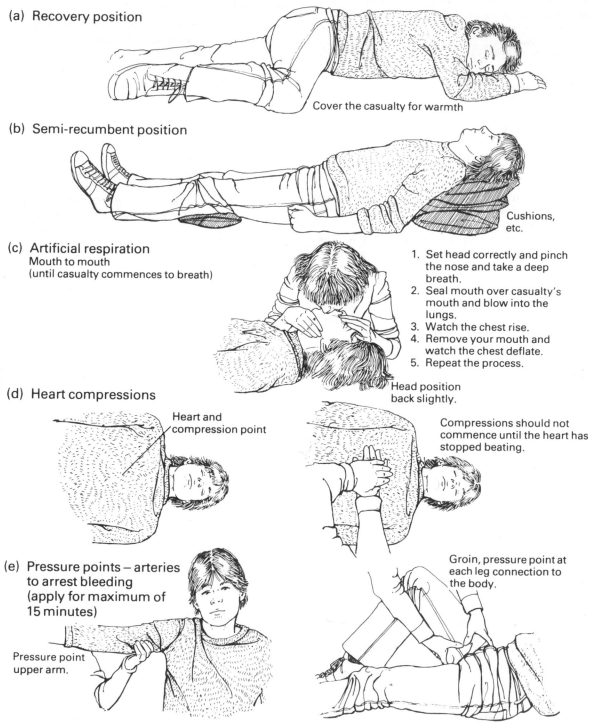

(a) Recovery position

Cover the casualty for warmth

(b) Semi-recumbent position

Cushions, etc.

(c) Artificial respiration
Mouth to mouth
(until casualty commences to breath)

1. Set head correctly and pinch the nose and take a deep breath.
2. Seal mouth over casualty's mouth and blow into the lungs.
3. Watch the chest rise.
4. Remove your mouth and watch the chest deflate.
5. Repeat the process.

Head position back slightly.

(d) Heart compressions

Heart and compression point

Compressions should not commence until the heart has stopped beating.

(e) Pressure points – arteries to arrest bleeding (apply for maximum of 15 minutes)

Pressure point upper arm.

Groin, pressure point at each leg connection to the body.

Fig. 1.3.1   First aid

Chapter 2
Supervisor's understanding of cost control

2.1 Site cost control

Cost control, or control of costs, is an essential objective of any firm. In order to achieve some measure of control over costs on-site the administrative requirements are such that suitably designed forms are necessary for use by those charged with the responsibility for collecting information which may be used by management in determining whether or not corrective action is required to a process or operation to achieve cost savings.

A cost control system should be so designed that it enables management to satisfactorily collect and produce information from which the monitoring of actual costs can be compared to the estimated costs. The information should be collected daily, but at least weekly, if the results for comparisons are to be meaningful. The earlier details can be discovered about a process which is costing more than was anticipated, the earlier a decision can be taken to rectify or modify the process in an attempt to reduce its cost. It is of little immediate use after a process has been completed to discover that it cost too much. The results, however, may be used for future projects to fix more realistic rates, or to ensure that resources are used more efficiently, perhaps with alternative techniques being adopted.

At the tender stage a contractor's estimator prepares estimates for carrying out various items of work shown in the bills of quantities, estimating being the anticipation of costs to the contractor for executing the work – excluding overhead costs and profit. Overheads are items of expenditure or costs incurred in running the main contractor's office, i.e. salaries, heating, lighting, rates, insurance, telephones etc., and the cost for such items must be considered if a contractor's true costs are to be eventually recovered.

A priced bill of quantities is retained by the contractor on the successful bid for a contract with a tender. It would then be used throughout the duration of the project work as a reference source, to compare the estimated unit rates with the actual, true unit costs – a unit being a measure such as: metre, square metre, cubic metre, item or number of work done by a particular trade, e.g. a square metre of brickwork.

It is essential to have sufficient levels of feed-back from sites for surveyors, from the Cost Control or Production Department, to calculate the actual costs for comparison with those initially allowed in the estimates. Any variances below, or in excess of, the estimated figures would be recorded so that:

1. Items of work showing excessive costs may be adjusted by corrective action at site level in the way the work is done, or, if the opportunity for action has passed, then rates may be increased for later contracts.
2. Items of work showing lower actual costs than estimated costs would obviously be in profit but may later be adjusted for future contracts, perhaps, to enable a keener, more competitive price to be submitted.

Controlling costs is a continuous occupation of management and causes the most problems in achieving. It is of little use waiting until a project is completed only to discover that a loss has been made. Monitoring costs daily or weekly would show up adverse trends. Counter measures may then be taken – using alternative materials allowable within the specified standards, alternative construction techniques, better motivation incentives, more plant, or even the replacement of the supervisor(s) with a more dynamic one.

Cost comparisons between actual and estimated costs must be made as early as the figures can be determined. This requires the close cooperation of the site manager and his staff and the head office staff.

The surveyor from the Cost Control or Production Department normally takes responsibility for generating feed-back from site to measure labour, etc., costs, and general efficiency. The documents from which information regarding costs may be gleaned are:

(a) material invoices from suppliers and manufactures;
(b) material issue sheets from main stores;
(c) material transfer sheets from or to other sites;
(d) plant hire firm invoices;
(e) plant issue sheets from own plant department;
(f) plant transfer sheets from or to other sites;
(g) site wages sheets for direct labour;
(h) labour-only subcontractor payments sheets;
(i) domestic subcontractor invoices/claims;

(*j*) nominated suppliers' invoices;

(*k*) daywork sheets;

(*l*) daily returns/site diary;

(*m*) foreman's weekly report (see Fig. 2.1.1);

(*n*) allocation sheets for labour, materials and plant (see Fig. 2.1.2);

(*o*) the contract progress chart;

(*p*) Bonus Surveyor's sheets;

(*q*) remeasurement figures from the quantity surveyor, and interim certificates;

(*r*) time sheets (if used).

A well-organised firm would prepare other documents (discussed later) to eventually enable a Cost Statement to be drawn up to show actual costs against budgeted costs resulting in either gains or losses (see Fig. 2.1.3), to show the profitable and unprofitable areas of production on a contract.

2.2 Methods and levels of controlling costs

Methods are obviously set up to suit the type and size of firm. Whether a business has just commenced or has been running for some time suitable systems should be in operation. Depending on the experience of the managers and their subordinates, systems of cost control may be utilised to enable results to be drawn up on a weekly or monthly basis or at the end of each contract period.

Cost control systems are generally organised to produce results for monitoring costs on contracts as follows:

1. At the end of a contract period – Contract Account or Contract Job Cost Sheet.
2. At interim periods during a contract – normally on a monthly basis, which analyses operations.
3. At more frequent intervals (daily or weekly) – applying unit costing on items of work and not whole operations.

1. Contract Account or Contract/Job Cost Sheet

This type of account or cost sheet is used within a small contracting set-up to calculate if a job was successfully executed by earning profits. Unfortunately, the results are determined too late for any corrective action to be taken in countering losses which may have been occurring. Nevertheless, the results may be used for future contract tendering where an estimator would adjust his/her unit rates in the light of the past experience. Management on the other hand would be looking to new ways for carrying out the subsequent contract work, and by appointing a site manager who is more cost conscious and knowledgeable about site organisation and motivation.

While the results on actual costs are somewhat belated, even a system as previously described is better than no system at all. It helps to ensure that further losses on similar work on future contracts are avoided, although this can never be guaranteed.

A typical Contract Sheet/Account can be seen in Fig. 2.1.3, which may be prepared by an accountant or by a manager/surveyor using simple book-keeping techniques. The system is to enter all expenditure for a contract on the payments or debit side (left-hand side), and the contract value and other credits on the receipts side (right-hand side). Unless further analyses are made information on material breakages, excessive usage, pilfering, etc., would be unknown. Similarly, plant and labour utilisation figures would be missing and levels of efficiency would go unchecked.

By studying the Contract/Job Cost Sheet one can see the level of expenditure and compare it with the receipts. It is already too late to take action on the areas where most expenditure occurred, i.e. labour and materials, other than to record the details for future jobs. A gross profit of £1000 was made on the contract but even this will eventually be eroded by the firm's general overheads, and therefore, it does not give an overall profitable picture. A true account of a firm's performance with a contract would not be known until a proportion of the cost of running the head office was deducted as expenditure from the gross profit to give net profit.

The overall position of a firm is given when the accountant prepares an Interim Trading/Profit and Loss Account which takes into consideration all the contracts being undertaken in a trading period and the overheads, such as the cost of running the head office (rates, salaries, heating, lighting, depreciation on capital items, insurances, etc.). This is outside a normal cost control system, however.

2. Monthly analysis

The monthly valuation certificate, prepared by the professional quantity surveyor (or firm's QS), may be used as a comparison with the actual costs for the month. Certainly, feed-back from site in the form of labour, plant and materials allocation sheets would have to be sufficient so that a summary of cost may be prepared which shows up adverse variances between actual costs and valuation costs; and where similar work still remains to be done, corrective remedies may be possible to prevent costs sliding still further into the red.

3. Unit costing

A unit of work in construction usually relates to materials which are removed or fixed in position, such as: square metre of brickwork, cubic metre of excavated material, etc. The number or quantity of units would be shown in the bills of quantities which are priced by the estimator.

Cost information entered on sheets and other documents, prepared generally at site level and fed back to head office, are correlated on to Cost Control Sheets (see Fig. 2.2.1) to determine the actual unit costs as a comparison with the predetermined unit rates. Comparisons at daily or weekly intervals would soon highlight adverse variances in cost on which management could take action to regulate if time permitted, but once again the information on costs would be used to determine future unit rates for future projects.

The objective of any management is to maximise profits at the minimal of costs. This is more achievable by the use of an efficient cost control system.

2.3 Total cost elements, break-even charts, etc.

In determining the total costs involved in production either at site or factory level one must have an appreciation of the financial terms relating to the elements which together make up these costs, such as:

1. Fixed costs.
2. Variable costs.
3. Direct costs.
4. Indirect costs.

Fixed costs (supplementary costs)

During production there are costs which remain fixed – which stay the same – irrespective of the levels of production. If the output increases or, on the other hand, is below the planned levels, the fixed costs will still remain the same. Example of items of fixed costs are:

1. At site level – on-costs (or site overheads):
   - administration staff salaries;
   - hutting hire charges and rates;
   - heating and lighting;
   - telephone charges;
   - supervisors' pay;
   - scaffolding charges;
   - fencing and other preliminary items.

These on-costs are, however, normally charged to the preliminaries section of the bills of quantities.

2. At head office level – overheads:
   - administration charges;
   - salaries;
   - yard wages;
   - directors' salaries;
   - insurances;
   - telephone charges, etc.

Fixed costs are not significantly affected even when a firm increases its output. The unit cost, on the other hand, would reduce by an increase in output, and would increase by a decrease in output. As a simple example:

Assume the fixed cost is £100 and the variable cost is £2.00 per unit.

If 100 units of work were produced the total costs would be:

$$\text{fixed cost} = £100$$
$$\text{and variable cost, } £2 \times 100 = £200$$
$$\text{total} \quad £300$$

Therefore, cost per unit $\dfrac{£300}{100 \text{ units}} = £3.00$

Now if the output increases to 200 units of work, total cost is:

$$\text{fixed cost} = £100$$
$$\text{and variable cost, } £2 \times 200 = £400$$
$$\text{total} \quad £500$$

Therefore, cost per unit $\dfrac{£500}{200 \text{ units}} = £2.50$

The more units to be completed the less effect the fixed costs have on the cost of production. (See Figs 2.3.1 and 2.3.2.)

Variable costs (prime costs)

These costs vary as production/output increases or decreases. Prime costs relate to the following types of expenditure.

- the cost of raw or other materials and components which are incorporated into a structure, e.g. bricks, concrete, timber, windows – including the cost of waste.
- site or factory wages actually expended in production and directly attributable to output, including overtime, bonus and other additional payments.
- national insurance – employers' contribution, for the employees involved in production.
- certain plant hire costs.
- fuel and power used for direct production, etc.

As output increases the variable costs will increase because more materials will be used and wages claims will be greater. Conversely, as output decreases variable costs will decrease.

It is therefore, with regard to the variable costs that particular attention must be paid to ensure no undue increases are incurred during production. Increases in variable costs should lead to a much greater increase in levels of output/production. Where there are decreases in output a corresponding drop in variable costs should occur. If an increase in output leads to a disproportionate increase in variable costs serious losses usually result, and it is with regard to such problems that businesses are advised to operate a sound cost control and monitoring system. (See Fig. 2.3.1 for the effect of variable costs on average costs.)

Direct costs

These forms of expenditure are directly caused by the production of units of work. In the construction of a building there are many areas of production – brickwork, carpentry and joinery, concrete, painting, landscaping, etc. In each the units of work vary. Cost control systems require that allocation sheets for labour, plant, materials, transport and subcontractors should be prepared to determine the actual costs and unit costs of each area of work. It is necessary, therefore, that only those items of cost attributable to each and every unit of work should be considered. The direct costs, being the first area of actual costs, revolve around the following:

(a) the cost of materials (direct material) used in an activity of work. Assuming that the activity was facing brickwork, the materials would be facing bricks, sand, cement, and lime or other additives;

(b) the cost of wages (direct wages) of craftsmen, apprentices and labourers engaged on an activity;

(c) plant hire charges where the plant has been used exclusively for the benefit of an activity (direct plant charges);

(d) other costs – subcontractors, etc.

B. B

FOREMAN'S WE

CONTRACT . . . .

WEEKENDING: . . . .

WORK IN PROGRESS

WORK ORDER	BUILDING OR SECTION	TARGET	% COMPLETE	TOTAL HOURS BROUGHT FORWARD	WEEKLY TOTAL

DETAILS OF VARIATIONS FROM SPECIFICATIONS, ETC:

QUANTITY

GENERAL REMARKS:

Fig. 2.1.1

LTD.

REPORT SHEET

PREPARED BY: . . . .

HEAD OFFICE USE ONLY

TOTAL TO DATE	TIME SAVED ON COMPLETION	HOURS	1 LABOUR	RATE	£	p.

MATERIALS USED:	QUANTITY	2. MATERIALS			
DESCRIPTION					
		TOTAL WEEKLY COST			

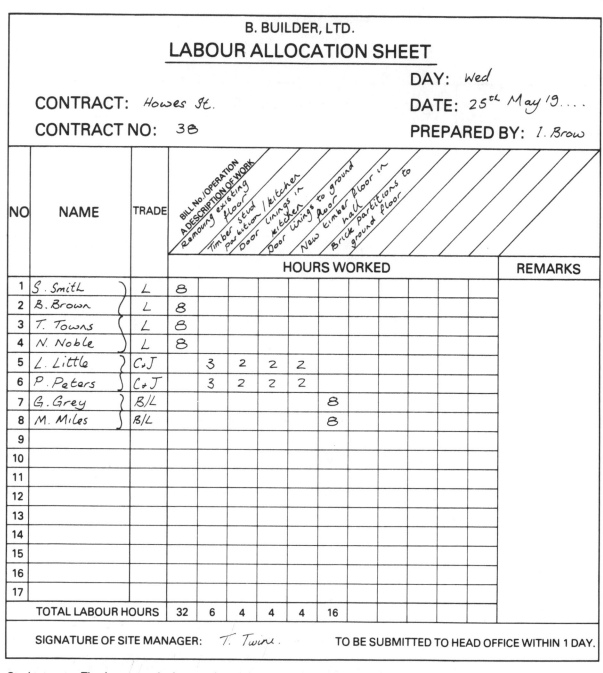

B. BUILDER, LTD.
LABOUR ALLOCATION SHEET

DAY: Wed

CONTRACT: Howes St.

DATE: 25th May '19....

CONTRACT NO: 38

PREPARED BY: I. Brow

NO	NAME	TRADE	BILL No./OPERATION A DESCRIPTION OF WORK Removing existing floor	Timber stud / kitchen partition	Door linings in kitchen	Door linings to ground floor	New timber floor in hall	Brick partitions to ground floor					REMARKS
			HOURS WORKED										REMARKS
1	S. Smith	L	8										
2	B. Brown	L	8										
3	T. Towns	L	8										
4	N. Noble	L	8										
5	L. Little	C & J		3	2	2	2						
6	P. Peters	C & J		3	2	2	2						
7	G. Grey	B/L						8					
8	M. Miles	B/L						8					
9													
10													
11													
12													
13													
14													
15													
16													
17													
	TOTAL LABOUR HOURS		32	6	4	4	4	16					

SIGNATURE OF SITE MANAGER: T. Twine. TO BE SUBMITTED TO HEAD OFFICE WITHIN 1 DAY.

Student note: The hours worked on each activity/operation or Bill item is shown allocated here. The totals can be transferred to a Summary Cost Control Sheet, if necessary, to enable calculations to be made as to actual labour costs for each item of work as a comparison with the estimated labour costs. A separate Allocation Sheet could be used for each trade and be prepared by the chargehand, etc.

Fig. 2.1.2

JOB/CONTRACT COST SHEET

CLIENT: S. M. All

CONTRACT: St Johns Street

CONTRACT NO: 39

DATE: 25th May 19......

PREPARED BY: S. Smith
(Surveyor)

	£	£
1.0 RECEIPTS CONTRACT PRICE	50 000	
ADD VARIATIONS/DAYWORKS	1 000	51 000
2.0 DEDUCT EXPENDITURE		
2.1 MATERIALS (From materials cost sheets) Includes: materials delivered direct from suppliers/manufacturers, materials from main store materials transferred from other sites, incidentals (site petty cash items), nominated suppliers less materials returned to main store and credited)	18 000	
2.2 (from plant cost sheets) Includes own plant charges, plant hire	5 000	
2.3 Labour (from labour cost sheets) Includes: direct labour, labour only subcontractors.	15 000	
2.4 SUBCONTRACTORS (from subcontractor cost sheets) Includes: domestic subcontractors who supply and fix, nominated subcontractors	10 000	
2.5 TRANSPORT (from transport cost sheets). Includes: direct charges, apportioned charges.	5 00	
2.6 ON-COSTS (from on-cost cost sheets) site overheads e.g. insurance, water, hutting, telephone, heating, etc.	1 000	50 000
3.0 BALANCE BEING GROSS PROFIT		£ 1 000

Student note: The nett profit is determined by deducting the desired percentage
normally allowed for head office overheads.

Fig. 2.1.3

B. BUILDER LTD.
COST CONTROL SHEET

CONTRACT: Hill Crest
CONTRACT NO.: 40

WEEK ENDING: 14th May
PREPARED BY: S. Smith
DATE: 17th May

Bill/ Op. No.	Operation/ work description	Quants.	Unit	Rate	Estimated total cost (value)	Labour Est. cost	Labour Actual cost	Plant Est. cost	Plant Actual cost	Materials Est. cost	Materials Actual cost	Actual total cost	Unit cost	Estimated/ actual cost variation + or −	Accumulative loss or gain on estimate	Comments
D2.2	½ Bk wall in Facings	400	m²	£11/m²	£4400	2600	2400	60	50	1740	1600	4050	10.13	+350	+350	8% in credit
D2.4	1Bk wall in Fletons	200	m²	£20/m²	£4000	2480	2100	20	30	1500	1350	3480	17.4	+520	+870	13% in credit
TOTALS					8400	5080	4500	80	80	3240	2950	7530			+870	

Fig. 2.2.1

Output of brickwork units (m²)	Fixed costs (£)	Variable costs (£)	Total costs (£)	Average costs per unit (£)	Marginal cost (£)
100	200	1 000	1 200	12	—
101	200	1 011	1 211	11.99	11
102	200	1 021	1 221	11.97	10
103	200	1 030	1 230	11.94	9
104	200	1 038	1 238	11.90	8

The increase from 100 to 101 units adds £11 to the total costs but the average cost is reducing. This means that while fixed costs remain the same in total the variable costs in this example are not rising in proportion to the output of units. Fixed costs per unit decrease with the increase in production.

Fig. 2.3.1   Example of how fixed and variable costs affect the average costs

Indirect costs

Costs which cannot be directly contributed in their entirety to an activity or area of work, but nonetheless, were incurred during a project are known as indirect costs, i.e.:

(a) supervisors' wages/salaries. The supervisor is called upon to direct more than one operation and therefore only a percentage of his/her time can be charged to an activity;

(b) storemen, site administrative staff, canteen personnel, etc., contribute a service to each area of production but not to any direct activity. Therefore, once again, the cost for each can only be apportioned to each activity;

(c) certain plant, e.g. hoists, cannot be directly charged to an activity but may have a proportion of its costs charged to each area of production;

(e) lubricants and consumable tools may only be charged as an indirect cost.

Break-even charts

Break-even charts are used as aids to management to show just at what points profits may be earned bearing in mind the effects of fixed and variable costs. There are a number of ways to illustrate such charts, the two shown in Fig. 2.3.2 are the simplest, with the second one being more realistic than the first, showing that it is the fixed costs which are not being covered by the sales in the early stages of production.

The information shown on the break-even chart in Fig. 2.3.2 assumes that the selling price of the units will remain fixed throughout, but management obviously make decisions which not only affect the selling price but also the fixed or, in fact, variable costs. The lines on the chart are generally never straight but are curved.

Marginal costs

Marginal cost is the extra cost incurred when the total output is increased from a planned level to a new level. Therefore, when a decision is taken to increase the volume of output from the original level the total costs would change; but it is the variable costs only which increase and create the change. Fixed costs remain the same. However, the fixed costs at substantially increased levels of output by necessity would change due to the depreciation of the extra plant or machinery which would be required to create the output increase. Also, extra salaried staff may be needed to cope with the additional administration. Agreements between unions and employers in many instances have changed many wage earners to salaried staff at site or shop floor levels, therefore creating guaranteed pay whether earned or not which may contribute to fixed costs.

Consider the following simple example for assessing marginal costs:

Units produced	1 000	2 000	3 000
Fixed costs (£)	1 200	1 200	1 200
Variable costs (£)	5 000	8 000	10 000
Total costs (£)	6 200	9 200	11 200

Example results

Output in units (m²)	0	100	200	300	400	500	600
Fixed costs £	1 000	1 000	1 000	1 000	1 000	1 000	1 000
Variable costs	0	500	1 000	1 500	2 000	2 500	3 000
Sales at £10 per unit	0	1 000	2 000	3 000	4 000	5 000	6 000

No 1.

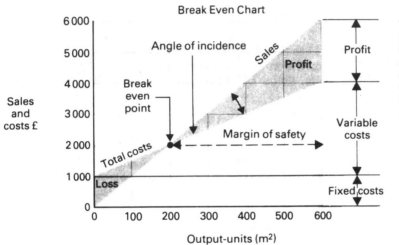

The preparation of a Break Even Chart is an attempt to compare the effects that both the fixed and variable costs will have on the sales. It also highlights at what point profits should begin to be earned.

No 2.

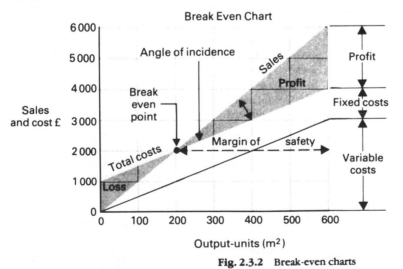

1. An alternative and more appropriate chart shows the fixed costs placed above the variable costs which indicates that below the break even point, on the left, the fixed costs are not being recovered completely while making sales.
2. The longer the length of the margin of safety the greater the chance of making a profit. The shorter it is means that if there is a small drop in productivity the profits will reduce considerably.
3. The bigger the angle of incidence the bigger the expected profit.

Fig. 2.3.2   Break-even charts

Average cost per unit

$(\dfrac{\text{total cost}}{\text{units produced}})$ £6.20   £4.60   £3.73

Margin cost of additional units:	—	£3.00 per unit	£2.00 per unit

The first 1000 units may be sold for £6.20 each (or more) which recovers the cost of producing the units. The next 1000 units may be sold for £3.00 per unit to recover the cost of producing the extra units. The remaining 1000 units may be sold for £2.00 per unit without incurring a loss, e.g.:

$$£6200 \text{ to } £9200 \quad = \quad \frac{£3000}{1000 \text{ units}} = £3/\text{unit}$$

$$\text{and} \quad £9200 \text{ to } £11\,200 = \frac{£2000}{1000 \text{ units}} = £2/\text{unit}$$

The whole concept of marginal costing rests on the accurate distinction between fixed and variable expenditure. An additional example on marginal costs is shown in Fig. 2.2.1. The total cost, £1200, is incurred while producing 100 units. Therefore the average cost is $\dfrac{£1200}{100 \text{ units}}$ = £12 per unit. The selling price must be in excess of this figure to earn profits. If one more unit is produced making a total of 101, the total cost is shown to be £1211. The average cost therefore is $\dfrac{£1211}{101 \text{ units}}$ = £11.99. The difference in the total cost per 100 units and 101 units is £11. This is the marginal cost. For a production of 102 units the marginal cost is a further £10, and so on. (Refer to Fig. 2.3.1.)

Application

Marginal costing provides a yardstick in the following situations:

1. In deciding prices during a recession, etc.
2. To compare the results for different contracts.
3. In assessing profits due to increases or decreases in sales.
4. To assess if it is suitable to sell below total costs or even the marginal costs perhaps for a limited period in order to retain the services of the skilled labour during a recession, or to maintain production when competition is at its fiercest. There comes a time, however when costs increase due to inflation, and losses will them become evident.

Chapter 3
Supervisor's appreciation of documents and references used by firms

3.1 Obtaining work and tendering

There are numerous ways by which a contractor can obtain work which are broadly divided into the following:

1. Those requiring a vigorous approach by the contractor's representatives who go out and seek land for purchase, and then arrange for the development of the land and its subsequent sale (speculation). Estate agents' and solicitors' services are generally used in both the initial purchase and subsequent sale of the land. Also, some form of market research should be undertaken to ensure that the right type of development takes place – housing, shops or industrial accommodation – to guarantee that the units can be disposed of profitably.
2. Those requiring the seeking out of tender invitations in newspapers, magazines or journals which obviously necessitates the assessing of the company's work load before making application.
3. Those requiring the advertising of the services offered by the contractor so that individuals and organisations may make approaches to have extensions, alterations refurbishments and other work carried out. Advertising may be done through newspapers, magazines, calling cards, direct canvassing, gimmicks, yellow pages of the telephone directory, names on vehicles, signboards on existing sites and head office, and by public relations exercises and promotion parties.
4. By reputation. Past satisfied customers may return to use a contractor's services, and they may also recommend to others the suitability of the company.

The whole of the above relates to marketing generally.

Tendering

A tendering situation does not arise where a contracting firm is to carry out its own speculative work. Prospective clients, however, need to have some indication of the expected costs of their proposed project. This may be conveyed to them in the form of an approximate estimate of costs by the architect/designer when tenders are submitted by contracting firms in competition with other firms. The lowest tender, with other conditions being equal, would be the successful one.

If the lowest tender was in excess of the architect's/designer's approximate estimate, and the client had budgeted within the estimate, there may have to be a reappraisal, possibly leading to the architect/designer having to look for ways to cut costs by reducing the size of the proposed structure, or by easing the specifications. Negotiations would generally take place between the client and the contractor who submitted the successful tender in an attempt to maintain the unit rates in the bills of quantities or to reduce them.

The methods of tendering (bidding) for work by contractors are outlined as follows:

Negotiated tendering
A contractor may be selected by a client because of previous successful performances, on the recommendation of others, or because of experience in a specialised field. Negotiations take place to resolve a price for doing the work, the form of contract to use, and to agree a programme for the work and the method by which the work should be carried out. Many of the medium sized local family businesses obtain their work by negotiated tender.

Open tendering
This traditional method of tendering is used by a client to achieve, in his/her mind, a completely free and fair way of offering work to local or other contractors. It may be used to show there is no favouritism within the business community. Because local authorities are accountable for public expenditure they must be seen to be getting value for money, and, therefore, many still operate such a system of tendering whereby the lowest bid from any contractor who cares to submit one would generally be accepted.

An advertisement inviting contractors to apply for tender documents is normally placed in a local or national newspaper, or trades journal. Each applicant may be asked to pay a deposit for the contract particulars and documents to deter them from frivolous application, and to recover some of the costs for the documents' production. In an open tendering situation numerous contractors may be interested and the documents produced for each could be stated as being an unnecessary added expense for the client.

On receipt of the documents the bills of quantities would be priced and tenders would be submitted by each interested contractor within a predetermined time limit, and the lowest tender would generally be accepted provided that:

(*a*) the contractor had sufficient resources to do the work (a performance bond is required in many instances by the client);

(*b*) the contractor was reputable, having completed similar work successfully in the past.

The open method of tendering requires that all the project particulars (design and documentation) are completed first.

Selective tendering

As an alternative to placing advertisements in newspapers, etc., a client could decide, with advice from the architect, on a select number of contractors (usually not more than six) whom he/she thinks would be most suitable for the proposed work. Letters of preliminary enquiry for invitation to tender would be sent to the initially selected contractors, and on receipt of favourable replies the appropriate tender documents would be issued and the contractor would then price the work as in open tendering. The advantages of selective tendering are:

1. It is a cheaper method than open tendering because fewer sets of tender documents are sent out and deposits therefore would not be required. It may not be necessary to insist on a performance bond because of the known reputation of the successful tenderer.
2. That proven performances by the selected reputable contractors means, generally, that a good service will be given.
3. That it eliminates the unscrupulous contractor.
4. That it saves time and gives a contractor a fairer chance to price the work more competitively with a better chance of being successful in a bid for the work.
5. That drawings and documents would be completed by the designer which assists the contractor in pricing the work accurately with less need to add-on for the unforeseen, thereby giving the client a fair deal.

Rotation tendering

To be included on a local authority list (usually by smaller specialist contractors – painting and decorating, etc.) means that ultimately as jobs become available the first few names from the list would be invited to tender for the work – the contractor with the lowest bid would be offered the contract. When another job becomes available the next few names on the list would similarly be invited to tender, and so on. This appears to be a fair way of sharing work out to local contractors.

Serial tendering

Once again this follows the principle of selective tendering. Between six and ten contractors may be invited to tender for work competitively using master bills of quantities for a first-stage contract. The initial tender offer for the first contract would then form the basis for a series of subsequent contracts. Further contracts would be offered later by the client of which the prices previously laid down in the master bills, and agreed, would be used after generally making allowances for inflation and fluctuations in costs (labour, plant, materials, etc.) since the award of the initial contract.

Where a client requires that a number of contracts be started and completed within certain time limits then serial tendering could be used. This system allows the contractor to participate with the architect in the design stages of subsequent projects. It also gives the contractor continuity of work, thereby allowing him/her to reduce costs to the advantage of both parties to the contract.

Performance and continuity tendering

Contractors are selected to competitively tender for work on the understanding that further contracts would be awarded to the contractor who submits the lowest tender, and provided that the performances on the first and each subsequent contract are satisfactory.

It must be appreciated that the subsequent projects need to be of a similar nature, as the first successfully completed one would determine the standards for the projects that follow.

Package deal tendering

Invariably a client who wishes to develop land, etc., selects two or more contractors and invites them to submit designs and layout drawings, with prices, of how the land in question should or could be developed. The contractor who submits the best competitively priced design with a back-up service to suit would be awarded the contract. There would be only two parties to the contract – the client and the contractor. The contractors who tender for the work have a considerable cost burden because, generally, they use their own design service department, and therefore the rewards on successfully bidding for the work must be substantial to justify the cost risk. The client, therefore, if he/she is to ensure that the contractors competing for the work are suitably interested and motivated to submit well-thought-out schemes at an acceptable price, should only invite between two and four to submit schemes and tenders.

Although there is no architect to look after the client's interests there are safeguards generally laid down in the contract document for dealing with alterations in design once work has commenced at site level, and for fluctuations, if necessary.

Prime cost contracts (cost reimbursement)

Where an early start is required on a project and, therefore, the early selection of a contractor is necessary before all the details have been finalised by the designer, this type of contract may be introduced. It allows for the contractor to be reimbursed for his/her prime costs (total costs incurred during the execution of the project, i.e. labour, plant and materials, etc., costs) and for a fee for managing the contract in the form of overheads and profit. This management fee is usually based on a percentage of prime costs, or as a lump sum on the estimated contract costs.

The tendering competition between selected contractors would be on the basis of the management fee – the lowest management fee submitted by a contractor would generally be sufficient for a successful bid for the contract.

Target cost and guaranteed maximum price contract

While this contract system is similar to the prime cost contract it is modified to provide an inducement to a contractor to control costs and complete the work as quickly as possible. Provisional target costs and target times are set, and by agreement, if the contract prime costs exceed the target cost, then the management fee would be reduced by an amount worked out by a formula. Similarly, where savings are made both the client and the contractor would share in the good fortune.

A maximum price (guaranteed maximum price) for the contract can be agreed so that the client knows the absolute limit of his/her expenditure for the project.

Tenders using bills of approximate quantities

These types of bills are used where drawings for a project are complete but some of the other details have not been finalised at the time of tender. The use of such bills of quantities gives an estimate of the work to be undertaken. Also, when the bills are priced, although the quantities are not complete, the unit and other rates included by a contractor in the approximate bills of quantities are used in determining the final payments made for work done.

As usual, in contracts of this nature, very careful remeasurement of work completed should be undertaken, recorded and be agreed with the client's representative at every stage of the work to ensure adequate claims for payment are made.

Two-stage tender

If the drawings, etc., have not been finalised contractors may be selected by the client's representative to tender for a contract after initial careful negotiations with each on how the proposed project should be undertaken. The selected contractors would then be invited to competitively tender for the work by pricing one of the following:

(*a*) approximate bills of quantities;
(*b*) bills of quantities of a similar past project;
(*c*) a fictitious/notional bills of quantities.

The contractor who submits the most favourable prices (tender) would then be expected to work closely with the designer to agree on an economic design and programme until a satisfactory solution to the client's needs are realised. A final tender would then be submitted by the selected contractor using the bills' prices previously outlined in the successful competitive tender.

3.2 Contract/project documents

Depending on which system of tendering is used, contractors who are interested in placing a bid for work would generally require to receive from the client or his/her representative most, if not all, of the following documents and particulars:

● production drawings;
● specifications;
● bills of quantities;
● tender form;
● appropriate standard form of contract (contract particulars);
● a list of nominated subcontractors and suppliers.

Production drawings

These drawings are produced by the architect/designer, and on major works other professional individuals or organisations could be involved producing details relevant to their specialism but on the direction of the architect/designer. These individuals could be:

Structural engineer.	Heating engineer.
Electrical engineer.	Ventilation engineer.
Mechanical engineer.	Acoustics engineer.
Communications engineer:	Drainage engineers.
The Post Office/	Other engineers:
Telecommunications.	Fire.
Personnel (lifts).	Security.
Goods (lifts, conveyors).	Refuse disposal, etc.
Computer.	
Gas engineers.	
Water engineers.	

The production drawings, which are essential for preparing bills of quantities, should be made available to contractors at the tender stage, and should give a clear portrayal of the magnitude of the project.

Specifications

This document is prepared to specify the expected quality of materials and workmanship and protection to be afforded to the works. British Standard Specifications (BSS) and British Standard Codes of Practice (BSCP) pertinent to each trade or area of work would be quoted within the specification. A contractor, by reading this document, would get a good idea of the standards expected by the architect and client and would determine the rates for each job depending on the degree of supervision necessary to achieve the standards laid down. Also, higher paid quality workers may have to be employed for certain operations which naturally would affect the pricing, as would the necessity to use better quality materials, fittings, components, etc. Specifications are either included as a bill in the bills of quantities, or are written in a separate document each of which should then be read in conjunction with the bills and other contract documents.

Bills of quantities

Assuming that bills are produced they should accurately represent the quantity of work to be undertaken on a proposed project and should be prepared meticulously by the quantity surveyor. If items of work are overlooked or quantities are incorrect in the bills, then the contractor would expect to do the additional work, etc., as variations. This is inconvenient and a nuisance because the extra work was not planned and causes delays and leads to increases in claims having to be made by the contractor's representatives.

Each type of work to be executed by different trades is shown in a separate bill of the bills of quantities. These range from demolition and alterations, excavations and earthworks, piling, concrete work, brickwork and blockwork, down to painting and decorating, drainage and fencing.

Tender form

This form is filled in by the contractor when submitting his/her tender figure to the architect/client. The tender figure includes the sum of the contractor's estimate for doing the work (labour, plant and materials) with additions for overheads and profit.

Standard form of contract

There are numerous standard forms of contract used in the building and civil engineering industries such as the following:

1. Joint Contracts Tribunal Forms of Contract (JCT), used normally where an architect is involved as a representative of the client:
    (a) the Standard Form of Building Contract – Private Edition, with Quantities. (Used for main contracts where there are drawings, quantities and specifications, etc.)
    (b) the Standard Form of Building Contract – Private Edition, without Quantities. (Used for small contracts where there are drawings and specifications.)
    (c) the Standard Form of Building Contract – Private Edition, with Approximate Quantities. (Used for contracts where some of the details are not finalised.)
    (d) the Standard Form of Building Contract – Local Authority Edition, with Quantities. (Used for main local authority contracts where there are drawings, quantities and specifications.)
    (e) the Standard Form of Building Contract – Local Authority Edition, without Quantities. (Used for small contracts where there are drawings and specifications.)
    (f) the Standard Form of Building Contract – Local Authority Edition, with Approximate Quantities. (Used for local authority contract where some of the details are not finalised.)
    There are various Supplements which may be used with the previously mentioned JCT Forms,

i.e.: fluctuations, sectional completion and the contractors design portion supplements.
    (g) the Fixed Fee Form of Prime Cost Contract of the JCT. (Used where all the details for a project have not been finalised but the client expects the work to be started early, and the contractor is to be reimbursed for prime costs plus a fixed management/profit fee.)
    (h) the Standard Form of Agreement for minor Building Works of the JCT. (Used for small contracts where drawings and specifications have been prepared.)
    (i) JCT Form with Contractors Design.
    (j) JCT Nominated Sub-Contract (Form NSC 1, 2, 3 & 4).
    (k) JCT TNS Forms (Tenders for Nominated Suppliers).
    (l) Form of Agreement between the Employer (Client) and Nominated Subcontractor, for use in conjunction with the JCT Standard Forms of Building Contract.
    (m) JCT DOM/1 and JCT DOM/2 Forms. The use of these standard forms ensures binding contracts between the main contractor and his/her sub-contractors. DOM/1 forms (with the Sub-contractor Conditions) are used where the contract between the main contractor and client is of the normal JCT 1980 type. DOM/2 forms are used when the main contract is of the JCT Contractors Design.
2. GC/Works/1 or GC/Works/2 General Conditions for Government Contracts for Building and Civil Engineering works (major and minor works forms).
3. Conditions of Contract and Form of Tender, Agreement and Bond for use in connection with Works of Civil Engineering Construction (better known as the ICE Conditions of Contract).
4. Form of Subcontract for use with the ICE Conditions of Contract (issued by the Federation of Civil Engineering Contractors – FCEC).

If one considers the Standard Form of Building Contract – Private Edition, with Quantities there are four main sections as follows:

(a) Articles of Agreement – this provides for the names and addresses of the parties to the contract to be shown; the site address, date of contract, etc.
(b) Conditions of Contract – this is divided into forty clauses which lay down the rights and obligations of the parties to the contracts.
(c) Appendix – this contains details regarding percentage additions on prime costs for dayworks, defects liability periods, dates of possession and completion of contract, liquidation and ascertained damages, periods of interim certificates, percentage retentions, etc.
(d) Supplemental Agreement – This is the Value Added Tax (VAT) Agreement. It makes provision for the contractor to recoup any VAT paid on materials, etc., from the client at interim periods.

A contractor's cash flow, expected during a contract, can be determined from the particulars given in the Conditions of Contract and Appendix, and from priced bill of quantities. Also, the contractor can plan the finances needed to run the contract and determine how cautious to be when pricing the work for a tender – especially if the contract is to be undertaken on a fixed price, or depending on which fluctuation clause is to apply.

A list of nominated subcontractors and suppliers

If a client or designer is to insist on certain subcontractors and suppliers then a list of their names, addresses and field of activity should be issued to contractors at the pre-tender stage. The main contractors can then make the necessary allowances to the sections of the bills of quantities which are affected by the nominations (allowances such as profit, general and special attendances). Also, later, when the successful contractor is awarded the contract, the nominated subcontractors and suppliers would be expected to sign a subcontract etc., Standard Form with the main contractor, agreeing to the contract conditions.

3.3 Project drawings

The number of drawings expected for any one project is difficult to determine. A thoughtful and efficient architect/ designer would produce a sufficient number of project drawings to enable all operations at site level to proceed smoothly. A less well-organised architect/designer who, perhaps, has fewer facilities available to him/her, may produce the minimum of drawings with the result that enquiries from the contractor's site manager will be constantly received with a loss of confidence and to the detriment of a good working relationship. The delays cause the site manager to become frustrated and may lead to claims for 'extension of time' caused by late instruction, which adds to the costs.

Where satisfactory thought has been given to the production of a set of drawings there should be an optimum number which gives sufficient details on each aspect of the proposed structure.

If information is shown clearly on a small scale drawing there would certainly be little achieved if the same information was drawn to a larger scale. This is unnecessary duplication, and because of the extra drawings produced additional storage space, codification and time to retrieve details would be required, leading to less efficiency at site level, and added cost to the designer. Sets of drawings are generally divided into the following groups:

1. Location drawings (general arrangement drawings).
2. Assembly drawings (detailed drawings).
3. Component drawings (detailed drawings).

1. Location drawings (see Fig. 3.3.1 (A))

This group of drawings shows information relating to:

(a) *The site and external works:* These include sizes of the site, trees, drives, drainage and other services, and the location of the site, i.e. key plans and block/site plans to minimum scales of 1 : 2500 and 1 : 1250, respectively.

(b) *The buildings:* These include dimensions of the parts and spaces (rooms) within, and show the shape of plans, elevations and sections to a minimum scale of 1 :100. Also, grids and reference points, door swings and setting-out dimensions would be included.

(c) *The elements:* These include the dimensions and positions of the elements of the building, i.e. walls, floors and roofs, etc.

(d) The references to larger scale drawings for assembly and components which are drawn to scales 1 : 50 to 1 : 10, and 1 : 10 to F.S., respectively.

2. Assembly drawings (see Fig. 3.3.1 (B))

These drawings show information relating to the assembly of parts of elements including the shape and size of those parts. Also, details of the junction of one element with another element would be shown (the jointing of a floor to the walls, or wall to a foundation).

Additionally, there would be details shown which refer, by cross referencing, to other assembly and component drawings.

3. Component drawings (see Fig. 3.3.1 (C))

These drawings show information relating to the elements and sub-elements of a building. The drawings are usually produced for components which will be made in a factory, workshop or work area away from the building, or to show the *in situ* assembly of component parts which cannot be clearly shown on assembly drawings. Component parts of a building may be doors, windows, reinforced concrete beams, etc.

A fourth group relating to a set of project or working drawings is the schedule. This groups of documents is not always used, but when it is, particularly on large projects, it shows a list of details of elements, components or other products which occur repetitively and in variety, and refers also to the part of the location drawings where each can be located. The schedule contains particulars of windows and glass, door and door frames, ironmongery, room finishes – floor, wall and ceiling, drainage, fittings and fixtures, etc.

Numbering/coding of drawings
The 'traditional method' of numbering/codifying drawings for storage and retrieval purposes, on or off site, depends on whether they were produced by the structural engineer or services engineer. The consecutive method of numbering has been, and is still, found to be quite suitable for small jobs. The details normally shown on drawings would be:

1. A short but concise **title** e.g. foundation details.
2. Name of designer (architect, structural engineer, etc).

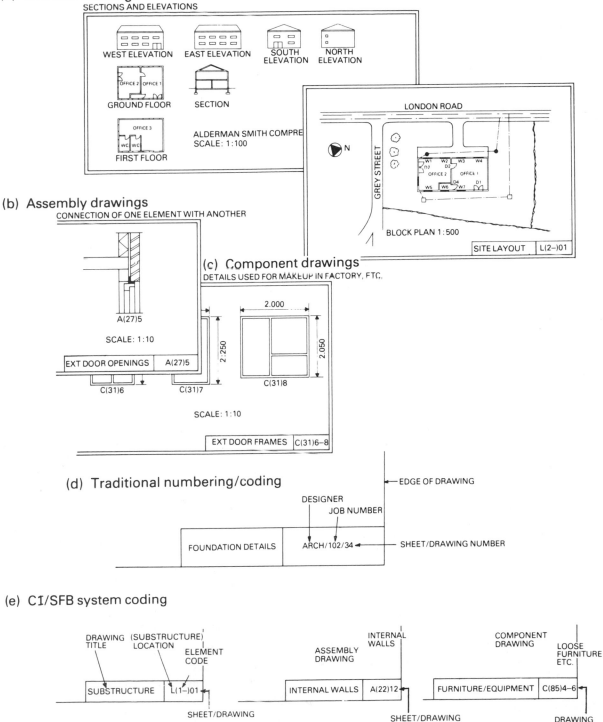

Fig. 3.3.1 Types of drawings in a set and their coding

3. Job number (could be different job number for the architect's office or structural engineer's office).
4. Sheet number (see Fig. 3.3.1 (D)).

There are many traditional variations to the above, and sometimes the site manager renumbers the drawings as he/she receives them from the different sources (architect, structural engineer or services engineers). If ten drawings were received from the architect and then an eleventh one came from the structural engineer – the drawing from the engineer would be numbered eleven, and so on, no matter from which sources the drawings came.

The Drawings' Register would be used to record the drawings received, stating:

(a) the number of drawings received;
(b) from whom the drawings were received;
(c) the date they were received;
(d) the number of copies of each drawing received;
(e) the date each drawing was amended or superseded;
(f) the scale of each drawing.

Separate Drawings' Registers may be used to record drawings received from each source – a Register for architect's drawings, one for structural engineer's drawings, and so on.

The CI/SfB system of codifying drawings is recommended for large jobs, and if each specialist designer uses it the drawings can be filed easily on-site by the site manager/supervisor. A set of project drawings are divided into groups (location, assembly and component), as described at the beginning of this section and in Fig. 3.3.1. For each drawing which is produced a suitable Title should be allocated which should be carefully chosen bearing in mind the reason for a particular drawing's production and what the designer is trying to communicate to the builder.

If the CI/SfB Project Manual is used by each designer the allotted code number can be extracted from Table 1 to be incorporated on the appropriate Group Drawing.

The details shown on the drawings, using the CI/SfB System, would be (see Fig. 3.3.1 (E)) as follows:

1. The Title of a drawing.
2. The Group letter (L = location, A = assembly, and C = Component).
3. The Element Code (from CI/SfB Project Manual). The Code is normally a number in brackets.
4. The Sheet/Drawing Number of the appropriate Group.

A typical page from a Drawings' Register can be seen in Fig. 3.3.4 (A).

Filing of drawings

The filing of sets of drawings should be in sequence using the CI/SfB numbering/coding as used by the designer. They should be filed in the groups as recommended by this system.

The storage of drawings on site may be by one or more of the following methods.

(a) laid flat in a drawings chest;
(b) rolled up and in racks;
(c) vertically hung between brackets;
(d) folded carefully as per BS 1192, and stored in suspended files in metal cabinets.

Letters (a) and (b) are the most commonly used, examples of which can be seen in Fig. 3.3.4 (B).

Drawings symbols

When reading drawings site managers and others should have an appreciation of the BS 1192: 1987 and other relevant British Standards which indicate the symbols and hatchings used to portray materials, elements, appliances, etc. The other British Standards are BS 308 for engineering drawings; BS 1553 for heating and ventilation; BS 1635 for fire protection installations; and BS 3939 for electrical installations, etc. (See Figs. 3.3.2. and 3.3.3.)

3.4 Reference sources on-site

In an attempt to maintain efficiency on-site the supervisor should consider how information should be stored and of what type. The information which is required regularly, day-to-day, is generally stored in the site supervisor's head. Information which is required less frequently may be recorded using a system which allows for immediate retrieval, taking seconds. Usually detailed, typed sheets which are pasted on to hardboard to protect them against damage are used showing information such as the more regularly required telephone numbers, instructions, rules or regulations (see Fig. 3.4.1). Other information which is required periodically may be stored in loose-leaf folders, lever arch files, on shelves or in drawers. Manufacturers' literature, technical data, other relevant documents, and even the telephone directories contribute as a back-up service for the site supervisor to use as and when the need arises.

Site supervisors are advised to collect suitable information for quick reference as soon as possible on the award of a contract. The information may be used to instruct, notify or advise individuals, and determine solutions to problems. The library of administration and other information is the memory of the site manager and other site staff, and invariably it allows prompt action and decisions to be taken for the efficient management of the site.

Although the following details are by no means a complete list they should be given consideration as a back-up to effective site administration (see also Fig. 3.4.1).

Telephone numbers and addresses

Project team
Head Office   Architect (and Clerk of Works or visiting engineers).   Quantity Surveyor.   Structural Engineer. Services Engineer.

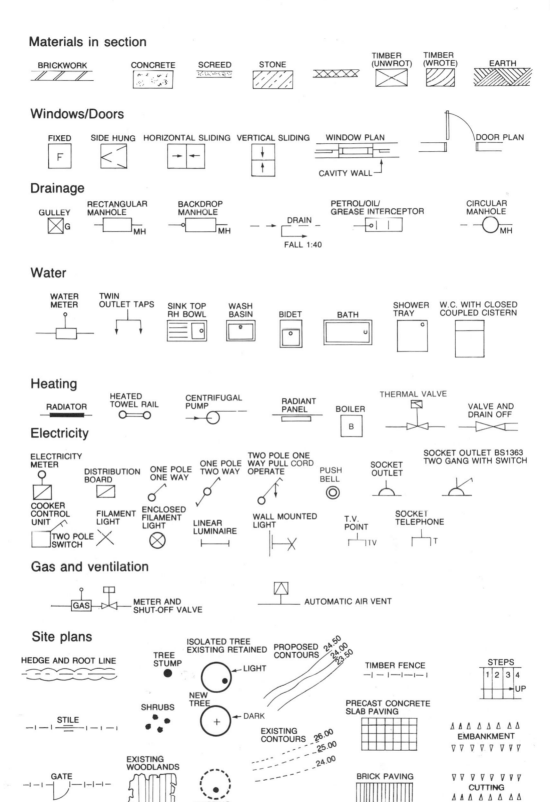

Fig. 3.3.2 BS Building Drawing Symbols

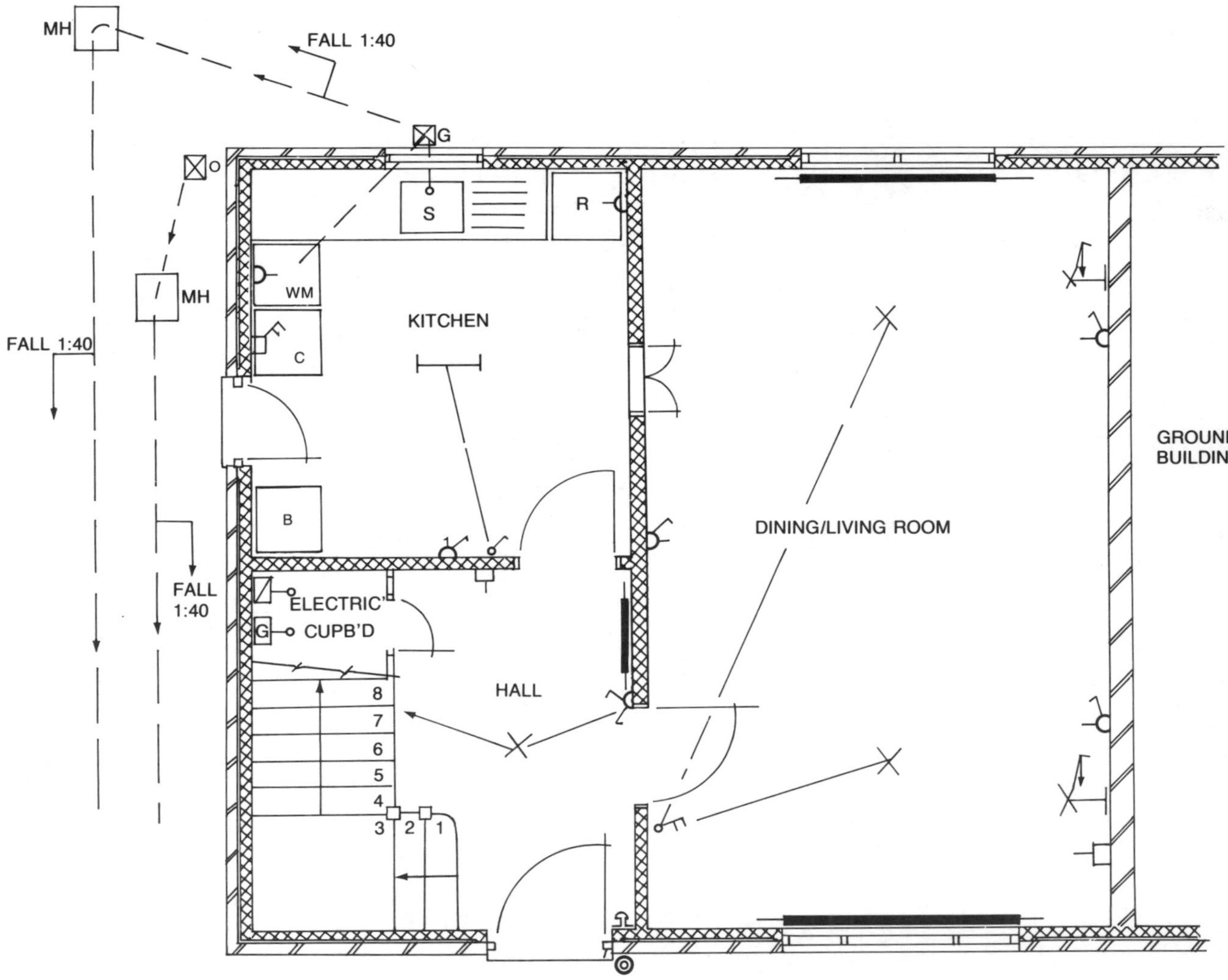

34

GROUND FLOOR
BUILDING PLAN

DINING/LIVING ROOM

KITCHEN

HALL

R

S

WM

C

F

B

ELECTRIC
CUPB'D

G

G

MH

MH

FALL 1:40

FALL 1:40

FALL 1:40

1
2
3
4
5
6
7
8

Extracts from BS 1192: Parts 1, 2, 3 and 4: 1987 and reproduced by kind permission of the British Standards Institution
Note: Other symbols recommended by the B.S.I. can be found in BS 308, 1553, 1635, 3939, etc.

Figure 3.3.3   Graphic Symbols for Construction
Drawing Practice (BS 1192).

Industrial relations
Full-time union officers of each union represented on-site. Local/Regional Council of the National Joint Council for the Building Industry (NJCBI).

Local authority
Planning Department. Building Control Department Highways Department. City Engineer. Environmental Health Department. Petroleum Officer. Police Station.

Health and welfare, etc.
Ambulance Authority and casualty department of the local hospital. Doctor. Fire Officer and Station. Factory Inspector. Environmental Health Dept.

Utility services organisations
Local Electricity Board office. Local Gas Board office. Local Water Board office. British Telecom

Employment
The Job Centre. Local newspaper – advertisements.

Testing
Independent Testing Laboratories. Colleges of Technology/Polytechnic/University.

Advisory bodies
Meteorological Office or Local Weather Centre. Concrete Society (CS). BEC (BAS Consec, etc.). NHBC. ROSPA.

Advisory library services
BRE. DOE/PSA. Professional institutions (CIOB; RIBA; RICS, etc.).

Research bodies
BRE. Construction Industry Research and Information Association (CIRIA). British Board of Agrément. TRADA etc.

Other bodies
The Building Centres. The Building Bookshop, London. Her Majesty's Stationery Office (HMSO).

Subcontractors
Domestic subcontractors. Nominated subcontractors. Labour-only subcontractors. Named subcontractors

Suppliers
Domestic suppliers. Nominated suppliers. Plant hire firms.

Documents

Contract documents
Bills of Quantities. Specifications. Standard Form of Contract. General Arrangement and Detail Drawings. Schedules.

Control documents
Method Statement. Programme and Progress Charts. Site Layout Plan. Site Diary. Drawings' Register. Site Structure Diagram. Company (Site Manager's) Handbook – policies and guide to dealing with problems on site.

Special documents
British Standards Specifications (relevant ones). British Standards Code of Practice (relevant ones). Working Rules Agreements (relevant ones). Construction Regulations – Lifting Operations Regulations, Health and Welfare Regulations, General Provisions Regulations, and Working Places Regulations. Also other relevant safety, health and welfare Regulations and Acts of Parliament. (See Chapter 10). A first-aid instruction book. Job guides/Instructions. Building Regulations.

Manufacturers'/Suppliers' literature
Price lists for material and plant. Other informative literature.

Text books on:
Technology. Management. Services. Surveying and setting out. Safety.

The Building Regulations

While some form of building control has existed for a number of centuries the Building Regulations of England and Wales were introduced shortly after the Public Health Act 1961, which empowered the then Minister of Housing and Local Government to prepare national regulations to supersede the separate Building Byelaws produced by each local authority. The Building Regulations 1984 apply at the present, but new Regulations come into force in 1992.

The regulations lay down rules to provide safe and healthy accommodation/buildings for the public. On approval being given to a designer's/client's design and application to build, a Technical Services Department's Building Controls Officer of a local authority would then expect notifications to be received from a contractor to inspect each stage of the construction work in the order as stated in the Building Regulations. The Building Controls Officer's inspection would be to ensure that the regulations were not being contravened by either the designer or contractor. Notification cards, provided by the Technical Services Department, are used by the contractor (see Fig. 4.4.1).

The Building Regulations lay down standards by which the materials to be used in structure must conform. The regulations also state the standards expected regarding: structural stability of a building, safety in fire, thermal and sound insulation, stairways, refuse disposal, open spaces

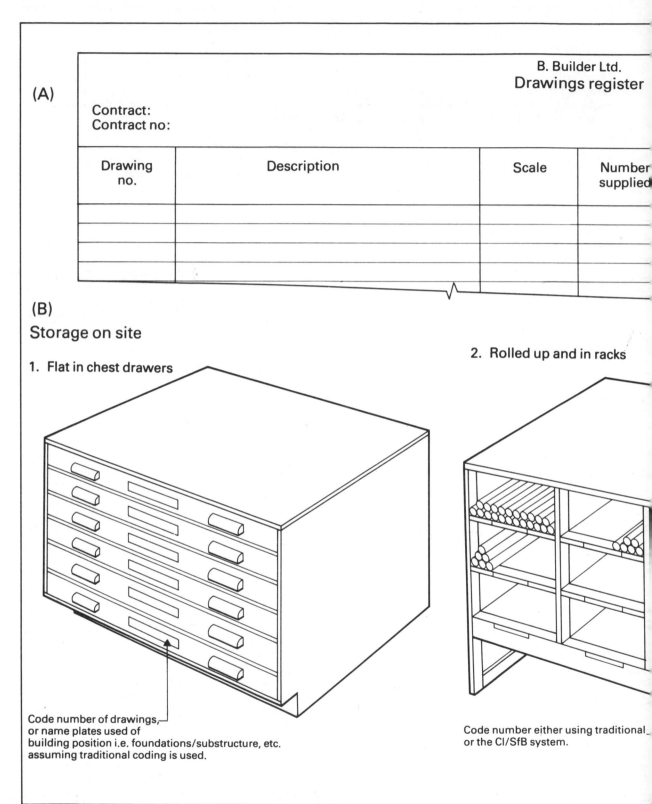

(A)

		B. Builder Ltd. Drawings register		

Contract:
Contract no:

Drawing no.	Description	Scale	Number supplied

(B)

Storage on site

1. Flat in chest drawers

2. Rolled up and in racks

Code number of drawings,
or name plates used of
building position i.e. foundations/substructure, etc.
assuming traditional coding is used.

Code number either using traditional
or the Cl/SfB system.

Fig. 3.3.4

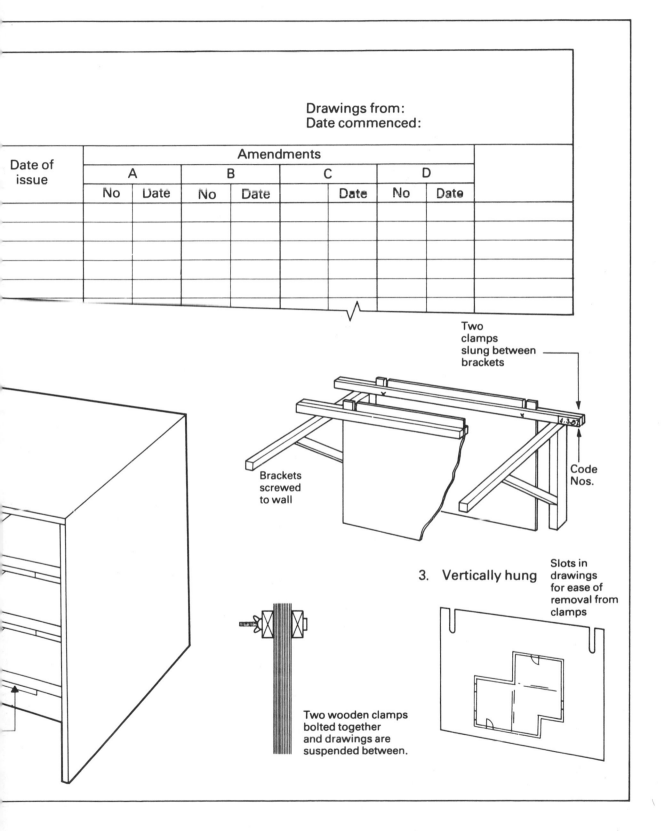

Date of issue	Amendments								
	A		B		C		D		
	No	Date	No	Date		Date	No	Date	

Drawings from:
Date commenced:

Two clamps slung between brackets

Brackets screwed to wall

Code Nos.

3. Vertically hung

Slots in drawings for ease of removal from clamps

Two wooden clamps bolted together and drawings are suspended between.

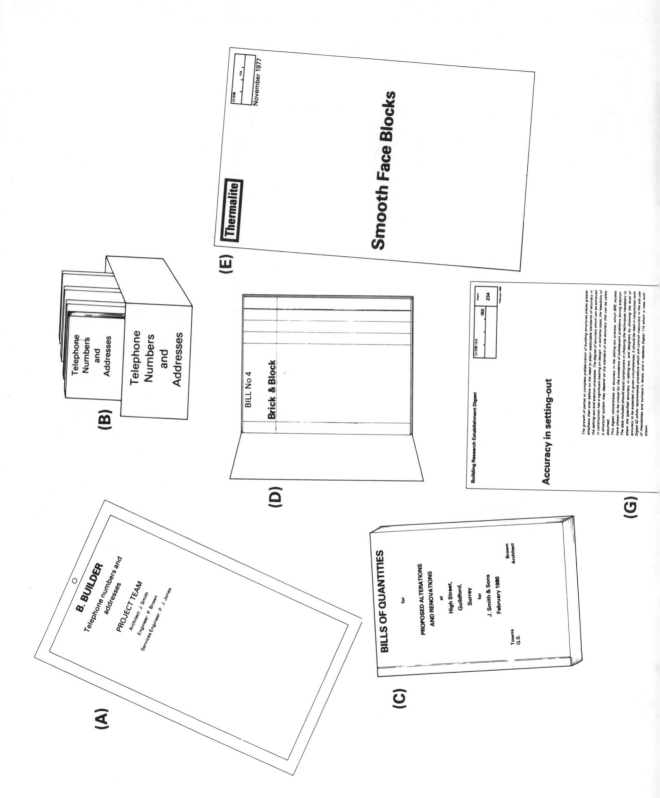

(A)

B. BUILDER

Telephone numbers and addresses

PROJECT TEAM

Architect: J. Smith
Engineer: P. Brown
Services Engineer: P. J. Jones

(B)

Telephone Numbers and Addresses

Telephone Numbers and Addresses

(C)

BILLS OF QUANTITIES

for

PROPOSED ALTERATIONS
AND RENOVATIONS

at

High Street,
Guildford,
Surrey

for

J. Smith & Sons

February 1980

Brown
Architect

Towns
Q.S.

(D)

BILL No 4

Brick & Block

(E)

Thermalite

Smooth Face Blocks

November 1977

(G)

Building Research Establishment Digest

234

Accuracy in setting-out

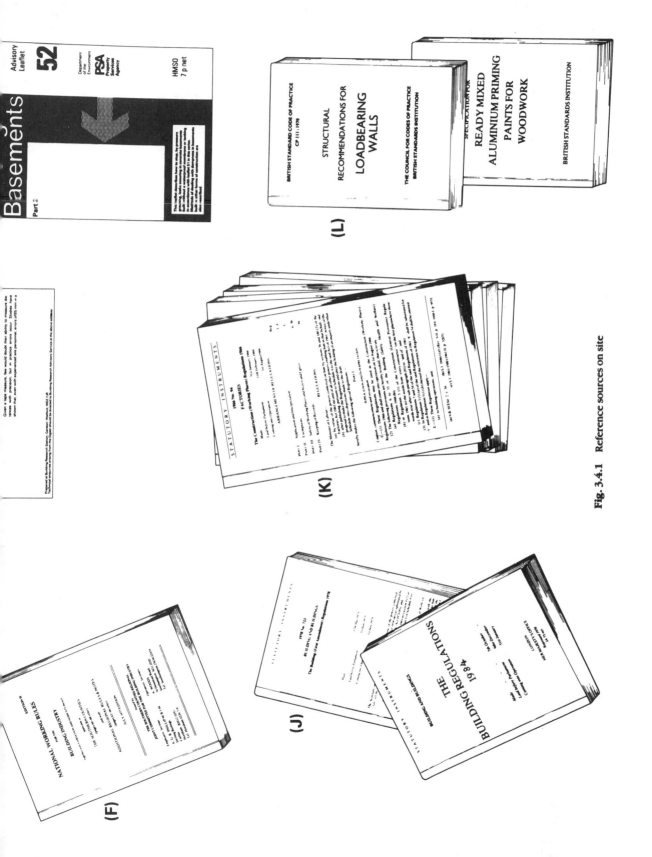

Fig. 3.4.1   Reference sources on site

around a building, and ventilation, chimneys and fireplaces, installation of heat-producing appliances, drainage and sanitation, etc.

The regulations can be obtained from HMSO.

Operatives employment conditions

Site managers should be conversant with the employment conditions of operatives/employees which are either laid down by the employer or, where agreements exist with the employees' trade unions, are detailed in the relevant Working Rules. Also, the most basic knowledge of the various employment legislation would help to prevent mistakes being made in dealing with personnel problems at site level (see Chapters 5 and 6).

The Working Rules Agreements are modified at regular intervals, however, as new legislation is introduced and to incorporate new agreements reached between the employers' and the employees' representative organisations. The Working Rules in the building industry are agreed by the NJCBI. There may be agreements at regional or local levels. The Working Rules for the civil engineering industry are agreed by the Civil Engineering Construction Conciliation Board (CECCB), whose members are representatives from the employers' and employees' organisations associated with civil engineering work.

The Working Rules Agreements for the building industry covers the following details:
● Wages; ● Working hours; ● Guaranteed weekly wage; ● Conditions of service and termination of employment;
● Extra payments for –
Discomfort, inconvenience and risk; Continuous extra skill; intermittent responsibility; tool allowance; special provision on servicing of mechanical plant; storage of tools; clothing;
● Overtime and holidays; ● Shift work and night work; ● Travelling and lodging; ● Trade Union recognition and procedures; ● Payments for absence due to sickness or injury; ● Grievances, disputes or differences; ● Register of Employers; ● Death Benefit Cover; ● Other relevant guides, etc.

Note: Similar arrangements exist within the Working Rules Agreement of the Building and Allied Trades Joint Industrial Council (BATJIC) which was created by the Federation of Master Builders (FMB) and the Transport and General Workers Union (TGWU) as a break-a-way Council from the N.J.C.B.I. which, it was believed, was dominated by the larger construction companies at the expense of the smaller companies.

Chapter 4
Supervisor's communications with those outside the site team

4.1 Communication with head office regarding site records and progress

It is beneficial to both head office and the site supervisor that there is a two-way communication to show how work is progressing; what problems have developed; what information is required to ensure continuity of flow of work; what materials and equipment are required to maintain progress, etc. Most of the communication flows from site because of the need to meet deadlines, while the remainder is to keep management informed of every reasonable detail so that they are as well placed to deal with an emergency as the site supervisor.

For the purposes of communications it is customary for someone to design suitable standard forms to be used in as many situations as is necessary and which everyone finds easy to understand.

Standard forms, used within an organisation, help to reduce thinking time and simplify clerical/administrative work. The simplest way to present information is to use a well designed form which enables individuals to be able to write down details without the need to think of the best way to present them.

Well-designed standard forms are used for the following, to: 1. inform. 2. notify. 3. request. 4. instruct. 5. advise. 6. collaborate. 7. report.

Well designed forms should:

(a) allow information to be easily illustrated;
(b) simplify typing or clerical duties;
(c) be designed to allow as much vital and relevant information to be presented as is necessary;
(d) allow for ease of reading and extraction of information;
(e) be designed to eliminate the unnecessary.

Standard forms enable all the various personnel within a firm to work more efficiently, and in each there is information which someone needs and from which he/she can improve or regulate conditions, etc., e.g.

● An estimator needs feed-back as a check on the rates originally worked out, and if costs are more than estimated the adjustments can be made for future use, or alternatively, management action can be taken to reduce costs by the application of different work techniques, construction and organisation.

● A plant manager would expect details of performances of plant and equipment. Plant, etc., which are inefficient due to break downs can be overhauled or sold off.
● A purchasing officer may wish to know what materials have been used and in what quantities to check if there have been excessive amounts acquired compared to those specified. Also it gives some control over wastage, pilfering and theft, etc.

Whenever forms are completed on-site there should always be a duplicate site copy retained for reference by the site supervisor or his/her staff. The following are but a few of the more important forms used at site level.

Site diary and daily report

Some firms have two separate documents – a site diary and a weekly report. The site diary is completed daily and is retained on-site for future reference, especially where queries arise or problems develop regarding promises made by individuals or firms, stoppages caused by disputes or extreme weather, dates of telephone calls, visitors to site, etc. A weekly report is an analysis by the site supervisor of the week's events extracted from the site diary, and is sent to head office for reference.

If, however, a combined site diary and daily report is prepared by the site supervisor which is sent on a regular basis to head office (loose leaf sheets used with carbon paper), there would be more than ample information for the personnel to work from. Also, by this combined daily system the site supervisor could dispense with the need to prepare a weekly report (see Fig. 4.1.1).

Material requisitions

See Fig. 4.1.2 and Fig. 7.1.3 for alternative designs for a suitable Materials Requisition Form. When materials are required on-site the site supervisor can call them forward by using the standard form shown. A minimum of two copies would be required: one to be sent to the purchasing department, the other to be retained on-site. The head office purchasing department would extract the details about the materials from the requisition form, and would confirm the delivery dates with the supplier (having already ordered key materials in bulk) on the award of the contract by the architect. Where a requisition is received from site of other materials the centralised purchasing department would be able to place orders obtaining the

42

B. BUILDER, LTD, MERROW, GUILDFORD.

SITE DIARY AND DAILY REPORT

Contract ....   Contract No: ....   Week ending ....
....   No. ....

Weather: Temperature   AM.
                                                PM.

Labour on-site			Labour required (state by which date)		
Own labour	Subcontractors	Nominated subcontractors	Own labour	Subcontractors	Nominated subcontractors

Overtime   Reason:

Man hours:

Job stoppages: Reason:

Man hours lost:

Drawings received   Required by   Information received   Required by

Visitors to site: Telephone calls   to:

                                                                  from:

Accidents: Were reports made?

VO or instructions   Plant required

                                           Plant received

Brief Report of work and progress and any difficulties, etc.

Signed
        Site manager ....   Contracts manager ....
Top copy to be sent to head office and the duplicate should be retained on-site

Fig. 4.1.1

	B. BUILDER, LTD, MERROW, GUILDFORD. ## MATERIAL REQUISITION	NO ...4.........
CONTRACT:	Wholesale Shop & Offices ...... NO 32...... London Rd, Guildford.	DATE 3rd Feb. 19....

BQ, AI, VO or DWG ref.	DESCRIPTION	QUANTITY
B Q 5/39 " 5/4 F " 5/4 c	Lintols - Dorman Long (0.950m) " - " " (1.950m) " - " " (1.200m)	7 No 2 No 1 No
B Q 7/5 G " 7/5 H	Boulton & Paul Standard Ext Door Frames (3NF) " " " " " " (3xF)	1 No 2 No
— — —	Half glazed hardwood doors (rebated) (5Y) Solid hardwood doors (rebated) (2P) Half glazed hardwood door (5x)	1 pair 1 pair 1 No

Delivery date/s:	All items to be delivered by 21st March 19...... except the doors these are wanted by 18th April 19......

Site manager's remarks about delivery restrictions, etc.	Diagram (if appropriate)
Buyer's remarks about additional cost recovery.	
	SIGNED ...... H. Helen. ......

Fig. 4.1.2

best rates and discounts. Sometimes, however, the site supervisor is given a list of suppliers from which to order direct. A pad of purchase order forms would then be issued to him/her. It is of course a recommendation that materials are purchased centrally – requisitions having to be made from site, which enables stricter controls on the use of materials, thereby eliminating the possibility of excessive ordering.

Advice of variations

Each firm should have its supply of standard forms called Advice of Variations with a pad being issued to each site for use by the site supervisor when he/she receives verbal instructions from either the architect or clerk-of-works.

A form is completed (in duplicate) by the site supervisor for each instruction received, and the top copy is sent to head office or is given to the contracts manager to enable written confirmation to be sent to the architect within seven days – as laid down in the Joint Contracts Tribunal Form of Building Contract. The architect should then issue an officially written architect's instruction within a further seven days. See Fig. 4.1.3. for the form which is sent by the site supervisor to head office.

Variation orders for subcontractors

If the main contractor is issued with an architect's instruction to do work or supply materials, etc., which differ from that which is laid down in the contract documents (bills of quantities, drawings, specifications) the instruction may also affect the services of a subcontractor. If so, a variation order for subcontractors would be completed and handed to the appropriate subcontrator for action to be taken. It would be prudent for the head office surveyor to issue the order, but sometimes the surveyor is site-based – therefore a copy would be retained on-site of the variation order submitted to the subcontractor and another copy would be sent to head office (see Fig. 4.1.4).

Daywork sheets

Daywork sheets are completed by the site supervisor each week that work is undertaken for the client and which is to be paid for on a daywork basis. Written instructions would have to be received from the architect for any dayworks which are to be carried out. The site supervisor should record on the daywork sheet the details of the work undertaken, the job and job number, the date of the work, the operatives' names and their hours worked, the materials and plant used. The clerk of work's or architect's signature should be obtained approving the details. A copy of the form (daywork sheet) should then be sent to the head office or (if site-based) surveyor for the total costs to be incorporated, and for submission for inclusion on the monthly valuations (see Fig. 4.1.5).

Dayworks are those works which are undertaken which cannot be accurately measured and priced beforehand, and nor are there any similar operations in the bills of quantities. It is work such as certain repair or replacement work. Payment is made from a fund (normally called a contingency sum) provisionally allowed in the bills of quantities for unforeseen work.

Depending on the size of the contract and type of construction firm, some or all of the following forms or particulars would be required.

1. Materials received sheets (see Fig. 7.3.1).
2. Materials/Plant transferred sheets (see Fig. 7.3.2).
3. Programme and progress reports.
4. Variation orders.
5. Subcontractor claims.
6. Valuation of work done.
7. Time sheets.
8. Wages sheets (if wages prepared on-site).
9. Bonus sheets.
10. Notification requests for inspection by Building Control Officer.
11. Notification requests to National House Building Council for inspections.
12. Notification requests under the Health and Safety at Work etc., Act 1974.
13. Details of disciplinary warnings to operatives/ employees (see Fig. 6.2.2).
14. Accident Reports and Statistics, etc. (see Fig. 10.2.3).

4.2 Letter and report writing, etc.

Letter writing

The purpose of a letter is to transmit information accurately, or to make enquiries which require an answer.

The writer of a letter is judged by the way he/she writes and presents the letter – an untidy letter portrays an untidy person; a vague letter a vague person. There is no excuse for letters being sent which have spelling mistakes or are carelessly presented. A dictionary should always be consulted if one continuously has doubts about spelling.

The more that one deals with the sending and receiving of correspondence the better and more confident one becomes at compiling clear and concise letters.

Before writing a letter
One must know:

1. What one wants to say.
2. The order in which one intends to say it.

Always be diplomatic, taking care in the choice of words, and be firm where required, or friendly and understanding. Rude letters only contribute to resentment on the part of the receiver leading to lack of cooperation. One should be polite at all times even though one is attempting to be firm and to the point.

Essentials of a business letter

1. Suitably designed and printed letter-headed paper should be used to portray a well organised concern and to minimise the typist's work. It would be unacceptable and a waste of time for a typist to type

B. BUILDER, LTD, MERROW, GUILDFORD.

ADVICE OF VARIATION

TO: B. Mitchell   NO: 5

FROM: H. Helen

CONTRACT: Wholesale Shop and Offices   JOB NO: 32   DATE 3rd March 19

London R.D.

Guildford

*I have been verbally instructed to carry out the following
as a variation to the contract by Mr J Joanne :*

ADDITION	OMISSION
Two extra electrical	
sockets with switches	
and wiring to counter	
area.	

Please confirm the instructions in writing to:

Mr J Joanne   Position: Architect

Signed: H Helen.
          Site Manager

Copies to: 1. Head Office
          2. Clerk of works
          3. Site.

Note:
This form should be sent to head office on the same day as instructions are issued.

Fig. 4.1.3

B. BUILDER, LTD,
MERROW, GUILDFORD.

VARIATION ORDER

No......*1*...

TO SUBCONTRACTOR
(NOMINATED/OUR OWN)   R.G. Bass Electrics Ltd.

Date *12th March 19......*

CONTRACT   Wholesale Shop and Offices,
London Rd., Guildford

Order No. .....*5*..........

Please execute and/or omit the following items as a variation on your subcontract.

Provide additional wiring and two extra sockets
with switches to the counter area as denoted
on the drawings.

Signed ......*B Mitchell*......
FOR B. BUILDER LTD.

Copies to	**Official V.O. No.** *5.*	**Price:**
R.G. Bass Electrics Ltd. Main Office Site		

Fig. 4.1.4

out the company's name and address every time a letter was prepared (see Fig. 4.2.1).

2. It is poor commercial practice to use abbreviated words when corresponding with others, although there are a few words which tend to be acceptable, e.g. ref – for reference, No. – for number, Ltd – for limited.
3. Use Our reference and Your reference, when necessary, to identify the compiler of the letter and typist.
4. Include a date. This would avoid the embarrassment of receiving a reply from a correspondent who commences his/her letter using the words – Thank you for your 'undated' letter. . . .
5. Show the name and address of the person to whom you are writing on the left hand side and above Dear Sir, etc.
6. The reference or subject heading should be underlined and preferably be on the next space below Dear Sir, etc., but in the middle of the page.
7. Start the letter with a salutation, i.e. Dear Sir, Dear Mr Blyth, Dear George, etc., depending on how informal or formal one wishes to be.
8. Introduction or first paragraph should state why the letter has been written.
9. The body of the letter should convey, with the minimum of effort to the reader, the information, problem or question which one has. It is essential therefore:
   (a) to be brief;
   (b) to use suitable punctuation and short sentences;
   (c) to cover the subject logically;
   (d) to set out paragraphs for each new idea or train of thought.
10. Use a suitable complimentary close. If the salutation is Dear Sir, the complimentary close should be Yours faithfully; Dear Mr Blyth – Yours sincerely; Dear George – Yours truly.
11. One's signature should be applied at the end of a letter, and always print or type the name underneath. If one is writing on behalf of a firm or company it is advisable to sign the letter as shown in Fig. 4.2.1.
12. If one encloses additional information or literature with a letter, one must state so in the letter. Also the word, 'Enclosure', should be typed/printed at a lower level than the signature, but on the left-hand side.
13. P.P. stands for 'Post Parenthesis' – which precedes additional information which is an afterthought once the main letter has been compiled. It is incorporated at the bottom left-hand corner of the letter (see Fig. 4.2.1).

Standard letters

These types of letters can be compiled for recurring situations. All that requires to be done is to type in the date, and the name and address of an individual. It saves compilation and typists' time.

Standard letters are used in countless situations, such as:

1. Acknowledgements of receipts of letters or enquiries from others.
2. Notifications to unsuccessful job applicants.
3. Warnings for misconducts by employees (see Fig. 6.2.2. and 6.2.3).

Reports

A report is usually a statement of facts and ideas presented for the attention of someone seeking information so that decisions can be taken.

Routine or recurring reports tend to be made by filling in specially designed standard forms, such as: daily or weekly site reports, site diaries, etc. Special one-off reports, however, need to be prepared by:

(a) observing the terms of reference;
(b) enquiring and collecting of information;
(c) arranging the material;
(d) drafting the report and editing;
(e) typing and checking.

Terms of reference

Before preparing the report one should obtain clear instructions from the person requiring the report regarding the following:

1. The kind of information required.
2. The amount of information required.
3. The date by which the report is required.

Drafting the report

Most organisations have adopted systems or layouts which are more suited to them depending on whether they are for internal or external use. Reports for internal use are usually on less formal lines than those produced for a client by a professional organisation.

Reports are basically outlined as follows:

1. The layout should show suitable headings with details of the subject under discussion clearly shown at the beginning. Always remember to use sub-headings and paragraphs to break the written work into easily readable information.
2. A contents section would assist the reader to easily locate the part of the report in which they are mainly interested, i.e.
   (a) Terms of reference.
   (b) Introduction and present situation.
   (c) Suggested improvements/alternatives.
   (d) Recommendations.
   (e) Conclusions (and Appendices if there are any charts or other documents, etc. which need to be added).
3. The report should be signed at the end by the person who drafted it (see Fig. 4.2.2 for an example internal report).

Editing. Always read through the draft copy carefully before submitting it for typing. Be a ruthless self critic and observe conciseness. Avoid awkward phrases and unnecessary words.

B. BUILDER, LTD.

CONTRACT
DATE

Description of work

Remove existing door and replace with
as V.O. 10 from 5 houses ready for hand
altering opening including and making good
cleaning up after   PER HOUSE

Labour : name	Trade	Hours	Rate	O/ti
A. Dunk	Carpenter	7	1 - 37	
F. Elliot	Bricklayer	6	1 - 37	
T. Blickett	Plasterer	2	1 - 37	
P. Taylor	Glazier	1	1 - 39	
C. Rakjohn	Painter	2	1 - 37	
G. Frobshom	Labourer	1	1 - 17	
				net la

Materials description	Quantity
75mm Cut nails	5 No
Carlite plaster	1 Bag browning 1 " finish
Emulsion paint	5 litres
Gloss paint	1 litre
100mm blocks	½ m²
Cement + Sand mortar	0.025 m³
	net mat

Plant description	Hours
Mixer	¼
	net

Signed. B Brown

Fig. 4.1.5

Rd oking B		DAYWORK SHEET		SHEET NO.	1

f glazed doors
lork includes taking out door,
twork, brickwork + plasterwork, and

OL IEY	Net labour		Additions				Total	
	9	59						
	8	22	Percentage = 120 %					
	2	74						
	1	39						
	2	74						
	1	17						
			net labour	25	85			
			percentage	31	02			
	25	85	gross labour	56	87	total	56	87
te	Net material							
ch		10	Percentage = 60 %					
each	4	00						
O	1	50						
0	1	00						
m²	0	75	net materials	7	48			
0m³	0	13	percentage	4	49			
	7	48	gross materials	11	97	total	11	97
te	Net plant							
0 hr	0	25	Percentage = 20 %					
			net plant	0	25			
			percentage	0	05			
	0	25	gross plant	0	30	total	0	30
				TOTAL			69	14
				VAT (10%)			6	91
				Total daywork sheet £			76	05

Directors:
W. SLATE
M. ORTAR
S. INK
B. RICK

B. BUILDING LTD.,
MERROW, GUILDFORD. GU1 5BY

ESTABLISHED 1880
TELEPHONE: GUILDFORD (0444) 5665878

REG: NO. 4224456 – ENGLAND V.A.T. NO: 77247631

OUR REF: GF/SB/8

YOUR REF: 28/B1/4

DATE : 15th November 19..

T. ILES, ESQ.,
J. PETER LTD.
DOWNLAND,
DERBYSHIRE.

Dear Sir,

Reference: Order No. Z584/21/-

With reference to the above order and my conversation with you on Wednesday 14th November 19.., I confirm my request for an earlier delivery of the roofing tiles from 17th March 19.. to the new agreed date of 1st March 19...
Thank you for your cooperation.

Yours faithfully,
for and on behalf of
B. BUILDER LTD.

H Heler

Site Manager

OTHER OFFICES IN WOKING, BYFLEET AND ALBURY

NOTE:
1. Enclosures or P.P. here.
2. The small 'f' in 'Yours faithfully'. Similarly there should be a small 's' in 'Yours sincerely', and small 't' in 'Yours truly'.

Fig. 4.2.1

Memoranda

This is a brief note of something which is to be remembered. In an office the most common 'memo' is in the form of a standard sheet, which anyone may complete when someone either from within or outside the firm is attempting to contact another person either by telephone or by having called in to see that person (see Fig. 4.2.3 for a typical example).

Circular

This is a special letter which is sent around the firm for a number of persons to read. Generally speaking it is generated from a higher level. It may be a single copy which has to be signed once it has been read by the intended individuals, or separate copies may be distributed to the individuals (see Fig. 4.2.4).

Notices

These are normally displayed on a notice board so that all and sundry can read, learn and inwardly digest. It is the cheapest way to reach everyone within the company, always supposing the majority of the employees read notices on notice boards. The boards should be in places of prominence. Notice boards are used for the following:

1. The display of safety protective clothing and equipment information.
2. Safety posters.
3. Accident statistics.
4. New contracts successfully tendered for.
5. To advertise vacancies whether of ordinary or supervisory levels (see Fig. 5.1.1).

4.3 Verbal communications and skills

All too often site supervisors have difficulty in communicating adequately with those from outside the work team, although they generally manage with those with whom they work either at subordinate or superior levels. Many of them have gained promotion from the craft ranks and their ability to communicate verbally is sometimes limited. They are unable to discuss site problems adequately, speak convincingly or win a point with the professionals or public authority visitors because they lack confidence due to their inarticulation and limited educational training, which inevitably leads to the loss of respect in the eyes of the site visitors.

Many site supervisors attend courses in further education establishments, and their firms offer sound industrial training through in-service courses. Many firms, however, are unable to allow their supervisors education and training time.

It is generally thought essential that site supervisors should have sufficient knowledge of construction technology, law, management skills, communication skills, industrial relations and safety to meet the challenges normally encountered on-site, but also to put them on a similar industrial footing as their counterparts – architects, engineers, quantity surveyors, etc., so that there can be some mutual understanding which leads to confidence and trust between the parties at site level. Training in this respect could be by in-company training, group training schemes; the Construction Industry Training Board; the Building Employers Confederation through the Building Advisory Service; business schools; and colleges of technology.

If the site supervisor means to succeed in the art of speaking (verbal communication) he/she must spend time reading to increase his/her word power. Care should be taken to speak accurately with individuals at all levels. Therefore, supervisors should read accurately, speak carefully, interpretate accordingly and use the correct manner in every situation.

A good site supervisor should have the qualities listed in Section 5.4 which altogether equip him/her for every eventuality on-site, particularly where communications are concerned.

The site supervisor should be an all-rounder, a trait which unfortunately only tends to develop in many of them later in life. There are, however, certain individuals who are born to lead, with most, if not all, of the qualities listed. It is, of course, now firmly believed that many individuals can be trained to achieve many of these qualities at training centres as previously described.

Barriers to effective speaking

There are numerous causes which lead to the breakdown in communications between individuals such as:

1. Personality problems.
2. Dislikes.
3. Misunderstandings.
4. Dullness and lack of intelligence.
5. Inarticulation and failure in projecting oneself.
6. Incompatibility.
7. Petty mindedness.
8. Carelessness in choice of words.
9. Laziness.
10. Poor power of persuasion.
11. Timidness.
12. Aggressiveness.
13. Inadequate knowledge of subject/technology.
14. Bad tastes.
15. Social background.
16. Tactlessness.

Site managers should attempt to overcome their own deficiencies or faults (as above), and should try to understand and ignore similar problems in others. Also one should not harbour grudges in business but should be conciliatory and outward looking.

B. BUILDER LTD.,   R E P O R T
MERROW,
GUILDFORD.

FROM: H. Helen, Site Manager.   No: 1/19....
TO: Mr Bigg, General Manager.   Sheet No: 1/
CONTRACT: Wholesale Shop and Offices,
           London Road, Guildford.   Date: 8th February, 19..

        SUBJECT: The recording of arrival and departures
                of operatives on-site.

      CONTENTS: 1.0 Terms of reference.
              2.0 Introduction and present situation.
              3.0 Suggested alternatives.
              4.0 Recommendations.
              5.0 Conclusions.

1.0   Terms of reference   To carry out a survey of the effectiveness of the present system of recording arrival and departure times of the operatives on-site, and to suggest and recommend alternatives bearing in mind the type of work undertaken by the company.

2.0   Introduction and present situation

2.1   It must be remembered that this company deals with the following personnel on-site: those of the nominated subcontractors and the company's directly appointed subcontractors, labour-only subcontractors and the company's own direct labour.
The labour mix on-site would be the determining factor as to which system of recording of arrival/departures would be used. The main control system chosen would relate, in the main, to the company's direct labour force - with a few labour-only subcontractor exceptions - the present system of all other subcontractors booking on to the site should still continue.

2.2   The system which has been recently in operation by the company on the sites is the one whereby operatives are entrusted to report to their immediate supervisors (foremen, changehands and gangers). This appears to have worked sufficiently well on the very small sites, but there are a number of weaknesses with the system, e.g.

2.21 It assumes that the supervisors (foremen, etc.) are honest and arrive early themselves, which in a number of cases is not so. If they arrive late they cannot generally check on subordinates.

2.22 If the supervisor is absent there may be a lack of productivity before the site manager or visiting contracts manager becomes aware of the situation.

2.23 Weak supervisors, through their inaction and lack of discipline, appear to condone absenteeism and lateness, especially where a subordinate with a stronger personality dominates within the gang.

2.24 Individuals are tempted to cover for each other or are in collusion with the supervisor (probably a friend).

2.25 There is no check on supervisors unless the site manager physically checks them himself. Because of pressure of work, however, this is not always possible to do.

3.0 <u>Suggested alternatives</u> The control of timekeeping of operatives on-site could be by one or more of the following systems. E.g.
3.1 reporting to the site manager/timekeeper on arrival and departure, who would then record the individual's name or works
number, and the time when the report was made.
3.2 individuals signing in and out on a time recording book kept
for the purpose. This could be open to abuse as operatives may
sign in their colleagues who they anticipate will be late.
3.3 the tally system - metal or plastic discs with individual's
works or allocated numbers imprinted on them. The tallies are
moved from one board to another by each operative who reports for
work. Once again it is a system which is open to abuse if not
closely supervised.
3.4 a clocking-on system. A works clock is installed which
accommodates operatives' works clocking on and off cards. Checks
should be made periodically to ensure early arrivers are not
clocking on late arrivers.
3.5 timesheets. Individuals are made responsible for the filling
in of their own timesheets to show what they have done and the time
taken to do it. Some would say that too much trust is given to
individuals with this method.
3.6 gang timesheets. Each supervisor is responsible for filling
in the sheets on a daily basis, each of which includes the names
of the operatives in the gang.
3.7 flexi-time. Clocking in or out, or signing a book with name,
date and time of arrival and departure. The idea is that individuals can come and go as they please provided that they remain at
work during the key time of, say, between 9.30 a.m. and 3.30 p.m.,
and that at the end of each week or month they should have put in
the normal expected total hours of work. Whilst this system
favours the employees, a reasonable level of control is maintained
by the employer of attendance times.
Flexi-time is difficult to operate on-site where gangs of workers
are involved and each individual in a gang relies on the other (a
concreting gang of eight would achieve little if at the start of
the day only two of the gang turned up in the first hour and most
went home one hour before the latest allowable finishing time). It
should however be given some consideration for other types of work.

4.0 <u>Recommendations</u> There must be some trust of supervisors.
They would never have been promoted to their positions unless they
had good qualities - honesty being one of them. They should be
trusted until they are found otherwise.
The system of clocking on can be used to give some control to the
site manager, with group or individual timesheets being used as a
back-up and as a check on the work done, who did it and when.

5.0 <u>Conclusion</u> I feel sure that the adoption of the two systems
recommended would be suitable for most sites and situations.

H. Helen.

MEMORANDUM	B. BUILDER, LTD, MERROW, GUILDFORD.

MESSAGE FOR

Mr _Lomax_

WHILE YOU WERE UNAVAILABLE

Mr _B Jones_

OF _Corby & Sons, Carlisle_

PHONE NO: _Carlisle 824444_

TELEPHONED	✓	PLEASE RING	
CALLED TO SEE YOU		WILL CALL AGAIN	
WANTS TO SEE YOU		URGENT	

MESSAGE: _He can supply the_
fittings you requested yesterday

DATE _23rd February 19....._ TIME _2.30 pm_

RECEIVED BY _H Lime_

Fig. 4.2.3

CIRCULAR   B. BUILDER LTD.,
                     MERROW, GUILDFORD.

FROM: S. Shirley, Contracts Manager.   No: 4/19..

TO: ALL SUPERVISORS   Sheet No: 1/1

                                  Date: 18th January 19..

SUBJECT: Misuse of Company Telephones

It has been brought to my notice that a substantial number of
employees are using the Company's telephones for non-essential
personal calls, sometimes with the approval of their supervisors.

All supervisors are reminded of the existing Company Policy
regarding the making of personal calls, which is:

> No personal calls should be made by subordinates unless
> they have the express approval of their immediate
> superior, and that where permissions are given, the
> calls should be in cases of emergency or great
> importance.

As the misuse of the telephone is incurring the Company in
considerable expense, this practice must cease, and every effort
should be made in the future, by those with responsibility, to
minimise the use of the telephone in this way.

Fig. 4.2.4

Those from outside the firm with whom one may have to communicate verbally while managing affairs on-site are:

1. The client.
2. The architect (or his/her representative, the clerk of works).
3. The quantity surveyor.
4. The consultants – structural/civil engineer and services engineers.
5. Manufacturers' representatives (nominated or ordinary).
6. Suppliers'/merchants' representatives (nominated or ordinary).
7. Subcontractors' representatives (nominated, ordinary and labour-only subcontractors).
8. Plant suppliers' representatives.
9. Local authority officers (planning, building control, highways, public health, engineer, fire).
10. Public undertakings' representatives (water, gas, electricity, British Telecom).
11. The factory inspector.
12. Amenity groups' representatives.
13. The public (complainants, colleges, schools, clubs, womens' institutes, etc.).
14. The police.

Each obviously requires a special approach whether it is verbally by telephone, or face to face. Also, the level of approach is important – with womens' institutes it would be friendly; with the police it would be helpful; with the building controls officer a serious and firm approach is necessary.

Verbal communications by telephone

There are three main types of telephones:

1. Internal systems.
2. External systems:
   (a) through the operator switchboard, or
   (b) direct (cordless or ordinary).
3. Radio:
   (a) direct dialling system.
   (b) relay through the headquarters of the firm.

Points to remember

Outgoing calls:
1. Collect ideas together and points to be raised and write them on a pad before dialling.
2. Dial the town code number first, then the directory number of the firm or person, and when someone answers ask for the department or extension number of the person you are trying to contact.
3. Be courteous and to the point (brief).
4. Use the correct tone, depending to whom you are talking (the tone for a friend would be different to that used when making a complaint).

Incoming calls:
1. If there is no operator use the correct salutation (good morning, etc). State your telephone number and firm's name to ensure that the person who has made the contact has the correct number and organisation.
2. Where an operator exists he/she will have received the call first and will put the call through to you; therefore, answer the telephone by stating the department and your name.
3. A pad amd pencil should be available beside each telephone, and there should also be a telephone memo pad for leaving messages for others if they are unavailable to callers. (See Fig. 4.2.3.)
4. Answer calls politely, concisely, and helpfully
5. Promises given should be promises kept – offer to call back if answers to queries cannot be given immediately.

It is good sound practice that registers for telephone calls and conversations are kept for both incoming and outgoing calls, in which the date, time, subject, organisation and employee's name are recorded for future references, particularly when someone wishes to query the receiver of a call.

Transferring calls: Sometimes a call is put through to the wrong department, and to enable it to get prompt treatment it may be transferred to the correct one by first apologising to the caller, stating that you are about to transfer him/her. First press the telephone button, dial the proper number required by the caller and on getting through replace the telephone which connects the caller to the correct receiver – this depends on the type of telephone system installed.

Meetings

References are made to types of meetings from time to time, and an appreciation of the meanings of each should be known, as follows:

1. *Conversation* – a talk or familiar discourse between two or more persons (individuals may make notes of what was agreed).
2. *Forum* – this is a discussion where the audience takes part, normally after a platform speaker introduces or speaks on a subject.
3. *Conference* – this is an infrequent gathering of many people with common interests to discuss business, exchange ideas on latest views, listen to papers being read.
4. *Discussion* – to examine in detail points or ideas, generally to arrive at an agreement.
5. *Seminar* – where a group of advanced students work in a specific subject of study under a teacher/lecturer.
6. *Interview* – a meeting to consider the suitability of a person or subject with a view to engagement, or a meeting with a famous or infamous person with a view to publication.
7. *Committee* – a group of people selected by a larger group or body who are charged with meeting together to control or recommend action.

8. *Debate* – where a large number of individuals gather to argue the case for or against a proposition which revolves around a current topic.

In each of these eight kinds, notes may be taken by individuals, but this does not mean that the meetings are formal.

All of the kinds of meetings listed above may be organised in one of two different ways: (*a*) informal meetings; (*b*) formal meetings.

Informal meetings

With informal meetings points are discussed generally without an agenda and without any formal minutes being taken as a correct record of what was said. The problems with these types of meetings are that there may be:

(i) no true record of what was discussed;
(ii) no proof of what was agreed;
(iii) no record of those in attendance;
(iv) no record of those who were supposed to attend but did not, or whether they sent their apologies;
(v) no proof of who agreed to do certain tasks. Perhaps the only advantages lie in the fact that persons can attend informal meetings knowing that what they say will not be published. Ideally only exploratory talks should be conducted along these lines in business or in sales pep talks by the sales director who wishes to inject some urgency for increasing sales into the sales staff.

The dangers of informal meetings are that what was discussed could be misinterpreted or misconstrued, and people may with afterthought, reflect on what they said and back down later, denying any agreements made.

Formal meetings

If discussions, debates, committee meetings, etc., are to be meaningful a proper agenda must be drawn up, and during the meeting minutes should be taken of discussions, motions or resolutions.

The following are necessary for formal meetings:

1. Plan and arrange the meeting.
2. Identify the committee or those who are required to attend the meeting, always remembering it is essential to have a chairman and secretary (or committee officers depending on the type of organisation and meeting being planned).
3. An agenda – prepared and distributed before the meeting by the secretary and chairman but with suggestions of points for discussions, sent by other committee members (attenders), being included in the agenda where necessary.
4. Minutes should be taken using the subheadings of the agenda which should be typed up quickly after the meeting for distribution to committee members for action and reference.
5. A set of rules for debate and a constitution (rules and regulations of the organisation).

Arranging a meeting

In Fig. 4.3.1 a comprehensive check list is shown for use by anyone who is about to call a meeting (chairman or secretary, obviously by agreement with the other officers). For project meetings many of the points shown in the check list would be inappropriate but, nonetheless, a list like it would be helpful to prevent some important point being overlooked.

Invitation, convening or calling notices would be sent to those who should attend the meeting; the committee members in the case of a committee meeting, perhaps with the Agenda printed below the letter as shown in the following section.

The agenda

For a project meeting Fig. 4.3.2 could be used as an agenda depending on what stage the contract work has reached and who or what firms are involved in the operations.

It may not always be necessary to call project meetings each month unless problems have developed or the architect is unsure of progress. Meetings for meetings sake should be avoided, and unless there is a purpose it is wasteful in time and effort for all concerned. Neither should invitations to attend a meeting be sent to those who cannot directly contribute to the meeting unless they wish to attend as observers.

Chairmanship

In committees a chairman usually holds office for a year or more, but at site level the architect and contracts manager could chair the monthly project meetings alternately. One must appreciate, however, that meetings held on-site can be of three types:

1. *Project meetings* – generally called by the architect who may expect the following individuals to attend: professional quantity surveyor, clerk of works, consultants, main contractor's representatives (contracts manager and site supervisor), nominated subcontractor and suppliers' representatives.
2. *Site meetings* – called by the contracts manager when required to review progress, etc., with representatives from subcontractors and suppliers, site supervisor and other site staff.
3. *Domestic site meetings* – called by the site manager (or sometimes by the contracts manager) usually on a weekly basis (or when required) to plan and maintain progress or to discuss other site problems. Those in attendance would be the foremen/supervisors, ganger, general foremen, the contractor's quantity surveyor, programmer and site engineer.

Ideally all the meetings should be minuted, so it is necessary to prepare an agenda which should be distributed prior to each meeting to those who need to attend. However, with domestic meetings or immediate command meetings the site supervisor/agent/site manager would make decisions after receiving information and advice

ARRANGING A MEETING

SITE MEETING, COMMITTEE MEETING, HEAD OFFICE DIRECTORS' MEETING, EXECUTIVE MANAGERS' OR UNION MEETING, A.G.Ms.

Check List

Are you a genius? If you are not sure then check your check list. Number your own priorities each time that you arrange a meeting.

Date?	Clock?	Ashtrays?
Time?	Lectern?	First Aid?
Place?	Lighting?	Projector?
Agenda?	Acoustics?	Telephone?
Numbers?	Water Carafe?	Blackboard?
Parking?	Tables/Chairs?	Power Points?
Catering?	Entertainment?	Pens/Pencils?
Reception?	Microphone/PAS?	Films/Slides?
Transport?	Heating/Ventilation?	Block and Gavel?
Name Cards?	Room Size/Suitability?	Paper/Note Books?
Seating Plan?	Nearest Doctor/Hospital?	Toilets/Cloakroom?
Record of Meeting?	Name of Manager/Caretaker?	Exhibition Facilities?

Checked your requirements?

Now check your written confirmations

Fig. 4.3.1   Arranging a meeting (site meeting, committee meeting, head office directors' meeting, executive managers' or union meetings, AGMs)

```
                        B. BUILDER LTD.,
                        MERROW,
                        GUILDFORD.
                        Tel. No. 592114


                                        7th January 19....

Dear (Committee Member),

     You are invited to attend the next project meeting to be
held at the new Wholesale Shop and Offices, London Road,
Guildford on Thursday, 18th January 19..., at 10.30 a.m.

                        AGENDA

1.  Notice convening the meeting.
2.  Apologies for absence.
3.  Minutes of previous meeting held on 4th December 19...
4.  Matters arising.
5.  Progress of work:

        6.1  Main contractor.
        6.2  Subcontractors:
             6.21  Nominated.
             6.22  General.
6.  Problems.
7.  Information required.
8.  Any other business.
9.  Date and time of next meeting.

                                Yours sincerely,


                                S. Shirley,
                                Contracts Manager,
                                B. Builder Limited.
```

Fig. 4.3.2

from his/her subordinates who are present at the meeting. The site supervisor would take notes of decisions, e.g.

(i) what is to be done next and why?
(ii) who is to do it?
(iii) when it is to start and by what time it is to be completed?
(iv) how is it to be done?
(v) where is it to be done?

In project or site meetings the chairman should:

1. Be knowledgeable of debating procedures.
2. Be able to control (lead and not be led) and ensure that dicussions and questions should be through the chair.
3. Be courteous and friendly, but firm.
4. Follow the agenda without deviation.
5. Allow each committee member to have his/her say without interruption.
6. Start and finish on time.

The procedure for conducting project or site meetings tends to follow similar patterns to those of more formal meetings. Instead of the need to have a quorum of attenders at the meetings on the site, as could be laid down in an organisations rules and regulations (the constitution), it is necessary that those who do attend have the authority and responsibility to make decisions and to take action.

The secretary

It is usual for the general foreman/site supervisor to be given the task of acting as secretary, perhaps alternating with the architect's representative or assistant.

The major part of a secretary's work is to record most of the discussions which take place at meetings. In a properly constituted committee the secretary would serve in office for a term of two or more years.

The secretary should know the rules for debate, and should prompt and advise the chairman, where necessary during meetings, having previously made all the arrangements for the meeting as outlined in Fig. 4.3.1. During the meeting the secretary would minute all the main points discussed and the decisions reached. The minutes would then be typed as soon after the meeting as possible, with committee members/attenders and other relevant individuals receiving copies immediately so that they can remind themselves of the resolutions made and the action individuals need to take to meet with the approval of the meeting.

The minutes (see Fig. 4.3.3) should follow the order of discussion as laid down in the Agenda (se Fig. 4.3.2).

Procedure in committee

Rules of procedure
The procedure usually follows custom and practice copied to some extent from Parliamentary procedures which have been derived over many hundreds of years.

The main differences in committee procedure, as compared with more formal meetings, are:

1. A general discussion usually precedes the moving of a formal motion.
2. Motions need not be seconded, although this is desirable.
3. Members may speak as often as they can catch the chairman's eye.
4. Members may remain seated when speaking.

Apart from these relaxations, procedure adopted at more formal meetings should be followed, for example:

(a) A quorum (the acceptable minimum number of committee or other members allowed under the constitution) must always be present throughout the whole of the proceedings (unless special provision has been laid down for the procedure to be followed in cases of emergency when a quorum is not present).
(b) Members should address the chair.
(c) Members wishing to speak should so indicate to the chairman and wait to be called.
(d) Coopted members who attend to express a view to give advice should not be allowed to vote.

Rules of debate

Standing orders
The Chairman's power and those of members are often defined in 'standing orders'. Standing orders should not be interfered with, but it may occasionally be desirable to suspend them; for example, to allow immediate discussion to a speaker to expound his case. To carry a vote on the suspension of standing orders normally requires at least a two-thirds majority of those present. Site meetings are conducted on less formal lines.

Motions
A motion must be proposed and seconded before it is discussed at more formal meetings. It may be put before the meeting orally or may be written and given to the chairman or secretary before the meeting. The rules of many organisations require notice of important motions to be circulated to members before the meeting. The chairman may have to refuse to accept motions unless the required notice has been given. As previously stated, at less formal meetings motions are proposed after a general discussion.

Amendments
Instead of speaking for or against a motion as it stands, a member may move an amendment. When an amendment has been moved and seconded it must be discussed and voted upon. If the amendment is lost, the meeting returns to the original motion and, after the proposer has replied, can vote upon it, unless another amendment is moved. If the amendment is carried however, the amended motion replaces the original motion. When an amended motion becomes a 'substantive motion' before the meeting, and must be put to the vote. If there is more than one

amendment proposed each must be voted upon in reverse order — the last one first, back to the first amendment.

Amendments should be framed so as to: (*a*) omit certain words. (*b*) insert certain words. (*c*) omit certain words and insert others.

Calling a member to order
The chairman may do this if a members' speech wanders from the agenda or point under discussion goes beyond the terms of reference of the committee, becomes unruly or is abusive to any committee member, attempts to speak out of turn or to speak twice to the same motion (generally only the proposer can do this). The chairman should also rule out of order anyone at a meeting who has no right to be there.

Calling the chairman to order
This may be done in accordance with standing orders. The normal procedure is for the chairman to vacate the chair and the vice-chairman or other officer to put the motion 'that the chairman's ruling to be upheld'. If this is defeated, the chairman's particular ruling is out of order.

Point of order
During one member's speech another may raise a point of order for one of the reasons given in calling a member to order (above). The chairman gives a ruling and the first speaker continues unless it is 'out of order' for him to speak further. The raising of points of order is often abused, and it is for the chairman to prevent such abuse.

Next business
At all types of meetings a proposal to pass on to the next business may be made. If the discussion is clearly un-rewarding and unlikely to lead to any useful decision, the motion 'next business' should be accepted, but the chairman should be careful not to allow abuse of this procedure.

Closure
At the end of a formal debate the chairman asks the proposer of the motion to reply to the debate. The correct time for this is at the end of the debate on the first amendment. If there is no amendment he replies at the end of the debate on the main question, before the final vote is taken. Once an amendment has become a substan-tive motion, the mover of the original motion may take part in the discussion on any further amendments. In replying the proposer must introduce no new matter. Sometimes a good deal of tact and judgement is required to decide when to close the debate; the general principle is that it should be closed when all the argument for and against have been stated.

That the question now be put
Debate may be closed by a motion moved by any member who has not spoken in the debate. The chairman can refuse such a motion if he considers the question requires further discussion. Sometimes it may be the duty of the chairman to move that the question be put from the chair.

The question 'are you ready to vote' will often be enough. The chairman must always be sure that the meeting is clear about matters on which it is asked to vote.

In more formal meetings, i.e. Annual General Meetings, Trade Union or Employers' Organisation meetings, etc., the rules of debate should be closely followed by all those in attendance and who have a right to attend.

Some important meanings (in addition to those previously outlined are):

1. Motion — proposition or proposal.
2. Question — motion under discussion.
3. Voting — show of hand, poll, ballot; usually on a question after some discussion — usually a majority vote carries the question.
4. Resolution — question which a meeting has voted in favour of.
5. Recess — a short interval for refreshment or to enable certain points to be discussed which would not be minuted.
6. Quorum — the minimum numbers of members who must attend a meeting for that meeting to be lawful under an organ-isation's constitution.
7. Adjournment — because the time allocated for a meeting has run out a proposal may be put to the chair that the meeting be adjourned until a future date to discuss the unfinished business.

4.4 Communications with central and local government personnel and others

Site supervisors must appreciate their responibilities re-garding communications with the various government departments, partially government-controlled bodies, and local authorities. Their responsibilities relate to answering queries when requested, but above all to noti-fying and advising the representatives of these bodies as and when required — usually to comply with legislation. The following is a list, with descriptions, of a site manager's collaborations and notifications with the authorities' rep-resentatives before and during the executions of projects:

1. Local authority

Planning officer
(i) This officer may visit the site to ensure that the siting and visible material choices remain unaltered from that which was approved on the original drawings. He is, however, more concerned with dealing with the client/architect, but there should be no work undertaken which requires approval from the planning officer until he/she gives permission, otherwise the contractor could be found as liable for infringement of planning law as the client or designer.
(ii) Where tree preservation orders exist licences should be sought from the planning authority before the felling

Minutes of the Project Meeting (or committee meeting) held
on Thursday, 18th January 19.. at 10.30 a.m. on the
Wholesale Shop and Office Site, London Road, Guildford.

Those attending:

     J. Joanne, Architect - Brown & Joanne Partners.
     S. Shirley, Contracts Manager - B. Builder Ltd.
     H. Helen, Site Supervisor - B. Builder Ltd.
     C. Clare, Clerk of Works.
     D. Haylock, Quantity Surveyor of Haylock Associates.
     M. Knight, P. B. & Pipe Ltd., Nominated Plumbing Contractor.
     R. G. Bass, Electrics Ltd., Electrical Subcontractor.

1. **Notice convening the meeting.**

   The meeting commenced at 10.30 a.m. prompt.

2. **Apologies for absence.**

   Mr Jones of Messrs. I. Ronmongers & Co. was unable to attend
   due to pressure of work, but had sent his programme of work
   as promised at the last meeting.

3. **Minutes of the previous meeting held on 4th December 19..**

   The minutes were read and agreed subject to the fact that
   under item 6 it was Mr Aberson of P. B. and Pipe Ltd. and
   not Mr Knight who agreed to commence the first fix
   plumbing in January.

4. **Matters arising.**

   Mr Bass, Electric Limited, said that he was successful in
   contacting his suppliers to arrange for additional electrical
   materials to be delivered in time for his employees to
   commence overtime working to complete the west wing in time
   for the plastering subcontractor in February.

5. **Progress of work.**

   5.1 **Main Contractor** - Mr Shirley, Contracts Manager, stated
   that now the bricklayers were working extra overtime the
   contract work will only be one week behind schedule by
   February. Mr Joanne, the architect, commented that his client
   would be pleased at this prospect.

   5.2 **Subcontractors** -

   5.21 **Nominated.** As the only nominated subcontractor involved
   so far was P. B. and Pipe Ltd., Mr Knight (their representative) stated that the work was slowing down due to the in-
   accessibility of some areas through which his employees were
   to feed pipes. Mr Joanne, the architect, offered to get

Fig. 4.3.3

advice on alternative positions for the holes from the structural engineer, and would issue instructions within three days.

5.22 <u>General</u>. - R. G. Bass, Electrics Ltd., reported that work was on target except for the west wing as previously discussed in item 4.

6. <u>Problems.</u>

The problem of too few main contractor's workers being available on the site was highlighted by Mr Clare, Clerk of Works, but as progress was being maintained he said that he was happy, but would still continue monitoring numbers.

Mr Shirley, Contracts Manager, said that a serious delay in concrete deliveries on another site had meant men had to be transferred for three days in December to improve output. He stated that he could foresee no problems similarly developing during the remaining contract period. Mr Joanne, the architect, said he was satisfied with the honest statement.

7. <u>Information required.</u>

Other than that discussed in item 5.21 no information was required because Mr Joanne, the architect, had already satisfied Mr Helen's request before the meeting commenced for particulars of the paved area in front of the east wing.

8. <u>Any other business.</u>

It was stated by Mr Helen, the Site Supervisor, that the new pavement cutter had arrived and that work was in hand.

9. <u>Date and time of next meeting.</u>

Thursday, 14th February 19.. at 10 a.m. was agreed as suitable, and an invitation was to be sent to Messrs Aluminium Windows Ltd. The meeting closed at 12.30 p.m.

Signed......................

S. Shirley (main contractor)

Copies to:

Members who attended,
Mr Jones,
Messrs. I. Ronmongery,
Messrs Aluminium Windows Ltd.
Head Office
Site file

of a tree or trees can take place. If the client wishes that trees should be removed proof of permission from the authorities to fell the trees would have to be checked first by the contractor's representative, to either ensure that preservation orders do not exist or, that if they do, the licence obtained is in order.

(iii)   Unauthorised advertising, particularly on hoardings, should be avoided by the site supervisor in certain situations unless a licence is first obtained.

(iv)   Before undertaking any demolition of buildings notifications should be sent for permission to do so to the planning authority by the site manager (this is done more usually by the client's representative) unless the contractor is himself/herself a speculator. Notification is required at least 28 days before the work is to be carried out.

Building control officer

(i)   A Commencement of Work Notice should be sent to this officer 48 hours before actual construction work starts on-site, by the site manager/supervisor or his superior.

(ii)   Inspection of certain stages of the construction work of the building is necessary by the building controls officer, and inspection certificates (originally issued to the building designer/architect who passes them on to the contractor) would be returned to the officer when and as the various building stages are ready for inspection (see Fig. 4.4.1).

(iii)   After visiting the site the building controls officer may find it necessary to instruct the site supervisor to alter or modify part or all of the work inspected of which the site supervisor would have to conform – usually after first notifying the architect.

Highways officer

(i)   A permit for a rubbish skip may have to be sought by the site manager when it is essential to deposit a skip on the highway (road, footpath or verge), but usually it is the skip owner who obtains the permit. A typical permit application form is shown in Fig. 4.4.2.

(ii)   An application should be made for a licence to open up part of the highways when it is necessary to do so (see Fig. 4.4.3).

(iii)   The fouling up of the highway with materials or dirt from lorry/dumper wheels may lead to a visit by this officer. Care should be taken therefore, to prevent the blocking up of road gullies, etc.

(iv)   A licence must be obtained for the erection of hoardings, gantries and even scaffolding which encroaches upon the highway. See Fig. 10.1.2 for an example of an application form.

City engineer

(i)   The site manager or other contractor's representative should first obtain permission from the city engineer to

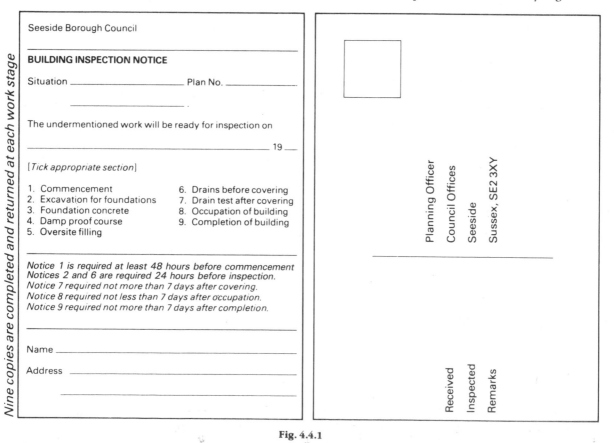

Fig. 4.4.1

BOROUGH OF MILDON

Directorate of Technical Services,
Works & Services Department,

Simpsons Road, Mildon, Surrey.

your ref   my ref

Telephone: 61 45231 6660

PLEASE RETURN IN DUPLICATE

Date ....

Dear Sir,

Application for Permission to Deposit a Skip or Container on the Highway
Sections 139   Highways Act 1980

Application is hereby made for the temporary deposit of a skip or container on the highway shown below and it is agreed that the skip or container will be removed on or before the end of the period mentioned or prior to this if police or the Council so request.

Any permission given will be subject to the provisions of Section 139 Highways Act 1980 regarding the removal of skips.

FIVE DAYS (excluding Saturday and Sunday) required for processing of application and POLICE notification by Council.

Name (Block Letters) ....   Signature ....

Address ....

....   Tel. No. ....

ADDRESS WHERE SKIP IS TO BE PLACED				
DATE OF START				
PERIOD OF DEPOSIT (Inclusive)				
LOCATION OF CONTAINER (Tick)	**ROADWAY**	**PAVEMENT**	**VERGE TARMAC**	**VERGE GRASS**
TRAFFIC (Tick)	Within 20 yards of Junction	Yellow Line Single	Yellow Line Double	
RESTRICTIONS	Parking Meter Area	Clearway		

SKETCH PLAN OF SITE

PLEASE RETURN IN DUPLICATE
N.B. Your attention is drawn to the provisions of Section 139 of Highways Act 1980 which are set out overleaf.
F. 184

Fig. 4.4.2

MILDON COUNTY COUNCIL

HIGHWAYS AND BRIDGES DEPARTMENT

FORM OF APPLICATION TO OPEN COUNTY ROADS

To the MILDON COUNTY COUNCIL

I/We . . . . of . . . .

in the Country of . . . .

(a) Strike out the operations which are not applicable

DO HEREBY apply for permission to (a)
    i. open
    ii. tunnel through
    iii. bore through
    iv. construct

(b) Name of Road. Mildon to

the (b) . . . . County Road

(c) Exact location of proposed trench, e.g. opposite house named "Downland"

at (c) . . . .

(d) Nature of Service required, e.g. Laying a sewer

for the purpose of (d) . . . .

and enclose Cheque, Postal Order, or Notes, to the value of

£ . . . . , as follows:–

	£	p
Deposit on account of the cost of reinstatement or/and charges incurred through supervision of tunnelling or other operations	_____	
* Premium for the insurance referred to in Condition No. 14	_____	
TOTAL £	_____	

• Please delete if not applicable

I/We, do hereby agree that the following Conditions shall be observed by and be binding upon me/us.

CONDITIONS

1. After I/we have commenced breaking open the surface of the road, I/we will carefully separate and keep the road surfacing, gravel and foundation in separate heaps and apart from the sub-soil, and will carefully return the same to the trench in the proper order in successive layers not exceeding 150 mm in thickness, and will thoroughly ram each layer providing at least three rammers to each filler, or will provide and use an approved ramming machine, and leave the surface of the road safe for the time being to the satisfaction of the County Engineer or his Representative.

2. I/WE will properly fence protect and guard the trench, opening, or other excavation and all materials, erections and obstructions placed in or on any part of the road at all times during the

[P.T.O.]

Fig. 4.4.3

make connections to existing sewers (a sewer is a drain which is maintained by the local authority).

(ii) Details of drains/sewer depths and positions are essential from the local authority engineer if there is any possibility of causing damages resulting from digger action.

(iii) Permission for temporary connections during construction operations should also be obtained by the contractor's representative.

(iv) Site supervisors should avoid depositing untreated waste which could contaminate the local authority sewage treatment plant (chemicals, etc.), and surface water drains should be protected against soil or concrete deposits.

Environmental Health Officer

Under the Control of Pollution Act 1975 a site supervisor should prevent unnecessary site noise, reverberations, smoke, dust or fumes from affecting surrounding occupiers of buildings. Also, other health hazards would be monitored, if requested by the contractor, by this officer.

Fire officer

(i) Fire certificates are required when using certain types of buildings as offices, etc. Under special conditions approval would be necessary for site hut usage, and precautions may have to be taken against the possible spread of fire (e.g. huts which may have to be situated on top of an existing building's flat roof which is undergoing refurbishment).

(ii) The officer may visit the site and the building under construction to check on conditions or processes.

Petroleum officer

(i) Licences are essential for the storage of flammable liquids, depending on the quantities to be stored.

(ii) Inspections would be made by the officer, where necessary, of any storage areas and storage structures to ensure proper conditions exist, and to ensure safety is being observed.

Rating officer

Notification of the use of site huts would be necessary where the huts are to be used for more than 12 months. The rates would not fall due until after the first 12-month period, but would be back-dated.

Careers officer

(i) A notification, on Form 2404, should be sent to the careers officer within seven days on the employemnt of a person under the age of 18 years (see Fig. 6.5.6).

(ii) Job vacancies, suitable for persons under 18 years of age, can be circulated to the careers officer who then notifies interested youngsters.

(iii) The careers officer may send a request for an employee under 18 years on-site, etc., to attend a medical. The site supervisor would normally allow the youngster time off to do so.

(iv) The careers office should be notified when a job vacancy is filled to save the time and effort of everyone and prevent confusion.

Job Centre personnel (Labour Exchange)

(i) Site supervisors or other contractor's representatives may circulate job vacancies.

(ii) Enquiries may be made from the Job Centre's personnel if there are suitable levels of labour available, especially when a new job is about to commence.

(iii) Notifications should be sent when jobs have been filled to eliminate further job enquiries and therefore save everyone time and effort.

(iv) Under the Employment Protection Act 1975, etc., impending redundancies of more than 10 employees must be notified through the Job Centre to the Secretary of State for Employment.

Police officers

(i) Site supervisors or other contractor representatives should be prepared to discuss problems such as suspected intimidation in labour dispute situations with police officers.

(ii) Approvals should be sought when deliveries of materials are to be made in congested streets. A police officer may be sent to direct traffic.

(iii) Parking problems generally would be queried by these officers unless prior approvals were given.

(iv) Complaints from the general public would be investigated by the police of contractor's nuisances.

(v) Notifications of payroll deliveries may be made as a security measure.

(vi) Site supervisors would be liable if they were responsible for the overloading of skips or vehicles, and may even face the prospects of having their own licences endorsed if found guilty of the aforementioned offence.

School Headmasters/Mistresses

(i) Job vacancy notices may be sent to these persons, or the contractor's representative may attend at careers days to help in recruitment of young persons.

(ii) Complaints could be sent by the site supervisor to the headteacher of constant trespassing by schoolchildren on the site.

(iii) Warning notices or posters of the dangers on certain sites may be circulated to schools when necessary.

Technical or further education college principals

(i) Notices of day release non-attendance of trainees/apprentices are normally sent by the principals to the personnel or training officers of businesses. However, some site managers have similar responsibilities to these officers, and non-attendance notices may be sent to site for disciplinary or other action to be taken.

(ii) Reports of progress or problems of trainees/apprentices would also be sent from the colleges.

(iii) References may be given by the colleges on individuals who may be applying for a job on the site being run by a site manager.

Health authority officers

(i) Notification is necessary to the Ambulance Authority, by letter, within 24 hours as soon as more than 25 persons are employed on-site. This is done by the site manager or other contractor's representative.

(ii)   Contact would be made with hospitals when an accident occurs which is serious, or injury results due to an unhealthy working environment.

2.   Government officers

Factory Inspector (Health and Safety Inspectors)

(i)   Notification, on Factory Form F10, of building operations or works of engineering construction which are to last for more than six weeks under the Factories Act 1961 should be made to the inspector.

(ii)   Notify cases of poisoning or diseases on Form F41 to this officer, and a second copy should be sent to the Employment Medical Adviser.

(iii)   It is necessary to notify the local factory inspector within 28 days when blue asbestos is discovered during demolition, etc., and before it is removed. (The earlier the better.)

(iv)   Notify dangerous occurrences – collapse of a crane, the shattering of revolving drums (grinding stones), etc., even though there are no injuries – Form F43B.

(v)   Other notifications are shown in section 10.4, e.g. explosions or fire caused by petroleum spirit, etc., and the use of offices, by persons on-site, of more than 21 hours and for more than a six-month period.

(vi)   Site supervisors should issue data when requested to the factory inspector, and should comply with Prohibition and Improvement Notices.

(vii)   The officer also visits the site to check inspection records, etc.

National Insurance and Social Security Officer

Query national insurance problems of employees.

Inland Revenue Officer

(i)   Communications may be necessary regarding a new starter's form P. 45 for PAYE.

(ii)   A form P. 46 is sent to the officer when a new starter fails to produce his/her P. 45.

(iii)   Querying details regarding subcontractors 714 1, P or C certificates (Individual, Partnership or Company).

Customs and Excise Officer

VAT returns – dealt with by head office accountant.

HMSO Employees

Government publications are normally purchased by contacting the above employees regarding PSA/DOE Advisory leaflets, and BSS and BSCP, etc., especially when a technological problem has to be solved.

Meteorological Officer (local centre)

Site supervisors may contact this officer at a number of local centres throughout Great Britain to obtain weather forecasts and temperatures, especially when certain external work could be seriously affected by the weather.

Advisory Conciliation and Arbitration Service (ACAS)

Site supervisors may be requested to give evidence in cases claiming unfair dismissal, etc.

3.   Utility services employees

Water authority officers

Permission should be sought either by the contractor's head office or site supervisor for the following:

(i)   A temporary water supply for the site (paid for through meters, or as a percentage of the contract sum).

(ii)   To make connections to the statutory bodies mains.

(iii)   To use water within a new building (generally the designer's responsibility).

Also details of the location of service/mains pipes should be obtained (position and depth) to prevent damages during excavation work.

Gas official

Similar permissions as those required for water, including details of location.

Electricity supplies official

Similar permission required as aforementioned, but, in addition, street lighting provisions would possibly be necessary.

Telecommunications employees

Similar applications to those required for water, gas and electricity.

When seeking details about the location of the various services, particularly underground, applications should be sent accompanied by block plans to each of the statutory bodies, to enable them to incoporate their services positions which would then be returned to the applicant. The positions of the services could then be marked out on-site from the returned block plans to safeguard against construction damage.

Chapter 5
Supervisor's ability to appreciate human resources

5.1 Recruitment and selection of operatives

Recruitment

The responsibility for recruitment of all personnel within a company generally lies with the personnel officer. However, when new operative recruits (joiners, bricklayers, mechanics, labourers) are deemed necessary the task of recruitment may be decentralised, and the site supervisor could be responsible for maintaining labour levels on his/her site, assisted at all stages by the foremen, gangers or chargehands. Should this be the case, then the supervisor must take cognizance of the company's personnel engagement policy outlined in guides known as Handbooks, examples of which are issued, by an efficiently organised company, to all supervisors.

The operatives, because of the casual nature of the industry – which leads to a greater turnover of labour than other industries – have to be screened carefully before being engaged by any efficiency conscious company. The recruitment and selection part of a supervisors' responsibility is critical to the satisfatory completion of construction work, more so than many of them realise.

Engagement of unsuitable operatives can lead to a reduction of output and an increase in accidents. If these workers are engaged, and are not detected until it is too late, eventually the site becomes inefficient and production drops – not to mention the cost to the company in terminating the employment of the inadequate workers caused by the protective legislation.

The process of recruitment can be an expensive one because of the time needed to replace an operative who has been transferred or has left the company, and by the cost of advertising and interviewing, when one considers the lost production time between the decision that a vacancy exists to the actual starting of work by a new operative, and the valuable time taken by the interviewer while conducting interviews.

If labour is scarce due to full employment within an area, good inducements must be at hand to entice the right kind of recruits to the site. This also depends on the use of proven successful techniques of advertising. Methods of advertising at hand to supervisors if necessary are:

1. The Site Board, facing towards the passing public, with details of jobs available within.

2. Information to the present work-force concerning vacancies and then details will, it is hoped, pass through 'the grapevine' to interested outsiders.
3. Advertisement card in local shop windows.
4. Details of vacancies sent to head office where an advertisement could be placed at the gates of the offices.
5. Advertisement through local, or indeed national newspapers.
6. Details sent to schools/colleges depending whether labourers or apprentices are required (usually a head office responsibility).
7. Requirements notified to local Labour Exchanges/Job Centres.
8. Vacancies telephoned to the Youth Employment Service.
9. An approach could be made to an Employment Bureau.
10. Approaches to local union officers (if this is within the policy of the company to do so).
11. As a last resort, advertisement for labour-only, or indeed ordinary, subcontractors.

In order to attract applicants who are experienced and trained to do the work, and to eliminate the possibility of receiving applications from the wrong type of applicants, a supervisor must undertake the following, albeit subconsciously:

(a) job analysis;
(b) job evaluation;
(c) job specification.

The above are really more specialised problems which the personnel officer is best trained to do. However, supervisors should have some ideas regarding them to appreciate recruitment difficulties, and expenses, so that where possible every endeavour should be made to retain the work-force under his/her charge to help keep recruitment costs to a minimum.

1. *Job analysis*
This entails discovering if a job really exists, and if so, to observe the particulars about the job; and to discuss details about the job with those who are doing similar work.

By studying the job certain factors may be highlighted to show what characteristics a person would have to be gifted with (if they have to be gifted at all) to make them suitable for the job, e.g. intelligence etc., or physically strong. It may be a dirty, noisy job which requires a special person to do it. On the other hand, it may require a self disciplined person who may have to work alone without supervision.

A universally recognised questioning technique which managers/supervisors are advised to adopt should therefore be used to determine most facets relating to the work.

(a) **What** has to be done?
(b) **Where** is it to be done?
(c) **How** is it to be done?
(d) **Who** does it?
(e) **When** does it need to be done?

By adhering to this analytical approach one can sometimes arrive at a better understanding of the work/problem in hand which helps in finding a solution.

2. *Job evaluation*

This means looking at the job to be done which has been determined by job analysis; and by studying the degree of skill, mental effort, physical effort, responsibility and working conditions required to undertake the job a picture emerges so that a job specification can be prepared and from it an advertisement to obtain the right personnel/operatives.

3. *Job specification*

This includes details gleaned from the job analysis and evaluation about the vacancy and the expected qualities of an individual to do the job. Additional benefits and fringe benefits as well as pay/salary, holidays and promotion prospects or other gains would be stated.

Naturally, if the job to be offered is standard, there should already be a job description in existence which could be used when preparing a recruiting advertisement. There may indeed by a filed copy of an advertisement which could be retrieved by the personnel officer for use when required.

Usually, if the job is of an unusual nature the procedure recommended previously could be adopted.

In most cases there is no need to determine, by evaluation, anything about the job, as rates of pay, etc., are already fixed by agreement with the operatives' union through the National Joint Council for the Building Industry, details of which can be found in the current Working Rule Agreements (WRA). There is, however, nothing to prevent the company from offering above the 'norm' payments and fringe benefits (provided there is no prices control and incomes policy in force by the Government at the time), if conditions dictate, e.g. shortage of the right kind of labour or where there are difficult working conditions etc.

Advertising vacancies

When using the previously described media for advertising a vacancy, careful wording of the job particulars is essential to prevent subsequent misunderstandings arising between the new employee and the company. For example, it is wise not to be specific when advertising for second fix carpenters/joiners, because if the second fix work runs out the employee could refuse to do alternative work offered. Effectively, the operative could claim that he/she has been made redundant, which due to this liability could cost the firm unnecessary expenditure. Therefore, titles of jobs must not be too restrictive in description but be broader in meaning to prevent the problems arising which have just been described.

Figure 5.1.1 (A–D) shows typical advertisements which have either appeared on notice boards outside some construction sites or in newspapers/journals. In Fig. 5.1.1 (A) the broadest meaning can be seen, but because it is not too specific a specialist or better graded carpenter may not apply as he/she would not wish to do just any type of carpentry (say, formworking), because some of this work can be dirty and rough. Figure 5.1.1 (C) is more specific, and is aimed at attracting quality workers. Many of the general quality joiners would not apply; thereby it is self screening, and makes the supervisor's task of choosing the right person for the job easier because of the smaller number of applicants which would result from such an advertisement.

Applications

Assuming that one or two applicants apply directly to the site and others send for application forms (which should be filled in and returned by the closing date specified – forms which arrive late should be rejected), the rule could be, and is so in some companies, that the site manager/general foreman, assisted by the appropriate trades foreman, ganger or chargehand, should conduct the final interviewing and selection of applicants.

Before an offer is made to a successful interviewee, references should be checked via the personnel department, particularly as this tends to be many companies' policy on engagement of personnel. It is also prudent to have those who apply directly to the site to fill in an application form, which would be processed in exactly the same way as the head office applicants.

It is of interest at this stage to appreciate the difference in the format of an application form used to recruit operatives and one for other personnel. The main difference being that while it may be thought necessary to enquire about an applicant's qualifications for clercial, supervisory, technical or managerial work, it is sometimes thought unnecessary in the case of most operative positions. Other differences may be appreciated by studying the two application forms in Figs 5.1.2 and 5.1.3.

Selection

Short list

There are different ways recognised in industry for the scrutiny of application forms before a short list of those to be called for interview can be prepared. In most cases this procedure can be dispensed with because of the small

TYPICAL SITE/SHOP ADVERTISEMENTS

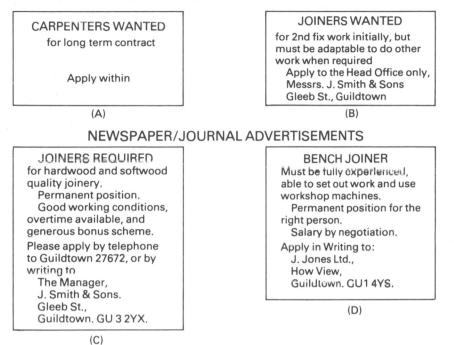

Fig. 5.1.1 Sample advertisements

number of applicants – all could then be called for interview. And, indeed, where the person applies directly to the site the interview could be conducted on the spot, subject to any company policy about the final engagement and checking of references and details. The site applicant should be made to fill in an application form so that his/her particulars can be retained for wage purposes and attitudes in the event of full approval being given for engagement by head office.

One method is to have a check-list with details printed down using information headings from the application form, with grading columns indicating whether the particulars given by an applicant comply with or come up to the standard required, as shown in the job specification. Those applicants who are given the highest overall gradings would be short-listed, the remainder should be sent a polite rejection letter, stating, if it is the company policy to do so, the reason for rejection. This could of course lead to complications if the wording is poor, so most companies take the simplest and easiest course by stating on the rejection letter/card that someone else has been appointed, and that should the unsuccessful applicant wish to apply again for a similar appointment in the future they should feel free to do so.

It is prudent to retain any application forms of rejected applicants for at least three months. They could then be used if someone fails to take up an appointment after being offered the job, or a complaint is received regarding discrimination of a racial, sex or, indeed, trade union nature.

Those who are successfully placed on the short-list should be sent polite invitations to attend for interview stating the time, date and place, or, if necessary, they could be telephoned to arrange a mutually agreeable date and time.

Interview arrangements

Arrangements should now be made to hold the interviews, and if they are to be formal interviews at site level, perhaps one could do the following:
- arrange for a room in which to hold the interviews.
- arrange adequate eating facilities to be provided.
- appoint someone to meet candidates – list of names issued.
- arrange tea if essential.
- appoint others to assist in the interviews, if necessary.
- see that standard engagement letter is available to type in successful candidate's name, etc.
- arrange a stand-in during own absence while interviewing.

If the interviews are to be informal, perhaps only a quiet room with a couple of seats could be booked.

Without the need to go into too much detail, the supervisor would do well to study the following brief

STUDENT NOTE: The particulars can be copied from this form onto the Operative's Personal Record Card.

B. BUILDER LTD., MERROW, GUILDFORD GU 1 2TZ.

TEL. 67676

APPLICATION FORM SITE/SHOP OPERATIVES

Name in full ....
Address Permanent .... Telephone no...
Address Temporary .... Telephone no...
Date of birth .... National Insurance no. ....

Occupation

Have you served an apprenticeship? ....When? ....
Have you attended a Government Training Centre? ....When? ....
Have you worked for this firm before? ....
If so, when? and where? ....

Give details of your last two employers

	(1)	(2)
Company's name	....	....
Location/site	....	....
Type of job	....	....
Occupation	....	....
Date start	....	....
Date finish	....	....
Reason for leaving	....	....
Pay	....	....

Next of kin
Name and address..
Telephone ....

Are you in good health?   Yes/No   Are you disabled in any way?   Yes/No

---

I certify that the particulars above are a correct record and agree that my previous employers may be approached for references.

Signed ....Date..

FOR OFFICE USE
Interviewed by...at..
Start date...Rate of Pay/Bonus ....
References ....Fares/Allowances ....
Letters Dated..

Fig. 5.1.2

BUILDING CONTRACTORS LTD.,
MERROW, GUILDFORD, GU1 1ZB.

REF......................

STAFF

PRIVATE

APPLICATION FORM

Application for position of ....

Personal particulars

Surname ....

Address permanent ....

Telephone no. ....

Address temporary ....

Telephone no. ....

First names ....

Age ....

Date of birth ....

Place of birth ....

Nationality ....

Marital status ....

Next of kin

Name and address ....

....

Relationship ....

Health: details of mental or serious illnesses in last 5 years ....

Education and qualifications

	Date		Subjects/examinations passed
	From	To	
Schools attended (age 11 onward)			
Colleges/universities			
Other courses			
Additional details			

Previous employment | In date order

Date		Name & address of employer and type of business	Position held and duties performed	Salary/ pay rates
From	To			

What date will you not be available for interview? ....

What date will you be able to take up the appointment? ....

Salary required? ....

Other relevant information

AN ADDITIONAL LETTER MAY BE INCLUDED

Signature .... DATE ....

Fig. 5.1.3

points regarding the function of selection interviews which are to:

(a) obtain and/or check factual information from completed application forms, if used;
(b) test orally the capacity of candidate;
(c) assess personality and intelligence;
(d) observe appearance and manner;
(e) state details regarding the vacancy;
(f) sell the job to the right candidate.

Planning the interview

It is necessary that interview documentation and other arrangements are made for the interview panel, e.g.

1. Copies of application forms prepared for panel members (if there is to be a panel), including the results of any enquiries for references.
2. Suitable setting – privacy, comfort, lighting, ash trays, etc.
3. Timetable for applicants.
4. Assessment forms, if necessary, with grading 1–5, under the following headings:
   - Appearance.
   - Manner.
   - Knowledge.
   - Ability.
   - Physical Ability.
   - Other Abilities.
5. Ensure that the panel members know which specialist questions to ask.

Conducting the interviews

Always conduct interviews as follows:

1. Start by being friendly (chairman).
2. Explain the job and its conditions.
3. Describe the firm.
4. Then keep the candidate talking by asking technical and other questions, but not questions which require a yes/no answer.
5. Invite questions from the candidate.
6. Record impression on Assessment Form immediately after the candidate leaves the room.
7. Be courteous and keep candidate at ease.

This ordinary 'selection interview' is the most common type used by the majority of firms throughout Great Britain, and is easily understood and accepted by everyone. But where special posts of, say, a management nature are offered, a different selection procedure exists in some of the more sophisticated companies and multi-national organisations. These procedures consider testing separately the personality, imagination, intelligence, aptitude and physical ability of candidates.

The interview can range from as little as one hour, to six or seven days – with each section being conducted by one individual or a panel, ranging from managers, psychologists to doctors. Naturally, for the ordinary run-of-the-mill site job it would prove too costly an operation and would be totally unnecessary.

Engagement

The candidate who satisfies most of the criteria for the job being offered (as recorded on the Interview Assessment Forms and by unanimous agreement of the interview panel – if there is one) would be told they have been selected, and if he/she accepted the offer engagement would be made in writing, subject, if necessary, to a medical examination.

Engagement of operatives should not normally be made on Wednesdays, Thursdays or Fridays, otherwise National Insurance Contributions will have to be made for that week, which is a costly burden to the company for, perhaps, only half a week's production from the new starter.

It would be advisable for the supervisor to ensure that when the new starter arrives for his/her first day at work that all documentation is dealt with either by himself/herself or by the site clerk. In cases when the employee has been engaged at head office their responsibility would fall on the employment/personnel officer initially. Nevertheless, a check must be made by the supervisor if most records are to be kept on-site, and to ensure a smooth induction into the company by the operative/new starter/employee. (Usually only applies on the larger contracts.)

Documentation can take the form of completion, for record purposes, the documents 1 and 2 and the issuing of documents 3 and 4 to the new employee, as follows:

1. Record of Engagement.
2. Personal Record Card.
3. Contract of Employment Statement.
4. Employee's Handbook.

The operative may have to prove membership of a recognised trade union where a 'closed shop' agreement exists between the company and the appropriate union.

Record of Engagement

This is a standard form (see Fig. 5.1.4) which is retained on-site and which is completed with details from the operative's application form. Copies are then sent to the Personnel and Wages Departments. One copy is retained on-site and should be signed by the operative as being a correct record and as proof of receiving a Contract of Employment Statement. See Fig. 6.1.2.

Personal Record Card

In similar manner to the above the relevant details are transferred from the operative's application form on to this form, which is retained on the site for reference purposes, and which is used for noting any changes in rates of pay, responsibility and warnings for misconduct – should it be necessary. Dates of warning and reasons may prove invaluable in the event of gross misconduct and dismissal later. See Fig. 5.1.5.

The Contract of Employment Statement is dealt with in section 6.1 and Fig. 6.1.2, and the Employee's Handbook is dealt with in the next section (5.2) 'Induction'.

B. BUILDER, LTD.,
MERROW, GUILDFORD,
GU1 2TX   TEL. 67676

ENGAGEMENT RECORD

Start date . . . .

Full name . . . .   Address permanent . . . . .

Date of birth . . . .   Address temporary . . . . .

Occupation . . . .   Works no. . . . .

National Insurance no. . .

Holiday stamps

No. affixed   Part 1 . . . . Part 2 . . . . Part 3 . . . . Part 4 . . . .

Hourly rate of pay . . . . Travel/fares . . . .

Registration no. – disabled person . . . .

I certify that the particulars above are correct and I have received a written statement under the *Contract of Employment Consolidation Act* and a copy of the Company's Employees' Handbook.

Supervisor's/Clerk's signature . . .   Date . . . .

Operatives Signature . . . .   Date . . . .

Separate copies should be sent to the Personnel Dept., Wages Dept., and one retained on site.

Fig. 5.1.4

B. BUILDER LTD., MERROW, GUILDFORD.
PERSONAL RECORD CARD – OPERATIVES

Surname . . . .   Forenames . . . .

Address permanent . . . .   Telephone . . . .

Address temporary . . . .   Telephone . . . .

Date of birth . . . .

National Insurance no. . . . .

Trade . . . .

Engagement date . . . .   At . . . .

Additional information (Allowances, expenses travel etc.)

. . . .

Courses attended

Date	Title

Transfers/rate increases

Date	Site	Occupation	Increase rate	Remarks/signature

Periodic travel

Show dates!

Sickness and absences

from	to

Warnings

Date	Type	Reason	Signature

Termination of employment

Date . . . .

Reason . . . .

Site . . . .

Assessment . . . .

Signature . . . .

Fig. 5.1.5

5.2 Induction and career development of subordinates

Induction

Induction means, in the first instance, an introduction of a new employee to important aspects about:

1. The company.
2. The job and employment conditions.

Regarding the company, the new employee/operative should be told, even the briefest terms, details relating to its:

(a) background, achievements and aims;
(b) organisation and leading personalities;
(c) general policies and objectives;
(d) amenities – sports club, etc.;
(e) importance it plays in the community.

Depending on whether a new employee is an assistant supervisor, leading craftsman, apprentice or labourer would determine whether this information is given by issuing a company general handbook or magazine, or by first showing a company film (with all other new employees attending), followed by a conducted tour around the various departments and ending with a talk by the personnel officer or general manager. Most organisations dispense with such formalities as being a waste of time, especially if the new employee is an operative (labourer or craftsman). This could be seen as a mistake especially if a company wishes to inbuild loyalties to it.

Supervisors are normally given the responsibility to receive new employees at the job location and to assist in helping during the settling-down period. It is, therefore, necessary to introduce the individual to the new work and its environment.

Induction could be a gradual process which would enable the new employee time to digest and fully appreciate each detail being introduced.

On the site a new employee may be given partial or the full treatment, particularly if he/she is a fresh apprentice, trainee or young labourer, and unused to construction work and the inherent dangers. On the other hand a fully fledged journeyman/operative would have had substantial construction site experience and is therefore accustomed to the procedures and conditions, so the induction period could be brief by simply issuing the relevant company guides and usual documents as laid down by successive legislation.

While some companies have elaborate systems for the induction of new employees with others only applying lip service to it, some consideration must be given to the following if management is to maintain some credibility in the eyes of its work-force, unions, and the law:

- The issuing of an Employee's Handbook.
- The sending of employees for Medical Examinations if called upon to do so by the company doctor.
- The giving of some form of instruction on the immediate aspects of safety (follow up later by showing films on hazards normally encountered on-site/in the shop).

This should be done by the Safety Officer.

- The giving of instructions on the work to be undertaken [this could be followed up later, if required, by sending individuals on an in-company training scheme, Construction Industry Training Board (CITB) course, or courses run by BAS (Building Advisory Service of the BEC).]
- The introduction of a new employee to the wages clerk, medical steward, site/shop steward and other necessary personalities.

Additionally, and perhaps immediately more important, inform the new employee of the following:

(a) site working hours (start and finish times);
(b) first-aid location points;
(c) welfare details: canteen, drying rooms, tool rooms and lock-ups, and other relevant rooms.
(d) bonus and other incentive scheme arrangements;
(e) fares and travel allowance, if necessary;
(f) the ganger, trade foreman and chargehand to whom he/she will be responsible.

It must always be remembered that new employees without construction experience are at greater risk than experienced operatives, and are not only a danger to themselves but are a danger to others as well. There should be a special emphasis on this due to the high number of serious accidents recorded each year in the construction industry.

Employee's Handbook

Apart from outlining details about the site procedures other points should be conveyed to the individual through the company's Employee's Handbook. This handbook is a printed guide to which, it is expected, the new employee will refer when personal problems arise. The handbook should state:

1. To whom wages and bonus queries should be directed.
2. To whom one notifies when there are changes in personal details, i.e. marriage, number of children, address, etc.
3. How to correspond with the company regarding any matter.
4. The need to sign for equipment, if required, from the stores, and the need to return on time.
5. The action required in the event of absenteeism, and who to notify or ask permission.
6. Grievance and disciplinary rules.
7. How to report accidents, and to whom, and any special safety points.
8. When instant dismissal would be implemented, i.e. for assault and threats, theft, malicious damage, serious breach of safety regulations, falsification of records for gain, and disobedience of reasonable instructions.

Follow-up

This process is to check that the employee has settled in, and to see what progress in the new job the individual has made.

A close watch is therefore necessary on all new employees (record kept by the supervisor), particularly in the early stages of starting for the company, so that if they are found to be unsuitable the earlier it is discovered the easier it would be to have their employment terminated, which keeps within the terms of the Employment Protection Act 1975, and Employment Protection (Consolidation) Act 1978 and 1979.

The follow-up also has the added advantage that it shows the new employee that the company cares about his/her welfare, and that its aim is to make him/her feel part of an important team. This leads to a sense of belonging and a happier working relationship, with a feeling of security, which, it is hoped, fosters loyalty to the company and pride in the job. Better relations create goodwill and trust on both sides. This is good strategy by those in responsible positions.

It is usual for the company's personnel officer to become involved where the termination of employment of an individual is recommended for one reason or another, especially if the employee has been employed for more than 52 weeks; but up to that time the supervisor, through the site manager, may be given the authority to terminate the employment of operatives within this period, especially in the case of serious misconduct: this would be outlined in the Supervisor's Handbook (if the company provides one).

Career development

The company would do well to consider each employee's career development: it could be a sound investment for the future. It could assist in the company's training scheme, if it has one, which is the normal way to ensure that the company retains sufficiently trained personnel for the type of work undertaken. Supervisors could be called upon to instruct on any special skills they may have, especially where there is an in-service training scheme in operation.

It is well for a supervisor to know the forms of training which are available to operatives, apprentices and semi-skilled personnel if judgement is to be made during the selection process. Also a request may be received from head office to choose suitable individuals for specialist training.

Apprentices

Module training should be used extensively for the training of apprentices instead of the adhoc arrangements of some firms, where apprentices are supposedly instructed and closely supervised by craftsmen on the rudiments of their trade, and means that too many apprentices remain working on mundane and monotonous jobs without someone attempting to arrange a scheme to ensure as many skills as possible would be covered in a 3 to 4 year apprenticeship.

Apprentices could start on a 'Module Scheme' devised say, by the CITB, which would be laid down in a Training Manual and which outlines each stage of work to be undertaken by apprentices in their chosen craft. A log book would be maintained by the apprentices of the work and skills covered. The log-books would be inspected periodically by a CITB inspector to ensure appropriate training, as laid down, was being given. Training would be supplemented by the apprentices' attendance at a technical college or training school/centre.

Technical college

Part-time, block release and YTS courses are offered at various colleges for craft apprentices in most trades. The part-time courses are well known. The YTS courses are comparatively new and need the full approval of the CITB, (which is the managing agent for the MSC or Training Commission). The courses are normally for a two-year duration. When the YTS apprentices complete their college two-year education and on-the-job work for the 'sponsor' firm/employer, they are eligible to attend day release classes to advanced craft level of the City and Guilds of London Institute, having already taken their ordinary craft level certificate at the end of the first two years YTS course.

Apprenticeships run for a three-year duration, which means that one could be a fully fledged tradesman by 19 or 20 years of age.

This system of training is now leading to a standardisation which at the least gives a new employer some idea of what level of work skills a fresh craftsman has achieved. It was not unknown for a young man previously to pass out of his time as a carpenter and joiner after spending a five-year apprenticeship assembling rough formwork and very little deviation on to better quality work; and young bricklayers being called craftsman, but in fact, doing nothing more worthy than laying bricks to straight walls on repetitive council housing schemes during their apprenticeships. Apprentices now have to sit skill tests to prove their worth.

CITB

This body is charged with the repsonsibility of maintaining the correct levels of trained personnel for the industry under the Industrial Training Act 1964. It has a number of training centres throughout Great Britain, Bircham Newton in East Anglia being one of the more important ones (see Fig. 5.2.1). Courses are offered in all skilled and semi-skilled disciplines relating to operatives and supervisors work, e.g. road building skills, scaffolding, paving, concreting, drain laying, steel erecting and steel fixing, plant operating and site supervision. Where it is considered that there is a shortage of key specialisms the Training Board will offer courses, and construction companies may send their chosen personnel to attend for instruction with certificates being awarded on completion. Persons who are being introduced to the industry for the first time could first attend up to one-year residential course at one of the various centres.

Fig. 5.2.1 CITB field areas

Companies who take the opportunity to send employees for training can offset the cost in doing so by claiming a 'grant' from the CITB.

Free residential places for training are given to selected pupils who are about to leave school and would like to enter the industry.

BAS

The Building Employers Confederation, through its Building Advisory Service, can be called upon by companies to run training courses for its members and others who are willing to pay for the service. Most of the courses offered by BAS are of a supervisory/management nature.

The Training Commission

The Training Commission is a Government-sponsored body and organises courses at its training centres throughout Great Britain to retrain individuals in an alternative skill to the one they may already have but the necessity of which is no longer required due to new industries replacing old ones.

The organisation can play a major role, particularly in providing finance for retraining in areas of high unemployment but where new industries are being enticed into local or regional problem areas.

Grants can be obtained for job training schemes (JTS) by persons wishing to train for work after a period of absence, due to unemployment.

If qualifications, training certificates or indentured certificates need to be inspected when selecting persons for a job, supervisors should at least know the types of qualifications which are in existence and the value to be placed on them, remembering that successful previous experience cannot always be substituted by qualifications and certificates alone.

Operatives/staff appraisal

This requires some consideration to enable promotions to take place from time to time. The morale of a company's employees must be sustained, and if new employees are aware of the possibility of internal upgradings and promotions by 'ladders of opportunity', the ambitious ones can only help to maintain the company's positions in society as well as assisting in its prosperity and competitiveness with other companies.

Status is an important aspect of life and where enhancement of status is stifled, alternative financial rewards, etc., must usually be offered.

Supervisors are called upon from time to time to keep records of subordinates' achievements, etc., so that when a vacancy arises within the firm existing employees should at least be given an equal opportunity to apply and be considered for the vacancy. Reports may therefore be called forward from the supervisors who should present them without being too blinkered to their subordinates' achievements and shortcomings. Too many people in

authority try to deny others on a lower level the opportunities they themselves have enjoyed, and take the view of 'push the boat out, I'm on board'. Fairmindedness and fair play should always be one of the many qualities of a supervisor.

Other ways one can qualify or train for industry's needs are:

- by qualifying through a college of Further Education to Higher National Certificate or Higher National Diploma.
- by attending a company's own training centre – grants obtained in some instances from the CITB.
- by attending a group training centre – financed by a number of companies.

5.3 Motivation of subordinates

Behavioural science is a study in the moods and attitudes of individuals to determine what it takes to motivate them generally, and how it causes them to act in their different ways. Through many years of research it is now understood that the behaviour, etc., of individuals depends firstly on one's social background, the environment in which one lives, the working conditions and challenges, financial incentives and relationships with others. These are synonymous with the way people are motivated in general life.

Motivation of individuals or groups in the work situation depends upon:

1. The ease in which one makes friends.
2. Being recognised and praised for good performance.
3. Job security.
4. The satisfactory level of financial rewards for the job being done.
5. Challenging and interesting work.

A good company personnel policy is one which lays down ways to create job security, and which endeavours to solicit the cooperation of everyone, from executives down to the operative level. A company can instil a sense of belonging in the minds of the operatives by organising such activities as:

1. The printing and distributing of news letters or journals to inform employees/operatives about the fortunes of the company, e.g. new contracts won, personality reports – marriages, births, promotions; and even job vacancies.
2. Social, recreation or sports clubs which encourage friendship and a sense of belonging.
3. Competitions, with prizes for : safety quizzes, suggestion schemes, new technical ideas to improve company performance and image.
4. Savings schemes.
5. Holiday trips or outings.

The fact that the company takes an interest in its employees socially, as well as being profit minded, shows it has a humanistic approach and care for the wellbeing of everyone within its framework.

The company

The responsibility for all aspects of personnel and their motivation lies mainly with the company and it should also look at additional benefits which could be offered to the direct labour force/operatives and staff as inducements to maintain harmonious company/employee relationships, such as:

(a) suitable wages structure, at least in keeping within the Working Rules Agreements and with a policy for maintaining regular earnings with the possibility of overtime when necessary;

(b) fair conditions of service or contract;

(c) suitable bonus structure with cost savings being paid to the operatives;

(d) improved public and annual holidays for longer service operatives;

(e) good promotion ladder of opportunities;

(f) good working facilities – safety, health and welfare arrangements organised by full-time officers, including a Welfare Officer;

(g) carefully selected colleagues;

(h) pension scheme (non-contributory);

(i) promotion to 'Staff' level for long serving, loyal workers;

(j) profit sharing;

(k) canteen facilities, with hot meals, where possible;

(l) transportation arrangements – workers' coach to site works;

(m) suitable flexibility of start and finish times should an operative require it for personal reasons, with the possibility of making up the time at a future date.

These facilities are offered by some firms, and with a little additional supervision work amicably.

The supervisor

The supervisor helps in the motivation process by having the long recognised qualities to appreciate human problems faced by different individuals during the various stages of life. He/she must therefore;

● have understanding, and be sympathetic.

● be a good disciplinarian but not a strict one.

● compromise, if necessary, and not be dogmatic – listen to other sides of an argument/suggestion.

● show a desire to help those with personal problems.

● have a good general attitude to the work, colleagues and subordinates.

● be a concise communicator.

Subordinate/operatives should be supervised and encouraged to work as a team, and should be made to feel part of an important group by enlisting their cooperation and help. Creativity should be fostered to give each subordinate some freedom to improvise and use their initiative throughout the work period.

The supervisor, although his/her goal is to increase productivity, where there is repetitive and boring work there should be an attempt at job rotation – with each operative being given the opportunity to do interesting work, and where there is a need, they should be placed in a challenging situation to suit their ability.

Harmonious relationships throughout ensures cooperation generally and where this relationship is threatened there should be prompt action by the supervisor to counter it. Some disharmony results when a worker fails to fit into a gang due to personality, laziness, etc; to recognise this and to replace with someone else by a discreet re-shuffle of individuals helps to relieve tensions and can assist in maintaining morale and team spirit, and, it is hoped, maintains the correct level of productivity at the right quality.

The supervisor should be fair, but firm, in dealing with a disciplinary problem. Behaviour of an operative which falls below the recognised standard requires a warning as laid down in the procedure given in the company's supervisors' handbook – verbally at first. A warning for consistently bad timekeeping, workmanship, etc., contrary to the belief of some people, has the effect of increasing morale and self-discipline, more so than if it was allowed to continue unchecked. Others would take the attitude 'well if he can get away with it why should we bother?'

Where workers are frustrated in their attempts to reach their self-assessed goal in life, they will cease to be motivated at work, whereupon the site and eventually the company suffer. Every effort must be made to encourage operatives to improve themselves in seeking advancement in status, and to increase their financial standing. In their quest to do more rewarding work and to elevate themselves within the company, assistance should be given where it is thought necessary. Transfers to other work or contracts should be arranged where possible for operatives, etc., who require it, and sound planning and forecasting of possible transfers to allow operatives on lower quality work, or lower status, to step up the ladder of opportunities to maintain their morale should be standard procedure. In the event of transfers it is necessary to ensure records, etc., are sent or are received on time so that there are no delays in wage and other payments.

Sound supervision both by management and site supervisors is necessary to ensure operations are planned in correct sequence, and materials, etc., are available on time, at the right position/location and correct quantities to create continuity of employment with the minimum of delays. Through this efficiency work can be completed within target times for bonus payments, which helps considerably for better production and motivates the operatives.

Extra payments must be made to key personnel with additional knowledge or skills and who are prepared to accept a level of responsibility not only for other's work, but for doing difficult or quality work themselves.

On the recommendation of the supervisor, faithful and good service should be recognised by the company and a scheme should be implemented to allow wage earners to transfer to 'staff', with all the inherent benefits, i.e. sick pay, pension scheme, additional public and annual holidays; concessions regarding purchases through the company; no loss of earnings due to stoppages, or even a company house/flat at a reasonable rent.

In winter in inclement weather some attempt should be made to create better site and working conditions by levelling or battening the ground to ensure ease of transportation and travel by operatives. Suitable cross-overs and protective screens to work points should be supplied to show the company's earnestness in dealing with the operatives' welfare and general working conditions. There are numerous other ways to help motivate individuals, i.e.:

1. Invite key operatives and those with recognised good qualities to attend in-company or other courses in special skills or management.
2. Ensure that the bonus clerk accurately measures the work done by operatives and be fair in assessments.
3. Treat persons as individuals, but at the same time insist on everyone observing the site personnel structure.
4. Instil cooperation by showing operatives, when possible, the confidence one has in them.
5. Prepare for replacement by foreseeing retirements and transfers, etc., by increasing subordinates' responsibilities and embarking them upon a suitable training programme.
6. By creating an atmosphere of congeniality to prevent antagonism.

In assessing whether operatives are motivated or frustrated one can look at the following:

(a) level of bonus being earned by each individual;
(b) quality of work produced;
(c) absences from work due to sicknesses, etc.;
(d) industrial action or disputes;
(e) labour turnover levels;
(f) levels of aggression, especially shown when orders/instructions are given.

While many supervisors, even under extreme working conditions, can get good results from their subordinates/operatives, generally the company personnel policy has the greatest influence on the motivation of individuals.

5.4 Leadership relating to work team performance

It is said that to get the best out of a workteam they need to be motivated. To be an influence in the motivation of individuals at site level the leader/supervisor must have extra qualities over those of the normal operative or member of a team. The old cliché of 'some people are born to lead and others are born to be led' is correct to a point. Under normal conditions many persons, although they are not natural leaders, excel; or by sheer personal effort find themselves, by either chance or by being in the right place at the right time, in positions of authority which is beyond their reasonable expectations.

Too often individuals find themselves in responsible positions above their capabilities, and resort to aggressive supervision to maintain their positions. This can lead to a breakdown in operatives' morale. Others quite naturally gain in status and reap the admiration from their subordinates through the examples they set in their approach to work, their intelligence or by being more generally suitable than others in being able to maintain a good level of work.

When a supervisory vacancy arises, many organisations are quite unprepared and immediately begin the search to engage someone who is suitable from outside the organisation. For very senior positions this is sometimes the acceptable practice. For a supervisory post, even though it is the first line of management, an organisation should be in a position to foresee possible vacancies arising and should plan, as do the better companies, for this eventuality. It can be said that only a limited resources firm is one which has one trained individual for each workteam. If assistants or understudies were appointed/nominated and were given periodic training it could be an automatic upgrading to the key position when a vacancy arises within a team, thereby leaving a vacancy which could be filled from outside for the lower or less responsible post. Some would say that this leads to inbreeding, and if a mistake was made in the selection of certain personnel the troubles for the company would only multiply; better to advertise all vacancies outside and within, and those within would have to compete with outsiders with the most suitable person succeeding. Certainly some grooming should be in evidence within a firm to maintain standards so that a vacuum created by a vacancy should never be allowed to remain for long, as 'the devil you know is better than the devil you don't know'.

There will always be a percentage turnover of supervisors and ordinary personnel through retirement and other natural losses. There should be a measure of selection of existing staff by observation, reports and follow-up procedures. Some individuals who are adjudged to be potential leaders would be sent on outside, or in-company, training schemes to maintain the existing standards; to keep abreast of modern developments, and particularly to uplift the individuals who deserve it and to show that the company is planning for the future and is looking after the individuals' future, by making them eligible for promotion later. Natural leaders will always be eventually recognised the longer they are employed by a company/firm.

It is essential to a company's survival that its leaders, i.e. managers and supervisors, have a number of qualities which make up a good, balanced personality. They must:

(a) be self confident.
(b) be technically knowledgeable.

(c) have a sense of humour.
(d) be tactful and helpful.
(e) be determined and enthusiastic.
(f) have power of persuasion.
(g) be courageous.
(h) have integrity.
(i) be good verbal communicators.
(j) be good listeners
(k) be good disciplinarians, but fair.

A good leader/supervisor should be one who knows his job technically, but also have managerial skills. In dealing with management problems he/she must have a humanistic and scientific approach. This is mostly learnt by adequate previous practical training and by attaining technical qualifications relating to his/her work. While there is no substitute for good practical knowledge technical and managerial studies help to supplement it.

Knowing the work and being able to give sound advice if asked for by subordinates, and being able to direct operations confidently at all stages – and with authority, is one of the most important criteria relative to a supervisors' leadership qualities, and gains respect from the workteam/subordinates under his/her control.

Mid-career training is thought essential for supervisors whether it is through in-company training schemes, or, as an example, through a scheme similar to the Building Employers Confederation/Chartered Institute of Building 'Site Management Scheme'.

Many practising site managers/supervisors who have little or no formal education/training can be sponsored by their companies to attend specially designed courses at some colleges of technology. The courses run mainly in the evenings for up to a period of three years, after which, if adjudged to be suitable, the successful supervisors are awarded a Diploma in Site Management (which can be used to gain associate membership of the Chartered Institute of Building). The subjects studied in depth and which will greatly assist the supervisors in their work are: planning and programming; legal studies; conditions of contract; safety; the Working Rules Agreement; communications skills; bills of quantities and specifications; industrial relations.

A good leader has many responsibilities. He/she is usually accountable to the immediate superior for any problems which may develop and therefore should have the ability to:

1. Secure individual interest by informing operatives of progress on-site, or, if necessary, progress relating to the firm generally.
2. Maintain the respect and loyalty from operatives by ensuring the correct level of discipline where necessary and securing correct financial rewards when earned.
3. Instil keenness by involving all concerned.
4. Listen and act sensibly and fairly to individuals' problems to prevent frustrations developing which could lead to disharmony.

5. Show sound judgement and not favouritism in dealing with grievances or disputes between individuals.
6. Encourage self respect and a sense of responsibility by operatives.

Leadership can be taught to some degree. When instructed on the many pitfalls experienced by supervisors, guidance on what one should do and not do in varying situations should be given. In being taught leadership supervisors can be guided on how best to deal with problems which frequently occur in day-to-day working, particularly on construction sites.

In studying psychology, etc., the supervisor can learn how to respond to the many situations, particularly those problems that develop where the many differing personalities are concerned which are usually found on a construction site.

A good leader should:

1. Be decisive.
2. Show self-control in awkward situations – not lose his temper, but be composed.
3. Be good at supervision, but not close supervision unless it is essential to quality control where the workers are below usual standard. A lack of trust shown by close supervision can upset the morale of most subordinates.
4. Be in a position to take control of any situation whether it is to solve a particular technical problem, or a human problem, by assimilation and validation of the information presented.
5. Be constantly monitoring progress in order to successfully complete the job in hand on time.
6. Be able to mediate in a conflict situation calmly and with courage.
7. Be able to read and correlate written and verbal communications.

A good leader must always be fair and in a position to train subordinates/assistants so that they can accept delegated responsibility when called upon to do so. In helping individuals to grow in themselves a trust is built up by respect; this can only help to make the task of supervision easier.

Supervisors work mainly under stress for most of their working day due to their numerous responsibilities. As leaders, operatives/subordinates expect them to be ready to deal with any crisis or problem. He/she must therefore be mentally alert to:

(a) hold meetings when necessary if a problem is unusual, and he/she must be competent to conduct the meeting;
(b) answer telephone calls with authority and precision;
(c) tour the site to check progress; answer general queries; deal with problems of safety; and to carry out inspections where necessary;
(d) answer mail and other written communications promptly.

Supervision/management by crisis is to be deprecated. A good leader, therefore, does not wait for a crisis to develop but should foresee it and 'nip it in the bud' to prevent it getting out of hand.

Decisions on any points, where possible, should be taken after generally discussing details with those concerned. A good leader is one who is reasonably flexible, but once a decision is taken he/she should do all in their power to carry it through.

A leader must create a suitable atmosphere by works and deeds. By setting good examples in conduct and one's affairs the supervisor/leader engenders his/her will into others, so that it inspires and motivates individuals to work willingly and constantly to the mutual benefit of all. The examples a leader could set would be to:

(a) arrive early on-site and leave on time at the end of the working day.;
(b) observe correct tea and other breaks;
(c) wear safety helmet and protective clothing when touring work places;
(d) show diligence;
(e) be polite to all and sundry;
(f) be precise in giving instructions;
(g) attend to individuals' problems promptly;
(h) take action on promises given;
(i) chase up details and queries quickly when requested;
(j) maintain a tidy site with neat layout of materials to reduce hazards;
(k) observe working rules agreements;
(l) observe safety codes;
(m) be able to work with a measure of autonomy;
(n) be prepared to delegate and trust subordinates;
(o) keep up to date with register, diaries and reports.

In general, practise what one preaches.

Leadership qualities are many and perhaps no one leader has all the qualities described, but to try is to perhaps succeed; to succeed without even trying can only lead to temporary success, with failure possibly following closely behind.

Good leadership qualities eventually assist in a young apprentice's or trainee's climb to the top of the status ladder (see Fig. 5.4.1).

5.5 Roles of trade-union representatives

The unions which are parties to the National Joint Council for the Building Industry (NJCBI) and which the employers recognise for negotiation purposes are:

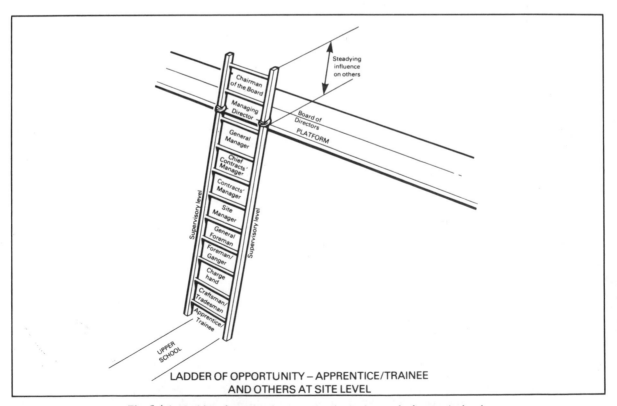

Fig. 5.4.1 Ladder of opportunity — apprentice/trainee and others at site level

1. The Union of Construction, Allied Trades and Technicians (UCATT).
2. The Transport and General Workers Union (TGWU).
3. The General and Municipal Boilermakers and Allied Trade Union. (GMBU).
4. The Furniture, Timber and Allied Trades Union (FTATU).

Some consideration must be given by the construction employers to the following unions, although in the main their members generally work for subcontractors in the construction industry service sections; it therefore would be the subcontractor's responsibility to conduct joint consultations with these unions:

1. Electrical, Electronics and Telecommunications/ Plumbers Trade Union (EET & PTU).
2. Engineering Workers Union (EWU).

Generally speaking, however, it is with the unions who are parties to the NJCBI with which the site supervisors would have to negotiate, if necessary, although any serious labour problems should be reported to the Personnel Department, and if the Industrial Relations Officer/Labour Relations Officer is attached thereto he/ she will take action immediately. Clearly the supervisor is able to assist in allowing any grievances to be dealt with promptly by the officer unless it is within the delegated responsibility of the supervisor.

The supervisor once again should have an understanding of the company's policy relating to industrial relations, which is part and parcel of the main personnel policy.

Types of union representatives

Part-time representatives on-site

Under the National Working Rules, which are agreements between the employers' and employees' unions, representatives of which meet under the NJCBI (or in civil engineering under the CECCB), the following trade union representatives may be found on-site, all of whom are unpaid, and will normally have worked on the site for at least one month to be eligible for recognition by the employer's representative, the supervisor:

1. One site steward for each trade or union.
2. Maximum of seven site stewards to form the site steward's committee.
3. A convenor steward and chairman of the site stewards committee who is appointed by the site stewards but normally works as an operative in the employment of the main contractor.

The site stewards are to be members of a union which is party to the NJCBI, and on being officially recognised as such by the unions and employers alike would be allowed the usual facilities to carry out their union duties.

Full-time officers

The full-time representatives of any recognised trade union (party to the NJCBI) may call at the site to conduct the affairs of the union. It would be expected that before venturing on to a site the union official must first seek permission from the site supervisor, who should not normally withhold his permission unless the official's business does not directly have any bearing on the site in question.

The supervisor should always check the credentials of the full-time official, who would be expected to be in possession of some form of identification. Problems have occurred where unofficial persons gained access on to the sites, leading to accidents and, in fact, unofficial industrial action. Certainly no site meeting should be allowed without a visitor first showing his/her credentials.

It has been known for certain sections of unofficial bodies to approach workers on-site without the knowledge of the official unions, and it has led to the disruption of work. The action has been mischievous and a nuisance, much to the regret of the companies concerned, and perhaps more so to individual operatives who's loyalties are torn between the unofficial and official bodies.

An official who in the normal course of his full-time union duties visits a site and eventually suggests to the supervisor that the company should deduct unions subscriptions from the union members on-site is entitled to do so, but the site supervisor should refer the official to the company's Personnel Officer/Industrial Relations Officer or Labour Officer – depending on which name is used within the company. A formal agreement can then be drawn up, making the company liable for the collection of union dues, which could then be passed on to the union local/regional headquarters or union's bank.

It is recommended that the site supervisor enters details of any union official's visits in the site diary, stating matters discussed or reasons for the visit, also the date, time and length of visit. Any other correspondence with the unions should be retained and a copy should be forwarded to the Industrial Relations Officer.

Usually after the appointment of a site steward the union concerned should send details of the appointment in writing to the site manager either directly or through his/her head office within seven days, otherwise the supervisor would be within his rights to withhold recognition of the steward. If there is a delay in receipt of confirmation from the union, the supervisor would be prudent to request confirmation by writing to the union, which can only maintain a mutual cooperation which leads to good industrial relations.

It has been known for some unions to object to the site operatives' apparent choice of a steward because of previous problems encountered, which may have led to poor industrial relations with a company. Recognition of a site steward chosen by site ballot/vote is not always automatic because of previous personality conflicts with the company or site supervisor, and thereby a site supervisor should normally contact the Industrial Relations Officer, giving reasons for the objection to enable him/ her to correlate all the facts before any official action (confirmation or rejection) is taken at site level, particularly as hasty action could lead immediately to an unnecessary confrontation situation. If there were reasons for not giving recognition there is a formal procedure for doing so through the Regional/National Joint Council for the Building Industry.

A superior is normally instructed not to recognise union stewards whose unions are not party to the NJCBI; but where there is an unusually large party of workers on-site belonging to alternative unions to those associated with the NJCBI, say, the EET & PTU, or EWU, the supervisor would be well advised to contact the Industrial Relations Officer before refusing acknowledgement out of hand. Under special conditions, as outlined, these unions and their representatives would be given recognition — although the subcontractors would generally deal with their own industrial relations.

Unofficial unions

Certain groups have formed which are outside the Trades Union Congress (TUC) guidelines and who act in an adverse manner to official unions. It is not unknown for these bodies to cause quite unnecessary hardships for employers, as well as the employees themselves, in pursuit of so-called workers' rights. The biggest problem with these groups — while attempting to gain better working conditions and renumerations which obviously benefit the employees in the short run — is that the negotiating methods adopted by them falls short of the codes laid down in the National/Regional Working Rules Agreement which have been compiled painstakingly over many years to the mutual satisfaction of workers and employers alike. Complete disregard for the codes turns the negotiating machinery in the building industry, which sets an example for other industries to follow, into a jungle of mistrust.

Site committee

If the site is large enough to warrant up to seven site stewards then it may be expedient from time to time for the convenor steward to call meetings to discuss important issues relating to site conditions. The site stewards, plus the convenor steward, would then be known as the Site Committee. The meetings may be called during normal working hours, and provided that the convenor steward requests approval for a meeting in an official capacity, then permission should be given by the site supervisor. It must however be understood that any site steward before attending a site committee meeting should seek permission from his/her immediate supervisor, and providing there are no critical operations being undertaken there should be no withholding of approval by any supervisor.

Sometimes a request is made from the convenor steward for the site committee to hold regular site meetings, then it may be suggested by the supervisor that a suitable time for the meetings to be held would be one half hour before lunch, so that if a meeting runs late it will encroach into the employees'/site stewards' lunch break and not

the company's. If meetings by the site stewards are to be held at irregular intervals to discuss important site issues, then the site supervisor should approve the meetings. Also, it is normal practice to allow the meetings to be held in the canteen. It must also be remembered that site stewards should be given paid time off to deal with his/her duties of office, and that the company telephone may be used for official business relating only to the union matters which affect the site.

The practice of the main contractor's site stewards representing subcontractors' employees, and similarly, the subcontrators' stewards representing the main contractor's employees is not normally tolerated, and the site supervisor should suggest that elections be held on-site to nominate site stewards for the company's employees, and similarly, site stewards for the subcontractors' employees; always bearing in mind the maximum of seven site stewards who are entitled to form the official site committee.

Site notice board

The supervisor's attention should be drawn to the misuse of the site notice boards by unofficial bodies/unions. As it is the company's property the supervisor should be alert to maintain its proper use for only officially recognised union literature, and literature and notices belonging to the company.

Because a site steward is also a site operative/worker he/she is not immune from dismissal for misconduct or serious breaches of company rules and possible redundancy, so long as the recognised procedures for warnings, or giving of notices, are served. It is usually the responsibility, however, of the Personnel Officer to deal with matters of termination of employment of site stewards, and the officer must be notified immediately by the site supervisor. The local or regional union officers in turn would be notified by the Personnel Officer of dismissals which are imminent; this is to ensure fairness in dealing with the site steward and to show there has been no victimisation because of the diligent way that union matters have been conducted by the steward.

Roles of the site stewards

The site stewards, who are unpaid part-time union officers, have many responsibilities relating to trade union matters and industrial relations, and although most of the duties revolve around the Working Rules Agreements there are other duties, some of which are included as follows: they
● act on behalf of the full-time officials within the WRA.
● attend Site Stewards meetings.
● call meetings of the operatives when necessary.
● inspect union cards and collect union subscriptions, if necessary.
● deal with redundancy problems affecting their brother workers.
● ensure that correct arrangements for apprentices/trainees are being observed, e.g. no overtime working on the days they are to attend day-release courses.

- assist the operative who has a problem with his supervisor to find a satisfactory solution.
- liaise with the convenor steward where a dispute occurs which affects more than one trade or union.
- try to recruit non-union members on-site.
- check on engagements of new employees in a closed shop situation.
- sort out problems relating to transfer of members from one site to another, particularly if the transfer requires the member to take up lodgings, or where there is an unnecessary long distances to travel to work to the new site.
- assist union members who need to communicate with the head office of the company.
- inspect health and welfare arrangements to ensure they are in order.
- ensure working conditions are maintained to a good standard.
- see that there are correct levels of remunerations for the work done.
- ensure bonuses, fares, travel and other allowances are maintained.
- act as a representative where a member is being warned for misconduct.
- see that members who are absent due to sickness, etc., receive their sick pay entitlements.
- check and report on accidents to the local/regional union headquarters.
- ensure the company applies in the correct way for over-time permits/permission.
- inspect safety equipment (helmets, etc.), to ensure that there are correct quantities to the right quality.
- attend courses on industrial relations (allowed to on full pay) when required, but the courses must be recognised by the NJCBI.
- be a member of the site safety committee and advise management on safety hazards.
- report to union local/regional headquarters from time to time.

One can see that if there is a large work-force on-site, a site steward would be a key operative, and some sympathy should be shown to the enormous task he/she has in looking after the union members' interests as well as having to do a satisfactory day's work.

5.6 Handling grievances and disputes

Human problems

Some individuals never appear to be satisfied and are obsessed with the desire to look for problems and complain constantly about everyone and everything. These disgruntled individuals create a certain amount of depression amongst those with whom they work and affect morale generally on-site.

Supervisors have to accept that there will always be individuals, such as those mentioned, in any society or work situation, with some sites having more than their fair share. One has to take a positive line of thinking in handling such individuals (which is not always easy, as

some are also abusive) by asking the question: Why? – Why do these individuals act or react like they do?

A number of causes which could have led to an individual's unfortunate personality may have generated from his/her personal background – from childhood, adolescence, adulthood or parenthood. An unhappy childhood due to many circumstances, e.g. broken home, battered baby, etc., could have contributed to the cause of the affliction. A serious illness (mental or physical) at one stage in a person's life, or a problem such as uncontrollable children or teenagers, could affect a person's whole outlook and make them react or complain and show disgruntlement with their surroundings at home and at work.

An unhappy marriage; concern about ailing elderly parents and debts would certainly depress most people and could contribute to the altering of the personality of an individual, albeit, temporarily.

One must have an understanding of human behaviour, and one needs to show patience with such persons who have problems as previously described. A certain amount of aggravation must be expected by managers and supervisors at some stage, and their personality should be such that aggravations must be accepted for what they are, although one should be on guard to prevent problems of aggravation going beyond the acceptable norm by being tactful, helpful, sympathetic, good humoured, firm, decisive – the normal attributes of a good supervisor. These forms of aggravation shown by some in their grievances tends to be the least serious encountered by supervisors.

The type of grievances which require a greater level of supervisory skill to deal with are those genuine grievances caused by a number of factors met with on-site, which are:

1. Poor working conditions.
2. Unhealthy working atmosphere.
3. Dangerous working areas.
4. Irresponsibility by other work groups/individuals.
5. Victimisation by supervisors, etc.
6. Personality clashes between an individual and gang or group.
7. Intimidation from other members of work team.
8. Lack of proper health and welfare facilities – clean canteen, poor food.
9. Lack of promotion opportunities.
10. Unfair level of discipline by a supervisor.
11. Inadequate level of supervision and control by supervisor to maintain work flow – materials and instructions not made available on time.
12. Low earnings (remuneration) – lack of overtime, etc.
13. Inadequate bonus – poor, unrealistic targets.
14. Stoppages by other trades or groups which affect others.
15. Poor methods used for doing work – work study helps.

Of these factors which causes an operative to be aggrieved, the most prevalent one is related to remuneration. Bonus targets are sometimes tight, and when one bears in mind that this kind of incentive scheme is to increase productivity and to allow the workers to earn up to 33 per cent more if they work at a regular and constant pace – wasting little time on tea and other rest periods by self discipline – there is little wonder at the attitudes shown by some operatives.

If one gang on-site earns excessive bonus compared to another gang whose efforts are just as vigorous, then this may lead to discontent and frustration with the system – so constant monitoring of bonuses is necessary by the supervisor in liaison with the bonus surveyor to retain a happy medium.

If bonus targets are not reached by certain diligent groups there should be an equitable way of adjusting the rates.

Some firms find it expedient to start on a tight target with a slackening off later in the light of experience and proof that the operatives' bonus earnings are low. Any increase in bonus payments has a better psychological effect on the work-force than if the bonus target set was too generous and there had to be a reduction for economy reasons. Workers do not readily accept reductions.

The basic rates of pay have for many years, it is argued, been too low in relationship to other key industries whose workers' skills and efforts fall short of the construction operatives'. It appears that to give the operatives a satisfactory level of earnings employers should offer a basic rate; a standing bonus; and a target bonus (plus overtime, travel allowances, etc.). The whole concept of bonus schemes in the first place was to increase productivity and to reward the workers for their increased efforts.

For the employers to offer standing bonuses and other fringe payments is to recognise the fact that operatives are underpaid on their basic rates. It would appear that a more realistic basic rate is called for. A typist's basic rate in many instances is in excess of a craftsman's/tradesman's and considering not only the skills of these operatives, but the mental and physical effort required at site level (not to mention the climatic problems faced), should lead to better rewards.

It is little wonder that supervisors find frustrations on-site and therefore are sometimes in conflict with various groups while representing the company, when in fact the problems stem from the management's inability to recognise their own shortcomings through an inadequate personnel policy, particularly relating to a decent wages structure.

While it appears to be good for all parties to have target bonus schemes there should be a measure of discretionary payments (operated by the site supervisor) where it is thought that a group's efforts are not being properly rewarded.

To offer overtime working is sometimes a way for an employer to appease the work-force to supplement the basic rates of pay and, really, only further highlights the low pay structure of the construction industry.

Stoppages

Although there has been, from time to time, stoppages on construction sites throughout Great Britain, on the whole the comparative percentage with other industries has been quite small (see Fig. 5.6.1); and where stoppages have occurred they have mainly been confined to the larger contracts. This does not mean that the industry is without its faults; it generally means that the workers are not united in the same way as workers from, say, factories. There is very seldom large concentrations of labour on construction sites and therefore the combination of operatives to air grievances tends to be limited. Therefore, although through the arrangements shown in the Working Rules Agreements (compiled by the NJCBI or CECCB) good industrial relations appear to prevail, and it can reasonably be said that serious stoppages do not occur because:

1. The industry is made up in the main of casual labour.
2. Labour-only subcontractors proliferate and are therefore self motivated.
3. There is a greater movement of workers between sites which prevents festering grudges worsening.
4. Fewer tradesmen/operatives belong to unions, particularly in the smaller firms.
5. Fragmented jobs and smaller jobs requiring three to four workers which cannot be seen as a combination of workers.
6. Some sites do not have union stewards and therefore the operatives have no representatives at this level.
7. Workers' earnings can be high, but only due to overtime/bonuses; and they do not realise that the basic rates are low. (Only when holidays are taken

Total days lost—yearly

1972	4 188 000 DAYS	1982	41 000 DAYS
73	193 000	83	68 000
74	252 000	84	334 000
75	247 000	85	50 000
76	570 000	86	33 000
77	297 000		
78	—		

Fig. 5.6.1 Total days lost in the construction industry due to industrial action (disputes and grievances). (*Reproduced by kind permission of the Central Statistics Office.*)

does a true picture emerge regarding wages, or if an operative is absent from work for some reason.)

The additional payments help to supplement low rates of pay and, therefore, the supervisor must be diligent in probing incentive schemes for the management if the workers request it.

It is a ridiculous state of affairs that some construction workers can earn up to £500.00 per week when others in the industry are lucky to earn one quarter of this figure. It is also an unsatisfactory situation that some employers are willing to pay above the norm when there is a shortage of labour, and yet revert to the basic rate when there is little work and labour is plentiful, although this bears out the 'supply and demand' principle expounded by economists.

The lower paid workers however may work for firms which look after the welfare of their workers better, and in place of high wages offer better holidays, good working conditions, non-contributory pension schemes, better quality of colleagues, social club and better quality work.

Site grievances

Domestic site meetings between site manager and his subordinates/supervisors can be useful for discussing not only output or shortages of labour, material, detail-drawings and subcontractor problems, but for discussing individual personalities who outwardly show they have a grievance, and to head off and prevent further resentment.

In listening to an individual's grievance, the problem should be studied in a similar manner to the problem solving techniques recommended to supervisors:

(a) listen to the aggrieved person's problem;
(b) listen to the other person's side of the story (if there is another involved);
(c) collect any other facts from witnesses, if necessary;
(d) study the facts and information;
(e) seek, if necessary, assistance from other supervisors;
(f) make a decision (fairly, and unbiased);
(g) state the decision and act on it.

Any problems of grievances outside the province of the supervisor, e.g. wage scales, promotions, etc., should be reported immediately to the personnel officer.

Where grievances develop regarding site duties, it is expected that a procedure should be outlined in a supervisor's Company Handbook.

It is also recognised that all supervisors should undergo some form of training by attending courses, lectures, etc., to prepare them for dealing with grievances, disputes or, if necessary, industrial relations. To have some knowledge of the many human problems encountered at site level prepares a supervisor for dealing with them.

As handbooks should be issued to supervisors as guides on how to handle problems on-site, so should an operative be given a handbook which explains how he/she should channel any grievances and how any communications with the head office should be made if he/she desires to do so.

Where a problem of dispute develops the supervisor would do well to remember the standard procedure as laid down in the National Working Rules Agreements (also shown in the Local/Regional WRA).

Small grievances can escalate into large ones if not correctly dealt with. Suggestion or complaint boxes could be a way round the problem of identification to ensure no victimisation. If, however, the supervisor maintains an impartial attitude and respect for the individual, then those with problems would feel free to air them openly without any fear of reprisal.

Individuals' grievances

The recommended way for an individual who has a grievance (no matter how trivial) should be verbally by first approaching his/her immediate superior, who should then endeavour to solve the problem by whatever means to arrive at an amicable solution. Failure of the immediate superior to solve the problem would entitle the aggrieved to approach the site supervisor/agent, accompanied by his/her union steward.

The course open to the aggrieved if once again there is no satisfactory solution at site level, would be to notify the full-time union officer (through the site steward), or the operatives' Regional Joint Secretary (joint, because of the application of more than one union). The full-time officer would next take up the case on behalf of the aggrieved operative with the company's management in order to try to resolve the problem.

Note: It is expected that at no time during the negotiations should there be any walk-outs, stoppages of work, go-slows, strikes, lock-outs or any other forms of industrial action.

Group grievances

Where there is a grievance by a group of operatives the union steward would be approached, if each member of the group belongs to the same union, to discuss a grievance with the supervisor. If the affected group members belong to more than one trade or union, then the convenor steward (site committee chairman) would take up the case with the immediate supervisor or site supervisor.

If the group(s) fail to resolve the problem at site level, then the union steward/convenor steward (as the case may be) should contact the full-time union representative of each union whose members are affected, or the joint union regional secretary, who should make representations to the company's management.

If there is no satisfactory solution to the problem at company level, the unions' representatives would process the grievance through the local/regional Conciliation Panel of the National Joint Council for the Building Industry under NJCBI Rule 8 (or similar if under the Civil Engineering Construction Conciliation Board).

Failure at regional level means proceeding to the National Conciliation Panel of the NJCBI, which, as in the local/regional panel, is made up of an equal number of operatives' and employers' representatives. This means that there must be a majority on both sides when a vote is taken. No party can out vote the other without the assistance of a member of the other party. This tends to mean that eventually when a decision is taken neither side loses face because the solution found is mutually acceptable.

A decision at national level would normally set a precedent for disputes of a similar nature; but a failure to resolve a dispute at national level would entail going to industrial arbitration.

There is at the moment a desire for the two bodies to the construction industry (NJCBI and CECCB) to merge, but as yet it appears that it will be sometime before the union takes place due to the smaller construction unions fears that they will lose some of their representative power to the larger unions.

Employers grievance

If an employer is aggrieved then the regional union secretary can be contacted in the first instance to try to resolved the problem. This procedure is then as before if no resolution can be found, through the employer's representative on the NJCBI.

In agreeing to such grievance procedures between the employers' and employees' representatives there is at present a limited level of industrial democracy.

Chapter 6
Supervisor's understanding of employment legislation

6.1 Engagement of people within the industry

On the engagement of operatives on-site and on completion of all documentation, as described in section 5.2, and to stay within the law and agreements made between the employers' organisations and unions through the National Joint Council for the Building Industry or the Civil Engineering Construction Conciliation Board as outlined in the Working Rules Agreements, the site manager may have the responsibility, in place of the personnel officer, to observe certain points regarding the engagement of personnel, and to take action where necessary regarding the conditions laid down in the various Acts of Parliament relative to employment; some of the Acts being:

1. Trade Union and Labour Relations Act 1974, and Trade Union and Labour Relations (Amendment) Act 1976.
2. Employment Protection Act 1975.
3. Employment Protection (Consolidation) Act 1978.
4. Employment Acts 1982 and 1990.
5. Sex Discrimination Act 1975.
6. Equal Pay Act 1970 (came into force 1975).
7. Race Relations Act 1976.
8. Disabled Persons (Employment) Act 1944 and 1958.
9. Social Security (Pensions) Act 1975.
10. Wages Act 1986.

Most of these Acts are inter-related and must be looked into collectively when trying to establish a procedure or point relating to the employment of persons or dealing with employees generally.

Subcontractors

Under the Tax Deduction Scheme for the Construction Industry introduced by a Finance Act of Parliament in 1975, and subsequently amended, when employing subcontractors Forms 714 I (individuals), 714 P (partnerships) and 714 C (companies) should be examined so that when payment is made for work undertaken there will be no need to deduct tax under Pay-as-you-earn (PAYE). If these forms are not produced by the subcontractors then it is the duty of the main contractor to deduct tax under the emergency code; it would then be up to the subcontractors to claim back from the Inland Revenue Office any excess tax paid under this system: heavy fines or, in fact, imprisonment could result of the main contractor's representatives if this duty is neglected. This system operates to reduce the £40 m. – 50 m. tax fraud each year by subcontractors and casual labour which started and gained momentum since the 'lump' labour system began to operate in the construction industry as far back as 1968.

Apprentices, trainees, etc.

Where the employment of a young person is involved, under the age of 18, the supervisor would have to undertake the following:

1. Notify the Careers Officer within seven days on the standard notice, Form 2404 under section 119A of the Factories Act 1961 (see Fig. 6.5.5).
2. Enter details of any young person employed on site on Form 36 of the General Register, under the Factory Act 1961.
3. Be prepared to allow a young person to attend for medical examination on the date and time specified if the Careers Officer deems it necessary, and on full pay.
4. Observe the recommended hours of work for those under 18 years of age, which is nine hours maximum per day and not more than 48 hours per week; remembering that 40 hours are normal hours. Neither must young persons commence work before 7 a.m. nor end work beyond 8 p.m.

In the last year of an apprentice's/trainee's three-year period of training a Holiday Credit Stamp Card should be started to enable the apprentice/trainee to receive holiday pay on graduating to a trained operative (tradesman/craftsman). A card would be started for young labourers on their commencement with the firm with the lower value of stamp being affixed each week (there is a higher value stamp for operatives over 18 years of age).

General operatives

On engagement various documents should be presented by the new operative to the site manager/supervisor so that wage payments and other remunerations can be made correctly, taking into consideration national insurance contributions, tax (PAYE) and holiday stamps; for example:

1. P45. (There should be two copies made out by the previous employer, one of which is then sent by the wages clerk to the Inland Revenue Offices.) A P46 would be prepared by the wages clerk for emergency coding for an individual who fails to present a P45. An emergency coding would then operate until the Inland Revenue Officer traces back the tax details of the new employee.

2. Holiday card. The card is divided into four parts (one part shown in Fig. 6.1.1) with credit stamps being affixed in turn each week by the new employer. This enables an operative eventually to take four separate weeks annual holiday (public holidays are allowed in the normal way, similar to technicians and office staff, without the need for stamping special cards as in the past). Two of the annual holiday weeks must be at Christmas and Easter.

Contributions for the cost of holiday credit stamps is entirely the responsibility of the employer – the employee makes no contributions. These stamps are purchased from a non-profit making company set up by the industry called the Building and Civil Engineering Holiday Scheme Management Limited.

These documents should be sent to the wages office for processing to enable a Tax Deduction Card to be prepared which will show an employee's earnings, national insurance contributions and tax (PAYE) payment. Also, wages sheets will be prepared and holiday cards will be maintained up-to-date.

Any extra payments for responsibility, etc., would be made by the wages department, particulars of which would be extracted from a copy of an operative's Personal Record Card. Overtime earnings would be noted from time sheets or clocking-on cards, and bonus from slips sent by the bonus clerk.

An employee is usually on probation for the first 52 weeks and if found unsuitable the employment could be terminated without the employer incurring redundancy payments or unfair dismissal payments provided that the correct periods of notice for termination of employment are allowed and that there was good reason, i.e. unsuitable for work offered, under the Employment Protection Act 1975.

Within 13 weeks of starting work an operative/ employee should be given a Contract of Employment or a Written Statement of his/her terms and conditions of employment where employed for 16 hours, or more per week (Employment Protection Act 1975). One must also bear in mind that part-time workers who have worked for eight hours or more per week for the firm during the past five years should also be issued with a Contract/Statement. (See Fig. 6.1.2.)

Under various employment Acts of Parliament a supervisor should be careful in his/her approach to particular types of persons under the following:

1. *Race Relations Act 1976*. There must be equal opportunities to reduce discrimination on grounds of colour or race.
2. *Disabled Persons (Employment) Acts 1944 and 1958*. Firms with not less than 20 employees have an obligation to employ a quota of registered disabled persons, the standard percentage being 3 per cent: to qualify one should be registered as such under the Act. The Secretary of State may designate certain classes of employment where this quota must be strictly adhered to, e.g. car park attendants and similar jobs.
3. *Sex Discrimination Act 1975 and 1986*. It is unlawful for an employer to treat women less favourably than men, although this also applies equally to men, especially when advertising a vacancy.
4. *Equal Pay Act 1970* (came into force 1975). If women are doing jobs similar in nature to those already being undertaken by men within the firm, they should also receive similar remuneration except if there were other considerations, e.g. responsibility, seniority, etc.

While employers have the right to employ whomever they wish, they should not deny employment to an individual who undertakes trade union matters (hard to enforce as can be seen from the outcome of various litigation outlined in the All English Law Reports).

6.2 Termination of employment

Due to numerous reasons it may be necessary to terminate the employment of an employee, but, as a supervisor, care should be exercised to stay within the law and the policy laid down by the firm/company regarding dismissals. Firms, although their prime objective is to make profits, should adopt a reasonable humanistic approach in dealing with individuals who become superfluous to requirements as, perhaps, in the future when business expands after a short depression, they may have to re-engage the same employees who were dismissed or made redundant previously in order to meet the demands made upon themselves. An uncaring attitude may deter workers from joining the firm because of past indiscretions. If poor impressions are allowed to permeate into peoples' minds regarding a firm's apparent total disregard for employees' feelings through its aggressive redundancy and dismissals policy, outsiders may be dissuaded from wanting to join the firm because security of employment, after all, is most employees' ideal. A sound personnel policy is a must, bearing in mind the various employment legislation which safeguards the interests to a point – of employees, and with it a suitable and equitable way of dealing with dismissals of individuals should the necessity arise.

Several Acts of Parliament make it necessary for all employers to abide by certain rules and procedures during periods of shortages of work or disciplinary problems.

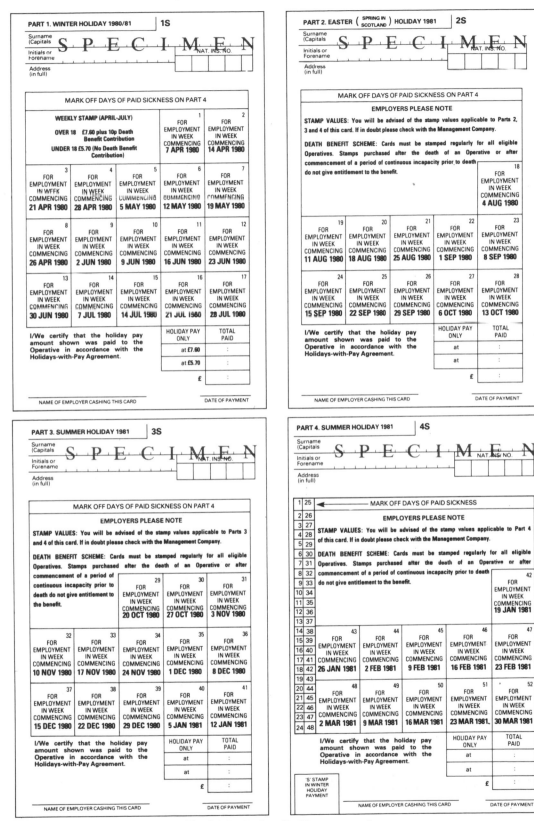

Fig. 6.1.1 Copies of holiday stamp cards

BUILDING OPERATIVES

CONTRACT OF EMPLOYMENT

WRITTEN STATEMENT OF MAIN TERMS OF EMPLOYMENT

Name of employer ....

Name of employee ....

Employment commencement date ....

Date of statement issue ....

Job title ....

Previous employment dates
  which count towards continuous service ....

....

....

Your rates of pay, hours of work, pay day, sick pay, annual
holidays, public holidays, holiday pay, disciplinary rules
and appeal procedures, grievance procedures, pensions and
pension schemes, periods of notice are provided for in the
following documents:

    1.   The Company's Operatives Handbook.
    2.   The Working Rule Agreement as approved by the
        National Joint Council for the Building Industry.

The above documents can be inspected by you on request, and any
future changes in the terms of Contract will be made to these
documents within one month of the change.

Separate forms exist for operatives who are not members of a union who are party to the NJCBI

Fig. 6.1.2

The important Act dealing with employment, etc., of personnel is the Employment Protection Act 1975, but even this must be read in conjunction with those other Acts already listed in section 6.1.

Supervisors should know exactly where they stand regarding the firm's personnel policy, and they should have a knowledge of legislation requirements relating to termination of employment (dismissals and redundancies), discipline and warnings for misconduct, etc. In addition, the Working Rules Agreements should be observed which contain details of agreements between the Employers' Organisations and Trade Unions, through the National Joint Council for the Building Industry or Civil Engineering, Construction, Conciliation Board.

Termination of employment and dismissals of employees from one type of employment may be due to many reasons, such as:

1. Redundancy (no work available).
2. Retirement.
3. End of contract agreements.
4. End of training periods (certain employees).
5. Maternity confinements.
6. Up-gradings.
7. Transfers.
8. Low productivity, gross misconduct, absenteeism, etc.
9. Notice served by operative/employee.
10. Lock-outs, strikes, go-slows.
11. Breach of contract.
12. Ill health.
13. Serious injuries.

The reasons for dismissals, or termination of employment by an employer should be recorded for future reference on Personal Record Cards. Also the reasons given by an employee for leaving may assist the firm in its efforts to reduce personnel wastage in times when high output is required. Earnings – flat rate, bonus and over-time – would determine the redundancy pay dismissed employees should be paid depending on the contract agreement, so careful recording of such must be preserved.

Engagements

On the engagement of personnel it is important that the supervisor maintains a close check on the suitability of the new employee, so that if problems develop within the Approval Period (2 years) and if the individual is lacking and is unable to do the job for which he/she applied (provided there was no shortage of labour), a termination notice should be served after the usual verbal and written warnings of unsuitability have been given. Provided that an individual's employment is terminated within the statutory 2 years, and the correct period of notice was given under the Employment Protection Act, the firm would not be liable for redundancy payments or additional payments for wrongful dismissal.

If a person, who is dismissed, has more than 2 years continuous service with the firm it may prove difficult to terminate his/her employment without incurring a penalty in the form of additional payments, under the various employment legislation. This could prove a heavy financial burden, particularly to a small business, although some of the added expense could be claimed back from the Redundancy Fund retained by the Government. Usually, the fact that an employee was not dismissed before the 2 years Approval Period means, in effect, the employee was suitable for the job he/she was employed to do. Therefore, under the Employment Protection Act 1975 some consideration must be given for an employee's loyal and satisfactory service in the form of a payment, and reasons for dismissal need to be given in writing; this document could be used if a dismissed worker decides to appeal against the dismissal through the Industrial Courts, and if the appeal is upheld, additional payments would have to be made by the employer to the dismissed worker, as previously outlined.

Periods of notice

Under the 1975 employment legislation the following periods of notice must be given unless both parties mutually agree to a payment being made in lieu of notice:

1. Employer to give:
   - (a) one week's notice if the employee has four weeks or more continuous service;
   - (b) two weeks' notice if the employee has two years or more continuous service;
   - (c) one week's additional notice for each year's continuous service up to a maximum of 12 years, making 12 weeks' notice.

   The WRA states further provisions for employees (operatives) whose unions are party to the NJCBI which are:
   - (d) two hours' notice during the first six days normal working days of employment expiring at the end of normal working hours on any day;
   - (e) one clear day's notice after the first six days (but less than four weeks) to expire at the end of normal working hours on Friday.
2. Employee to give:
   - (a) one week's notice (unless there is a contract between the employer and employee to do otherwise);
   - (b) same periods of notice as employer is expected to give in (d) and (e), under the WRA.

Time off to seek work

Under normal conditions where an employee has been given notice terminating his/her employment in a redundancy situation, reasonable time off from work should be allowed in order to seek employment elsewhere, provided that the employee has worked for 2 years or more in continuous service. In some instances employees may only need one period of absence to attend for an interview and be successful, while in others it may entail two or three visits to seek work; this is normal and should be allowed on full normal pay. Those who are not covered are employees working less than 16 hours per week, unless they have been employed continuously for more than five years and their weekly hours remain in excess of eight hours, under the Employment Protection Act 1975 and Employment Act 1982.

Unfair dismissal

Employees can be dismissed due to many reasons but the most usual ones relate to the following:

1. Conduct – as outlined in the National Working Rules Agreements (for operatives) and Employment Acts.
2. Capability – inadequate due to attitude, training, experience, physique and mentality.
3. Redundancy – as described in the Employment Protection Act 1975.

Employers and supervisors should be careful in their approach to dismissals, and usually procedures are laid down by all personnel-conscious firms for those with authority to observe them.

There are many acceptable reasons due to the Employment Protection Act 1975 for certain dismissals to be unlawful, and which, if proven in an Industrial Court could be at the expense of the employer: five examples are given below.

1. *Closed shop dismissals*
If, where a closed shop agreement exists or is implemented by an employer in agreement with a trade union, an employee refuses to join a recognised trade union, the employer may be advised by the union to dismiss the individual. The employer under these conditions would not normally be liable for paying compensation to the dismissed person, unless the person was dismissed because he/she refused to join a union due to religious beliefs, or other recognised reasons, under the Trade Union and Labour Relations Acts 1974 and 1976 and Employment Acts 1980 and 1982.

2. *Pregnant women dismissals*
The rights of pregnant women are safeguarded (Employment Protection Act 1975), which means that they may not be dismissed due to their condition. They must, however, have been continuously employed by the employer for at least two years to qualify; they also have the option of safeguarding their employment after their confinement, which means the employer may have to keep a woman's job open for her.

3. *Medical suspension*
If, after being inspected by the Employment Medical Adviser or doctor, an employee is suspended on medical grounds because of contacting some disease or has been exposed to a dangerous substance, he/she is still entitled to a normal week's pay for up to 26 weeks, and should not return to the same type of work until advised to do so by the doctor (only applies if employee has been employed continuously for four weeks or more). The employee

should still be prepared to do other reasonable work (less dangerous or hazardous) to qualify for pay if called upon to do so by the employer, after the doctor/medical adviser clears the employee.

4. *Short-time working*
Employees are entitled to guaranteed pay if laid off for a short time because there is no work or work has been interrupted by a trade dispute of which the employees are playing no part.

5. *Trade union activities*
It is foolhardy for an employer or a supervisor to terminate the employment of an individual because of trade union activities. Apart from paying compensation for unfair dismissal if the individual successfully appeals through an Industrial Court, extra compensation may have to be made if the court rules for the reinstatement of the dismissed employee and the employer refuses.

Industrial tribunal applications

In the case of unfair dismissals an individual may apply to the Central Office of Industrial Tribunals (COIT) within 3 months of the effective date of termination of employment. Application forms are obtainable from local Job Centres. The COIT will then send a copy of an application to the Advisory, Conciliation and Arbitration Service (ACAS). Conciliation Officers will try to assist the parties to reach a voluntary settlement without the need for a Tribunal hearing.

If a settlement between the parties is not reached when assistance is given by ACAS officers, the employee's complaint will then proceed to a full Industrial Tribunal hearing. The Tribunal will determine whether the dismissal is unfair or not. The remedies open to an employee where unfair dismissal is proven are:

(i)   reinstatement – job retained/given back to employee;
(ii) re-engagement – alternative job found by employer;
(iii) compensation – if (i) and (ii) are impracticable.

This procedure is generally the last resort in the construction industry because grievances can be taken up by the unions through the conciliation machinery of the NJCBI or CECCB at local or national levels.

Redundancy

The following can be expressed as a redundancy situation regarding employees:

(*a*) employer intends to or has ceased to carry on business;
(*b*) employer intends to or has moved the business from the place employees contracted to work;
(*c*) work of a particular kind has diminished or has ceased to exist at a particular place.

Lump sums can be claimed as redundancy pay by those who are to be, or have been, made redundant as above.

Large firms have to find the redundancy money for payments to redundant workers, while firms with less than 10 employees can claim some of the payment back from the redundancy fund retained by the Government.

Notification of redundancies

Firms which are contemplating redundancies should consult any trade union whose members are to be affected, and they should be prepared to discuss ways of implementing the redundancies, if in fact redundancy appears to be the only solution available by the employer due to: lack of business; changes in mode of production; changes in types of work produced, leading to, perhaps, more skilled men and less unskilled ones being required. Firms must also notify the Secretary of State for Employment (through the Job Centre) of any planned redundancies which may affect more than 10 employees.

Minimum period of consultancy with unions (where more than 10 employees are involved under the Employment Protection Act 1975)

1. If 10–99 employees are to be dismissed or made redundant at one establishment over a period of 30 days or less, then the period of consultancy is to be at least 30 days before the first dismissal takes effect.
2. Where 100 or more employees are to be dismissed, etc., over a period of 90 days or less, the consultancy period should be at least 90 days before the first dismissal takes effect.

Employers must supply the following details in writing, to the unions whose members are affected:

1. The reason for the firm's dismissal proposals.
2. The numbers and description of employees to be dismissed.
3. The total numbers overall of employees of such description at the establishment.
4. The proposed method of selecting employees for dismissal.
5. The proposed method of carrying out the dismissals and the period over which dismissals take effect, always taking into account previously agreed procedures with the unions (see Fig. 6.2.1).

If the employer does not give correct 'Notice of Intent', and the unions appeal to an Industrial Tribunal and this is upheld, then a Protective Award could be allowed which would be as follows: normal pay each week allowed through the protected period to each employee to be made redundant which, if necessary, commences from when the first employee would have been made redundant. The maximum length of the protected period would be:

1. *90 days* where 100 employees are to be made redundant over a period of 90 days or less.
2. *30 days* where 10 or more employees are to be made redundant over 30 days or less.
3. *28 days* where fewer than 10 employees are to be made redundant.

Where an employer fails to give the required notice to the Secretary of State a fine could be imposed.

Under the Wages Act 1986 the Government's redundancy fund is not available to firms who employ more than 10 employees unless they are in serious financial difficulties.

Rates for redundancy

Rates for redundancy, according to the Redundancy Acts, depends on the length of continuous service each individual has served with a firm. An employee/operative is entitled to a redundancy payment after serving at least two years with the same employer as follows:

1. For each year of reckonable service between the ages of 41 to 65 years (male), 60 (female) – one-and-a-half weeks' pay.
2. For each year of reckonable service between the ages of 22 to 41 years – one week's pay.
3. For each year of reckonable service between the ages of 18 or under 22 – half-a-week's pay.

Redundancy pay is in addition to the recognised periods of notice expected under the Employment Protection (Consolidation) Act 1978.

The employer must observe the following when making payments for redundancy to redundant workers:

1. Provide written statement of how the Redundancy Payment has been formulated.
2. Written statement of why redundancy occurred and why they have been dismissed.
3. Deduct no tax from Redundancy Payment.
4. Redundancy payment is reduced by $\frac{1}{12}$ if redundant worker is over 64 years of age (male), 59 (female), for each month up to retirement age of 65 and 60 years respectively.

Those not eligible for redundancy payments in the construction industry are:

(*a*) wives or husbands of the employing persons;
(*b*) employees with fixed term contracts;
(*c*) those without two years' continuous service;
(*d*) those who normally work for less than 16 hours per week (although someone working 8 or more hours with 5 years continuous service would be eligible).

Selection for redundancy

The principle of last in first out does not have to apply when considering redundancies, although in the main this would be the case, particularly as it would be extremely costly to dismiss employees with many years service; however, selection could be made on the following basis:
1. length of service. 2. age 3. working ability 4. qualifications. 5. experience.
6. personality 7. travel distance (nearest to job best). 8. any special aptitudes (capabilities and suitability, skills). 9. disciplinary and attendance record.

Special Note: The name of the full-time union representives should be typed in, and the appropriate union should be underlined.

NOTIFICATION OF PROPOSED REDUNDANCIES

FROM: B. BUILDER, LTD.,
MERROW,
GUILDFORD GU1 2QB
TELEPHONE: 6262

To: ....

1. THE UNION OF CONSTRUCTION, ALLIED TRADES AND TECHNICIANS.

2. THE TRANSPORT AND GENERAL WORKERS' UNION.

3. THE GENERAL MUNICIPAL BOILERMAKERS AND ALLIED TRADES UNION

4. THE FURNITURE, TIMBER AND ALLIED TRADES UNION

5. THE ELECTRICAL, ELECTRONICS AND TELECOMMUNICATIONS UNION/

   PLUMBERS TRADE UNION.

6. THE DEPARTMENT OF EMPLOYMENT.

Notice is hereby given, in accordance with the Employment Protection Act 1975, that the following is a list of redundancies proposed at the site stated:

Site and Location: ....

Types of Employees	No. to be Dismissed	Total No. Employed	Dates of Dismissal	
			First dismissals	Last dismissals

Reason for redundancy: ....
....

Method of selection for redundancy: ....
....

Method of termination of employment: ....
....

Please note the names of the person nominated by the company to whom proposed redundancies may by discussed: ....

Name: ....

Address: .... Telephone No. ....

Position: ....

Signed: .... Date: ....

Internal copies to: ....

Fig. 6.2.1

Firms should have a contingency policy for dealing with redundancies should the situation arise.

Dismissals

Where dismissals of individuals takes place due to redundancies, misdemeanours, etc., they are entitled to a 'Written Statement', if they have been employed for 52 weeks or more, stating why they have been dismissed (if they request one). It is important, therefore, that personal records are maintained on each employee to check on general conduct, warnings given and other relevant details should they be necessary in the future (see Operative's Personal Record Card – Fig. 5.1.5).

Within the industry where agreements exist between the trade unions and employers' organisations, and where operatives are prevented from working in a pay week because of weather, etc., guaranteed pay should be received by the operatives. If, however, in the second week work cannot be provided by the employer as a temporary measure, the operatives are advised (under an agreement with the employers) to register as unemployed persons until work recommences. This temporary lay-off would not affect the continuity of employment of the operatives unless the lay-off continued indefinitely; redundancy pay could then be claimed depending on length of service.

Supervisor's responsibilities

Employee's first week of employment
During the first six days of their employment, operatives are entitled to two hours' notice which should expire at the end of normal working hours on any day. Where the operatives are found to be unsuitable dismissals at this stage cause few problems. Site supervisors should therefore maintain a close check on new operatives, through their immediate supervisors, to ensure that they are suitable for the work they have been engaged to do

Transfers
Operatives on-site who are required for transfer to another site should be told verbally by the supervisor, and they should be given as much information as possible about the new site. Where an operative refuses a transfer he/she should be ordered to the new site – in writing – stating the Working Rule 12, new site address, date to be transferred and the occupation to which he/she will be employed. If the operative still refuses to transfer he/she should be informed of his/her breach of Contract of Employment and the appropriate notice for termination of employment should be given.

Disciplinary procedures
Where an operative's attitude or behaviour becomes unacceptable because of absenteeism, poor workmanship, bad timekeeping or low level of output, the supervisor is advised to do the following:

1. *Give verbal warning:* for the initial warning ensure that the site supervisor talks to the offending operative about the problem with a witness present. The operative should be warned that his/her behaviour or attitude must improve if continuous employment is to be maintained. Where possible, evidence should be provided of low output, absenteeism, etc., and the operative's personal record card should have recorded on it all relevant details by the person giving the warning.

2. *Give written warning:* this is a final warning if the verbal warning has little or no effect. It should spell out that if the operative does not improve his/her ways then termination of employment would be inevitable. For the record, itemise the details about the operative which have been objected about in the past. It may be necessary to send the warning by registered post if the operative is absent from work. The problem can, however, be passed over to the Personnel Officer or Labour Relations Officer if this is within the policy of the firm. Details of this written warning is entered on the operative's personal record card and a copy of the warning letter is sent to the personnel department, the site union steward and the full-time union steward. (See specimen warning letter, Fig. 6.2.2.)

3. *Dismiss:* the dismissal of the site operative is generally the main responsibility of the site supervisor if the conduct, etc., of the offending operative does not improve substantially relating to prior warnings regarding possibly:

(*a*) poor performance (low productivity);
(*b*) poor timekeeping (arrives late, prolonged rest breaks, finishes early);
(*c*) poor attendance at work (absenteeism).

See specimen letter, Fig. 6.2.3.

Dismissal of trade union stewards
Stewards of trade unions are not entitled to special treatment because of their position under the Working Rules Agreement, but where dismissals are imminent, discuss the details with the firm's Labour Relations Officer/Personnel Officer before serving notice.

Gross misconduct
It is usual first to suspend an operative if there is a case of gross misconduct so that the case can be discussed properly and to prevent a decision of wrongful dismissal being given which could cost the firm dearly should a dismissed operative successfully appeal to an Industrial Tribunal. On suspension an operative should be ordered off the site pending an enquiry.

Only the most serious offences are classed as 'gross misconduct',

1. Physical assault or threat.
2. Theft of company or other employees' property.
3. Malicious damage to company property.
4. Serious breach of safety regulations.
5. Refusal to comply with reasonable instructions.
6. Removal of company property without permission.

Sample letters for Low Productivity

```
B. BUILDER LTD                      NO:
MERROW                            DATE:
GUILDFORD GU1 2QB                  SITE:
TEL. 6262


                    FINAL WARNING

Dear
After being verbally warned on          of your low productivity and
general performance you have not shown a significant improvement
according to our records of output. This is a final warning and it is
expected that if no improvement is made your employment with the
firm will be terminated.


                         Yours faithfully,



                         Site Supervisor
```

Fig. 6.2.2   Sample letter for low productivity

Sample dismissal letter

```
B. BUILDER LTD                      NO:
MERROW                            DATE:
GUILDFORD GU1 2QB                  SITE:
TEL. 6262



                        DISMISSAL

Dear
As you have made no significant improvement in productivity or
general performance since you received a written warning. It is
regretted that you are now given          notice to terminate your
employment with the firm on

                         Yours faithfully,



                         Site Manager
```

Similar standard letters can be drafted to deal with unacceptable
timekeeping and absenteeism including prolonged absenteeism.

Fig. 6.2.3   Sample dismissal letter

Investigations: the site manager/supervisor should conduct an investigation by calling in witnesses, the main one being the suspended operative's immediate supervisor. The operative should be allowed to defend him/herself and can be represented by a union steward later when all the facts have been collated. The site supervisor should, where necessary, consult with the contracts manager and personnel/labour relations officer before a final decision is taken, particularly if the only course of action appears to be dismissal. Records should be kept of any relevant details relating to the hearing given to the defending operative in case an appeal is made to an Industrial Tribunal.

Notification to Wages Department and Personnel Officer should be in good time where dismissals/termination of employment are to take place so that the following can be prepared and sent to the site for issue to those who are leaving the employment of the firm:

(*a*) wages, including bonus, overtime, etc.;
(*b*) P45, two copies;
(*c*) Holiday Stamp card – stamped up to date;
(*d*) dismissal letter (not in cases where the operative has tendered his resignation in order to work elsewhere).

Where mutually agreed, these documents could be posted on to the individual by registered post.

If an operative wishes to leave on his/her own request, the reasons for leaving should be forwarded by the supervisor to Head Office to help management assess any changes which could be made in the future to improve employment conditions. The usual reasons given for leaving are:

1. Not enough overtime.
2. Wages and bonuses are inadequate.
3. Welfare conditions poor.
4. Dissatisfaction with type of work.
5. Physical conditions on-site poor.
6. No prospects of promotion.
7. Domestic reasons.
8. Excessive need to travel.
9. Fares and other allowances inadequate.
10. On medical grounds.
11. Promotion elsewhere.
12. Personality clash with supervisor.

6.3 Discrimination and protection of employment

For too many years, it is argued, certain sections of society have suffered discrimination in employment due to conditions which were either not of their own making, or because of personal decisions they had taken relating to societies or organisations they had joined and behaviour therefrom which may not have been acceptable in many individuals' eyes. Examples of discrimination by employers relate to the following:

1. Sex discrimination.
   (*a*) females not acceptable for certain employment;
   (*b*) women becoming pregnant while in employment;
   (*c*) single persons being more readily acceptable by employers than married persons.
2. Discrimination against any active trade unionist.
3. Discrimination of persons with public duties to perform.
4. Racial discrimination.
   (*a*) having different coloured skin to that of the majority of individuals;
   (*b*) originating from a different country whose race and culture are at variance with the acceptance country (England, Scotland, Wales and Northern Ireland).
5. Discrimination against those with homosexual tendencies instead of heterosexual ones.
6. Discrimination against those who have contracted a disease at work (see section 1.3).
7. Discrimination against disabled persons (see section 6.1).
8. Other discrimination cases:
   (*a*) those who belong to special organisations or political parties;
   (*b*) middle/older age discrimination;
   (*c*) part-time workers;
   (*d*) the parents of one-parent families;
   (*e*) individuals who did not attend a public school or other special schools (old school tie network);
   (*f*) individuals with a criminal record.
9. Discrimination against those who have asserted their rights against their employers by appealing under the Sex Discrimination Act 1975 and 1986 where they have had a grievance.
10. Personality discrimination.

Supervisors are held responsible in the same way as their employers and should refrain from any kind of discrimination which, because of the very act, could undermine any respect the subordinates may have for them. This does not imply, however, that the supervisor has to relax his/her firmness in dealing with individuals who may be in the minority class, but it is important to remain impartial with a measure of detachment to all the work-force under the supervisor's control.

Particular care must be exercised when advertising employment vacancies and suitable wording of advertisements must be maintained to prevent anyone from translating them into discrimination of any kind. However, a firm which employs five persons or less would be exempt from observance of this rule in advertising and from many rules laid down in the Sex Discrimination Act.

Sex discrimination

Both the Sex Discrimination Act 1975 and 1986 and the Equal Pay Act 1970 attempts to safeguard the interests of certain employed individuals, but particularly females.

Under the Sex Discrimination Act it is unlawful:

(*a*) for an employer to treat women less favourably than men;

(*b*) to treat a married person less favourably than an unmarried one;

(*c*) to treat a person unfavourably after they have asserted their rights by appealing, under the Sex Discrimination Act, against their employer and were either successful or unsuccessful in their appeal.

It must be noted, unfortunately, that it is one thing having various legislation dealing with security of employment, but it is another to see that it is enforced by parliament.

Under the Equal Pay Act 1970, which incidentally came into force in 1975, the purpose for its introduction was to eliminate discrimination particularly where women were concerned regarding pay and other terms of contract (overtime, bonus, piecework payments, holidays and sick leave entitlements). It seeks to achieve this purpose in two ways:

1. By establishing the right of the individual woman to equal treatment when she is employed in similar work to that of a man, and by job evaluation (a term used to grade jobs) to be assessed fairly in comparison to the work done by men.
2. To remove discrimination in collective agreements, employees' pay structures and statutory wages orders which contain any provisions applying specifically to men or to women.

The Equal Opportunities Commission was set up under the Sex Discrimination Act to deal with complaints by individuals and to conduct legal proceedings against suspected offending employers at the Industrial Tribunals.

There are, of course, certain types of employment which are strictly retained for men, and there are those which are strictly retained for women, e.g. washroom attendents for men or for women, etc. ACAS would naturally try to encourage a settlement before the opposing parties reach the Industrial Tribunal stage.

Pregnant women should not be dismissed due to their condition because they would be entitled to a 're-instatement order' or 'compensation' if they successfully appealed against the employer for wrongful dismissal through an Industrial Tribunal; the employer would then be liable for either. Even after her confinement a woman would be entitled to her job back provided that she had:

(*a*) been in continuous employment for a minimum of two years with her employer;

(*b*) worked until 11 weeks before the expected birth; and

(*c*) informed her employer that she intended returning to work after the child's birth (she can return at any time up to 29 weeks after the baby is born).

If the woman's job is to be taken over by someone else, that person must be told the job is a temporary one.

During the first six weeks of confinement a woman is entitled to nine-tenths of her normal pay each week, less any maternity allowances payable from Social Security (under the Social Security Act 1975). The employer can claim a full refund from the Maternity Fund retained by the government to which a small contribution is made weekly through the employer's social security contributions.

It is unlawful, in employment, to treat a married person (male or female) less favourably than an unmarried one, under the Sex Discrimination Act 1975 and 1986.

Supervisors should understand that while they and their employers must obey the law, they must also attempt to see that others' acts do not force them into positions where criticism can be levelled their way because of condoning fellow workers' (subordinates) victimisation of other workers and, trade unions' victimisation of other workers – especially where any action taken is against the law. The labour relations officer should be notified where the unions are involved, and a low profile should be maintained by the supervisor especially if all the facts are not to hand: impartiality is the key word in situations such as these.

Employees are entitled, under the Trade Union and Labour Relations Act, to join any independent trade union without the employer or supervisor victimising them; neither should there be any victimisation because an operative carries out lawful duties on behalf of his/her trade union.

Discrimination of persons with public duties to perform

The type of discrimination relates mainly to the refusal by an employer of an employee to reasonable time off work during working hours to perform a public duty. Under normal circumstances time off should be allowed in the same way as employees who are trade union representatives are granted time off to conduct trade union affairs, particularly when the duties relate to serving as a:

1. Justice of the peace.
2. Member of a local authority committee.
3. Member of any statutory tribunal.
4. Member of a governing body of an educational establishment of a local education authority.
5. Member of a water authority.
   (See Employment Protection Act 1975.)

Racial discrimination

Colour, religion, culture or creed should not normally affect a person's position or prospects within a firm.

Individuals should be given equal opportunities under the Race Relations Act 1975, and should be selected for work which they are qualified to do, and should have work conditions and terms of contract not less favourable than others. See Trade Unions and Labour Relations (Amendment) Act 1976.

Employers are, however allowed, under the Act, to maintain a balance between racial groups in their employment, and may refuse employment to an individual on these grounds.

Discrimination against those with homosexual tendencies instead of heterosexual ones

It is really no defence to dismiss someone who is a homosexual, unless by their very actions or deeds at work fellow workers object to their presence. The employer must heed their objections and take action if absolutely necessary. The objections, however, would not normally be valid if the objectors are not directly affected by the actions of the homosexual under Common Law.

Other discrimination cases, etc.

Most of the cases mentioned are protected in the normal way by the facilities shown in the various employment legislation, as listed in section 6.1. Where discrimination takes place regarding the types of employed persons stated here, there is no recourse under the Employment Protection Act 1975, Employment Acts 1980 and 1982.

Vacant jobs should be available to all individuals provided that their qualifications, experience and other relevant personal details meet with the job specifications. Certain personnel officers and employers, however, attach great importance to the type of school one should have attended, or what one's father's position is, for various employment positions. These problems do exist in employment and are difficult to overcome because of natural divisions which exist in the society in which we live. Social status, superior thinking and personal interests create barriers between certain types of individuals, which is their choice and are quite within their rights so to do.

A highly educated person could be discriminated against in the type of employment a much less able person could undertake, mainly because a supervisor or employer suspects that a short stay is envisaged by the individual.

Those with criminal records should not normally be in danger of dismissal as operatives on construction sites provided they are diligent in their work; but persons may find themselves at risk if trust is the keyword, in the type of employment where money or valuables are to be handled.

Members of one parent families should be treated ordinarily as should those who belong to special organisations or political parties. Part-time workers are protected to a certain degree under the Employment Protection Act 1975 depending on number of hours worked per week and length of service.

Because of an employee's age a firm may appoint someone else to take on some of the work normally undertaken by the older person. Provided there is no salary drop or serious undermining of one's position there can be no cause for complaint. If, however, one's job (including responsibilities) are seriously affected a claim for redundancy could be made due to the change in conditions and contract of employment.

Personality discrimination is a difficult problem for an individual to overcome; it can happen in any type of employment, and where it can be proven that an employer is deliberately aggravating the situation to force an employee to leave, a wrongful dismissal claim can be made, and redundancy pay awarded.

6.4 Trade union and labour relations

All site supervisors require an outline knowledge of the law relating to trade unions and labour relations. It assists the supervisor to appreciate situations relating to the employment of individuals on-site and how best to approach negotiations with trade union representatives if and when the need arises.

The immediate document to which a site supervisor would refer in dealing with a query on labour relations would be his/her handbook issued by the company. The second document, the details of which should be fully understood, would by the appropriate Working Rule Agreements. Neither of these documents give the full picture for dealing with every eventuality on-site; therefore, it is essential to have an appreciation of the appropriate sections of legislation covering trade unions, etc., such as: the Trade Union and Labour Relations Acts 1974 and 1976 – which deal entirely with trade union matters, showing what is allowed and not allowed where disputes or other problems develop between the employees on one side and the employers on the other. Other Acts contain clauses dealing with trade union matters, such as: the Employment Protection Act 1975, Employment Protection (Consolidation) Acts 1978 and 1979 and the Employment Acts 1980 and 1982, to mention but a few. Each refers to particular points which may affect individuals, and how trade unions can make provisions to redress any situation which appears to be unfair and which affects their members.

Employers' organisations

These relate to organisations of which employers are members, which negotiate on behalf of the employers with the trade unions regarding matters covering employment, conditions of work, rates of renumerations, holidays, training and safety health and welfare. The Building Employers Confederation and Federation of Master Builders are two such organisations.

Trade unions

These are organisations where employees combine to negotiate with the employers' organisations to improve working conditions, etc. They are divided into two classes under the Trade Union and Labour Relations Act 1974:

1. Dependent trade unions.
2. Independent trade unions.

Dependent trade unions
The dependent type can be listed, provided they satisfy the requirements, by the Certification Officer under the

Trade Union and Labour Relations Act. These types of unions could be set up by and even dominated by the employers, or the employer has some direct influence over them: the more usual being Staff Associations or types of local unions with most employees within a firm being members. The members within these unions are not, therefore, what could be described as staunch trade unionists and are not given the same safeguards by the various Acts dealing with trade union and labour relations as are independent trade unions.

Independent trade unions

These types of unions are independent from the employers' control, examples of which are: UCATT, TGWU, GMBU, etc., and have no difficulty in getting 'listed', which entitles them to a Certificate of Independence with all the benefits which are inherent.

The listing of trade unions and employers' organisation is by register which is maintained by the Certification Officer for anyone's inspection.

Independent trade unions usually have a central national headquarters which is financed by the members' subscriptions and interest from shrewd investment. Each union is divided into regions, each of which employs full-time trade union officials. These regions can also be sub-divided into local area branches still with, perhaps, a full-time official in charge. Where shops or contracts exist within a local area part-time shop or site stewards may be nominated or voted into office, but operate unpaid, but obviously being able to claim for expenses incurred while carrying out official business. Usually, if the site is large enough, an agreement exists through the WRA for one site steward to be appointed for each trade, and where it is thought necessary a site committee can be set up with a maximum of seven members. From the seven members a convenor steward, or site committee chairman, would be appointed who should normally be in the employment of the main contractor (see section 5.5).

Under the Trade Union and Labour Relations Acts 1974 and 1976 trade unions are given protection from liability in tort: in other words, any action taken by an individual or organisation against an independent trade union in the following situations would fail in a court of law (although the Employment Acts 1980 and 1982 further limit the protection given to unions).

1. A claim for damages caused by any union member during a trade dispute.
2. A claim for damages to trade caused by walk-outs, work-to-rules, go-slows, withdrawals of labour, unwillingness to work, and sit-ins (or work-ins) and through unofficial strike actions.
3. A claim for intimidation by members of a union (separate charges would have to be brought against individuals here).

A trade union would be liable, however, in the following situations:

1. Where a trade union official carelessly injures some-one by running them down with his/her car while on official union duty.
2. Where the union itself is negligent or causes a nuisance which leads to a disease, physical or mental injury to persons.
3. Where strikes are called by a union without balloting members.
4. Secondary picketing or picketing in numbers greater than 6 at each work entrance.

The Employment Act 1980 lays down strict rules on unions relating to disputes procedures.

Trade disputes refer to disputes:

1. Between workers and employers.
2. Between workers and workers.
3. Concerning a worker's trade union membership.

Under 'Bridlington' the TUC has made arrangements within its ranks for each union to refrain from the poaching of membership, and where independent unions exist within a place of work the dominant union of similar category workers would be the one with which the employer would have to negotiate directly; for example, the FTATU, when the majority of woodcutting machinists in a machine shop belong to this union, and the remaining minority woodcutting machinists belong to another, perhaps smaller, union, negotiations on conditions of work and remunerations, etc., would be conducted between the majority union and the employer, and the agreements would generally be binding on the minor union members.

Closed shop

Under the Trades Union Labour Relations Acts 1974 and 1976 Employment Protection (Consolidation) Act 1978 and Employment Acts 1980 and 1982, individuals may terminate their membership of a trade union, provided that they give suitable reasons for doing so. Where a 'union membership agreement' (closed shop) exists, however, the termination of membership by an individual would most certainly affect his/her security of employment particularly as the agreement is to preclude non-union members from the workplace. Main features of a closed shop agreement are:

1. 80% of those entitled to vote approved such an agreement in a secret ballot or 85% of those who voted.
2. The class of employee covered may only relate to a particular trade or skilled workers (usually, however, it relates to all employees).

There are exceptions which can be made, under the law, which are:

1. All existing non-union members within an organisation.
2. Employees over a certain age (to be agreed).
3. Employees with religious beliefs which forbid other such organisation membership.
4. Employees who are conscientious objectors.
5. Employees who pay a contribution in lieu of membership (this may not be insisted upon).
6. Employees in the class covered by a closed shop agreement may be required, or allowed, to join an

independent trade union which is not party to the closed shop agreement.

Disputes

Where a dispute arises between independent trade unions and employers organisations employees are entitled under the law to withdraw their labour, unless there is a written collective agreement between them which includes a no-strike clause, and this is also incorporated into the employees' contracts. This does not generally affect operatives of the construction industry.

While on strike peaceful picketing is within the law, but the intimidation of individuals is not, and, therefore, the police would take serious view of the use of threats and assaults on anyone by union members.

If a firm locks-out its employees (refuses their entry to work) because of work-to-rules, go-slows, etc., which is seriously affecting production, employees cannot claim unfair dismissal if the employer later offers to take on the employees as from the resumption of work.

Supervisors must note that in cases of dispute there is a recognised procedure for processing the problems through the machinery laid down by the National Joint Council for the Building Industry, or the Civil Engineering Construction Conciliation Board in the appropriate Working Rules. Failure, at national level, to find a satisfactory solution to a dispute could mean that ACAS could step in and would refer the problem to the Central Arbitration Council (CAC) if it failed to get the parties to resolve the problem.

6.5   Health and welfare

In employing individuals precautions must be taken to ensure their health is safeguarded against the many hazards encountered in construction work which may cause, if allowed, long lasting damage to health or even early termination of life.

Health

Health safeguards can be divided into those which protect in the main the following areas of the human body: skin, eyes, ears, bone, lungs, muscles.

Skin

Dermatitis is a most prevalent problem to operatives' skin caused by the skin's contact with numerous harmful substances, with different persons being more allergic to chemicals/materials than others. Bricklayers can be affected (and plasterers) by the cement/lime/plaster powder or, in fact, by the wet mix coming in contact with the skin, particularly the skin on the hands. This has, particularly in the operatives' earlier years in the trade, caused dermatitis and has forced many individuals to seek employment in a different industry or trade. The irritation of the skin which results from such materials inevitably causes this disease, especially if the skin is not washed reasonably regularly — although too much washing can lead to the

same problem, as the skin's natural protection is prevented from forming by constant washing. Stone masons or paviors are also particularly vulnerable to this disease due to stone, dust and cement, etc.

Cleanliness is of paramount importance in combating the dermatitis problem but many operatives in their attempts to maintain clean hands from dust, etc., use substances which aggravate the situation even more, e.g. mould oil, petroleum spirit, these substances being preferred to remove grease and dirt more than others, when, in fact, 'care' should be the keyword; therefore specially manufactured hand cleansers should be provided by the employer and their use encouraged.

To reduce the possibility of infection, either appropriate barrier creams should be used, which should be applied to the hands before commencing the type of work previously mentioned, or gloves should be worn which are chosen to suit each individual's preference, bearing in mind the type of work to be undertaken. Gloves are made from leather, plastic, rubber and even cotton.

By providing the necessary protective gloves where heavy, rough work is undertaken, and other protective clothing to guard against skin infection when using certain chemicals (creosote spraying), the Health and Safety at Work, etc. Act 1974 would at least be partially observed.

Eyes

Protection of Eyes Regulations 1974 and the Protection of Eyes (amendment) Regulation 1975 makes provisions for the shielding of the eyes where the production process could be harmful to the eyes. Spectacles, goggles, masks or shields must be adequate for the work to be undertaken. It is particularly important to protect the eyes against such processes as:

1. Oxyacetylene work.
2. Arc welding.
3. Woodwork turning.
4. Metal turning.
5. Stone chipping.
6. Grinding operations.
7. The use of lasers (see BS 4803:1972, guide to the protection of personnel against hazards from laser radiation).

There are different types of eye guards, from the types which give protection from flying foreign matter and dangerous bright lights, to processes which could cause burning of the eye tissues.

Proper lighting of work areas also helps to protect the operatives' eyes and thereby reduces eye strain and fatigue, in addition to reducing the possibility of accidents.

Ears

Where certain noisy machines, plant or processes are used operatives are then subjected to noise bombardment which, if allowed to continue unchecked without some remedial measures, causes gradual damage to the eardrums which may never be rectified. The Control of

Pollution Act 1974 covers many points regarding noise control on construction sites, as does the BS 5228 Code of Practice for Noise Control on Construction and Demolition sites.

Continuous noise leads to fatigue and, thereby, accidents. There are therefore a number of ways of protecting workers, such as:

(*a*) using quieter tools/machines (silencers added);
(*b*) isolating noisy machines from other work areas;
(*c*) jacket mechanical tools/machines;
(*d*) providing screens around machines, etc;
(*e*) resting machines on rubber pads to reduce reverberation;
(*f*) providing ear muffles to workers at most serious risk (operators of machines).

Bones

The most obvious problems associated with bones caused by either constant poor working conditions or unfortunate occurrences are: arthritis; slipped discs; dislocations and broken bones; bone rot.

Arthritis: This occurs within individuals who continually work in wet or cold weather conditions without adequate protection or facilities. The precautions which are recommended apart from providing alternative work during bad weather would be to provide the right quality and quantity of protective clothing (warm clothing when necessary, e.g. donkey jackets, gloves), rubber or other boots. If drying and warming facilities are made available they add to the operative's comfort and wellbeing. The provision of hot meals on-site, reasonably priced, contributes not only to a healthier outlook for all, but adds to the morale on site.

Slipped discs: These can occur in the most unusual circumstances but generally through falls, carrying of heavy weights and wrenching oneself while working in awkward positions or situations. Safety would be the keyword here, but in addition to safety precautions heavy weights should be lifted by proper plant or apparatus. A visit to a physiotherapist would generally be recommended where an individual suffers a slipped disc.

Dislocations and broken bones: The supervisor should ensure safe working conditions and methods to prevent such occurrences.

Bone rot: A particularly distressing problem could result where individuals have to work in compressed air, conditions which exist in certain tunnelling, caissons and diving operations. Although there are stringent safeguards in existence control during work in these situations is not foolproof. Because of poor supervision and lack of knowledge of compressed air working otherwise fit operatives can be affected by the 'bends' or even gradual rotting of the bones, and if the rotting takes place at the bone joints

it leads to a most serious deterioration of an individuals' health which usually cripples him/her for life.

X-ray examinations are therefore necessary at regular intervals, under the law, of individuals who work in compressed air. It is also a recommendation that longer decompression times should be observed for workers ending their shifts from compressed air working.

The Work in Compressed Air Regulation 1958 (and various codes of practice) and the Diving Operations at Work Regulation 1981 must be observed carefully by employers of operatives who work in such conditions.

Lungs

There are many operations which, because of the materials and substances involved, could lead to infection of the lungs and which may result in death. The materials are too numerous to mention, but special reference is made to the following and the problem they cause to the lungs:

Dust: Silicosis is a disease which is unpleasant to say the least. It is a contamination of the lungs caused by inhalation of excessive quantities of stone dust (in particular) over many years. The dust solidifies, which then prevents the lungs from inflating or deflating correctly, which subsequently makes breathing difficult. The types of operatives most usually affected are: stone masons, bricklayers and those who undertake jobs requiring the use of grinding machines, etc. (Some miners suffer from pneumonicosis.)

Where there are such processes which emit dust of any kind protection could be afforded by:
1. Damping down material with water to prevent the dust from rising.
2. Providing protective breathing masks or pads depending on the volume and density of the dust.
3. Providing breathing suits and apparatus particularly where the dust is harmful and in large quantities.
4. Providing exhaust vents or vacuums and collecting equipment to certain dust-generating machines (as stated, for example, in the Woodworking Machines Regulations 1974).

Fibreglass: This is a particularly unpleasant material to work with as the small needle shaped pieces of the glass, if released, tend to float in the atmosphere in which the operative works. Breathing masks and even goggles could be provided.

Asbestos: This is dangerous substance, which, if carelessly used, can lead to one of the worst industrial diseases, 'asbestosis' or cancer. Blue asbestos is of the most dangerous and is not now used; unfortunately it is still in existence in old buildings, and before demolishing or altering buildings care should be exercised to remove the material before work commences, and if found, protective suits and breathing gear should be worn by the operatives, and observation of the Asbestos Regulation 1969 and Asbestos (Licensing) Regulations 1983 is absolute.

Gases and fumes: Most operations using burning techniques contribute to the output of gases or fumes, and if precautions are not taken to ensure there is adequate ventilation of the work area, damage to one's lungs is inevitable. When the provision of ventilation is difficult to attain then breathing suits should be provided to prevent the operatives from being inflicted with the disease known 'welders' lung'. Particular attention is therefore needed on sites which use the following processes:

1. Arc welding.   2. Oxyacetylene cutting
3. Brazing.   4. Soldering.   5. Lead welding.
6. Zinc welding.

Combustion machines, if used indoors, causes carbon monoxide poisoning when care is not taken to ventilate the work area. It is advisable to provide an exhaust lead to the outside. Space heaters use up oxygen, therefore, to prevent workers being starved of oxygen in confined spaces ensure provisions are made for ventilation.

When work has to be undertaken in deep pits and trenches, also in headings and sewers, a close check should be made at commencement of work and at reasonable intervals to ensure there are no toxic gases or fumes present.

Painting, polishing, creosoting or the use of petroleum rubberised solutions in confined spaces can be injurious to health because of the fumes given off by these substances. Windows should be left open during, and after, application to provide ventilation, and the drying process will also be accelerated.

The most common ailment contracted by individuals who have continually been exposed to the aforementioned processes is bronchitis.

Muscles
Where operatives need to strain their bodies in carrying out operations and perhaps at the same time work in extreme cold or wet weather, the eventual outcome is to perhaps suffer one or more ailments, such as: lumbago, sciatica, muscle strain and back pains, which results in lost production time. Site management could help to alleviate the working conditions by using better techniques (plant for easing the burden on the operatives) to eliminate the heavier work, and to give in-company instruction on correct manual methods of lifting or doing special strenuous work. Work study helps to find easier methods of doing various operations.

Pneumatic drills or hammers can cause blood circulation problems to the hands and are dangerous. Where individuals report personal problems of this kind while using such equipment, they should be given alternative work.

Where it is essential to conduct non-destructive tests on welded joints to steelwork, or to test the strength of *in situ* concrete, both the Ionising Radiations (Unsealed Radioactive Substances) Regulations 1968 must be observed.

Inadequate training and supervision can be the cause of individuals suffering from unnecessary exposure to doses of radiation.

Hot weather, while enjoyable to most operatives, causes, if operatives are careless, sunstroke and the indi-

viduals should be instructed not to expose themselves unnecessarily to the sun and should be encouraged to take plenty of liquid.

Food preparation
Where food or drinks are prepared on-site the Food Hygiene (General) Regulations 1970 contains provisions to help prevent illness from contaminated food; it makes reference to the following points:

(*a*) cleanliness of premises (canteen);
(*b*) clean, reliable cooks with clean habits;
(*c*) clean utensils;
(*d*) observance of correct temperatures at which food should be kept to prevent bacterial growth.

Welfare

Contractors (employers) have to comply with the Health and Safety at Work, etc. Act 1974, and, when considering the welfare of their employees on-site, they should observe the Construction, Health and Welfare Regulations 1966. If there are office personnel on site the Offices, Shops and Railway Premises Act 1963 must be observed.

There are many facilities provided on-site, depending on the numbers employed there, which ensures the welfare arrangements for the workers is adequate. These facilities, where necessary, can be then offered by the main contractor to subcontractors, and provided that a record is made in the register (Form 3303) the factory inspectorate would be satisfied. A certificate (Form 2202 – Part A) is issued to those subcontractors who have agreed to use the main contractor's facilities, as proof of the agreement. If construction work is to be undertaken in a client's factory and the factory is equipped with health and welfare facilities to Factory Act standards, a contractor could arrange with the factory owner/directors to use such facilities, provided that the appropriate registration and certification is made (see Fig. 6.5.2).

In deciding how many employees are employed on-site or in the works to satisfy the Constrution (Health and Welfare) Regulations 1966, one must consider not only the directly employed operatives but also the subcontractor's employees where there are to be shared facilities: one considers the anticipated maximum number.

The facilities normally provided on-site and which are described in the Construction Regulations are dealt with under the following headings:

1. First-aid boxes, etc.   ⎫ refer to the Health and
2. Ambulances.   ⎬ Safety (First Aid)
3. First-aid rooms.   ⎭ Regulations 1981.
4. Shelter and accommodation for clothing and for taking meals.
5. Washing facilities.
6. Number of sanitary conveniences and other requirements.
7. Protective clothing.
8. Safe access to places where facilities are provided.

1. First-aid boxes and kits

At least one first-aid box must be provided on site regardless of the number of employees. On a large site it may be necessary to provide more than one box to give employees ease of access when required and the positions should be clearly marked. First-aid kits would be supplied to employees working singly or in small groups which are isolated from the first-aid box central positions. The minimum contents of first-aid boxes and first-aid kits are shown in Fig. 6.5.1 A and B.

In the Health and Safety (First Aid) Regulations 1981 references are made to the following individuals who would be expected to deal with accident victims and ill persons on site or in the workplace.

(a) Occupational first-aider − an employee who has undergone additional training to that of a first-aider, particularly where special hazards exist in the workplace.
(b) Trained first-aider − one who is trained and has qualifications as approved by the Health and Safety Executive.
(c) Suitable person − one who holds a first-aid certificate issued by St John's Ambulance Association, St Andrew's Ambulance Association or the British Red Cross Society (First-Aider).
(d) Appointed person − where less than 50 employees are employed there is no statutory duty to have a first-aider. Employers must therefore ensure that a reasonably intelligent person is appointed to deal with any emergency (calling the ambulance, etc).

There should be a first-aider where 50 to 150 employees are employed at work, with an additional first-aider for every subsequent 150 employees. Occupational first-aiders should be available if special hazards exist.

2. Ambulances

Notification should be made to the Ambulance Authority within 24 hours as soon as more than 25 workers are employed on-site, giving the address of the site, nature of the work and the anticipated completion date. There should also be provided on-site a stretcher and someone should be appointed to be available to summon an ambulance when needed; this person should be instructed on how to make emergency or radio calls (name of this person should be posted in convenient position on-site).

If it is impracticable to arrange for emergency telephone, etc., calls, a suitable motor vehicle should be kept available on-site.

3. First-aid rooms

Where more than 250 persons are employed on site there should be a suitably staffed and equipped first-aid room. The room should be:

(a) under the charge of an occupational first-aider and names of first-aiders should be displayed;
(b) used only for first aid;
(c) accessible for stretchers and of suitable size;
(d) fitted with first aid material (as in Fig. 6.5.1) and the following:
    (i) a sink with hot and cold running water,
    (ii) drinking water,
    (iii) paper towels, soap and nail brush,
    (iv) smooth-topped impermeable work surfaces,
    (v) clean garments for use by first-aiders and occupational first-aiders,
    (vi) clinical thermometer,
    (vii) a couch with pillow and blankets;
(e) heated, lighted, ventilated and cleaned regularly;
(f) equipped with a telephone or similar communication link.

4. Shelter and accomodation for clothing and for taking meals

The following must be provided on sites by an employer:
(a) adequate shelter from inclement weather;
(b) accommodation for personal clothes;
(c) warming facilities and drying arrangements for clothes where more than five operatives are employed on-site;
(d) drying facilities should be laid on where practicable if five or less operatives employed on-site;
(e) accommodation for protective clothing;
(f) drinking water (marked DRINKING WATER).
    Shelters should be kept clean and not used as stores.

5. Washing facilities

Washing facilities should be provided when operatives are to be employed on-site for more than four consecutive hours.

If the contractor employs more than 20 but not more than 100 operatives on-site and the work is to last for more than six weeks, the following facilities are to be provided.

(a) wash troughs, basins or buckets;
(b) soap and towels, etc;
(c) sufficient hot and cold (or warm) water.

Where the work is to last for more than 12 months and there are more than 100 operatives there must be:

Four wash basins for the first 100, and one extra basin for each additional unit of 35 operatives (fractions of 35 rounded up or down) with soap, towels and hot water provided as before.

When operatives are employed using red lead compounds (paint, etc.) the aforementioned facilities should be provided for every five operatives using this dangerous compound, but with the addition of nailbrushes.

6. Number of sanitary conveniences and other requirements

If sufficient urinals are also provided there should be one sanitary convenience (WC) for every 25 operatives where there are up to 100 of them employed on-site, and

A.	FIRST AID BOX MINIMUM CONTENTS					
		NUMBER OF EMPLOYEES				
ITEM		1–5	6–10	11–50	51–100	101–150
a. Guidance card.		1	1	1	1	1
b. Individually-wrapped sterile adhesive dressings.		10	20	40	40	40
c. Sterile eye pads with attachment.		1	2	4	6	8
d. Triangular bandages.		1	2	4	6	8
e. Sterile coverings for serious wounds (where applicable).		1	2	4	6	8
f. Safety pins.		6	6	12	12	12
g. Medium-sized sterile unmedicated dressings.		3	6	8	10	12
h. Large sterile unmedicated dressings.		1	2	4	6	10
i. Extra-large sterile unmedicated dressings.		1	2	4	6	8
j. Sterile water or saline in 300 ml disposable containers, where tap water is unavailable.		1	1	3	6	6

NOTE: the box should be soundly made to protect the contents from dampness and dust and clearly marked with a white cross on a green background.

B. *FIRST AID KIT MINIMUM CONTENTS.*
The contents vary depending on the hazards expected to be encountered:
   a. Six individually-wrapped sterile adhesive dressings.
   b. One sterile unmedicated dressing (100mm × 80 mm).
   c. One sterile triangular bandage (if unsterile, include a sterile covering.)
   d. Six safety pins.

Figure 6.5.1

one additional WC for every unit of 35 operatives over 100. Example: for 140 operatives there needs to be five WCs (having accounted for four WCs for each 25 operatives up to 100, and one extra for units of 35 operatives over 100. Units less than one half of 35 are disregarded).

The sanitary conveniences must be separated by partitions and should be reasonably accessible on-site from all workplaces; doors should have fasteners and may not open into work/mess rooms; cubicles should be ventilated. Naturally, separate conveniences should be provided where women are employed in the same ratios as previously described.

7. *Protective clothing*
When there is a possibility that operatives could be affected by inclement weather protective clothing must be provided by the employer.

8. *Safe access to places where facilities are provided*
Safe footpaths should be provided for access to (and from) the shelters and conveniences and should so be maintained.

It is important to know that the Minister is empowered, firstly, under the Factories Act 1961 and, secondly, under the Health and Safety at Work, etc. Act 1974 to prepare safety, health and welfare regulations when it is deemed absolutely necessary because of the ever changing conditions met with in industry.

The following is a list of statutory registers, forms, certificates and notifications which needs to be maintained or sent to various statutory health bodies or firms under the health and welfare legislation.

1. Register and Certificates of shared welfare arrangements.
   (*a*) Form F2202 Part A – issued to subcontractors.
   (*b*) Form F2202 Part B – retained by main contractor (see Fig. 6.5.2).
2. General Register (F36) for buidling operations and works of engineering construction:
   (*a*) Certificate of Fitness for Persons Under 18 (Part 2);
   (*b*) Persons Taken into Employment When Under 18 (Part 2) (see Fig. 6.5.3);
   (*c*) Cases of Poisoning or Diseases (Part 4) (see Fig. 6.5.4).
3. Written reports to H.S.E. under the Reporting of Injuries and Dangerous Occurrences Regulations 1985.
   (*a*) Report of an injury or dangerous occurrence F2508;
   (*b*) Report of a case of disease F2508A (see Figs. 6.5.6 and 6.5.7).
4. Notice of Taking into Employment or Transference of a Young Person, Form F2404 (see Fig. 6.5.5).
5. Notice of Employment of Persons in Offices, Form LSRI. (Where office is to used for more than six months and more than 21 hours per week.)
   Under Offices, Shops and Railway Premises Act 1964.
6. Notification (by standard letter) to local Ambulance Authority within 24 hours as soon as more than 25 operatives are employed on-site (state address of site, nature of operations and probable completion date).

109

Factories Act 1961

Construction (Health and Welfare) Regulations 1966

Form approved by H.M. Chief Inspector of Factories

Certificate of shared welfare arrangements
made under regulation 4

Part B   F2202

(To be retained in the Register)

Name or title of Employer or Contractor providing the facilities

Address of Site

Name of Employer or Contractor for whom facilities are provided

Facilities provided	Shelters and accommo-dation for clothing and taking meals. (Reg. 11.)	Washing facilities (Reg. 12.)	Sanitary conveniences (Reg. 13.)
Whether facilities provided (Yes/No)			
Date arrangements began			
Date arrangements ended			

Signed

For and on behalf of

Name of Employer or Contractor providing facilities

Date

Fig. 6.5.2

PART 2

Persons taken into employment or transferred to work subject to the Factories Act when under 18

Surname (1)	Christian name or forename (2)	Usual residential address (3)	Date of birth (4)	Date of taking into employment or transfer (5)	Careers Office to which notice of taking into employment or transfer was sent (6)	Date of despatch of notice (7)	Date of leaving employment (8)

Fig. 6.5.3

Part 4

Cases of poisoning or disease

Address of operations or works where person affected was employed (1)	Date of Notice (on F41) to		Name of person affected (4)	Sex (5)	Age (6)	Precise occupation (7)	Name of Disease (see notes on page 36) (8)	Remarks (9)
	Inspector (2)	Appointed Factory Doctor (3)						

Fig. 6.5.4

112

Department of Employment   F 2404

FACTORIES ACT 1961, EMPLOYMENT MEDICAL ADVISORY SERVICE ACT 1972

NOTICE OF TAKING INTO EMPLOYMENT OR TRANSFERENCE OF A YOUNG PERSON

Section 119A of the Factories Act 1961 requires an employer not later than seven days after taking a young person under the age of 18 into employment to work in premises or on a process or operation subject to the Factories Act 1961 or transferring a young person to such work from work not subject to that Act, to send a written notice to the local Careers Office.

NAME OF OCCUPIER   ....

ADDRESS OF FACTORY OR PLACE OF WORK (*If construction industry and the young person has been taken into employment, or transferred, to work on a particular site, the address of SITE should be given*).

....

....

....

DATE OF TAKING INTO EMPLOYMENT/TRANSFERENCE   ....
(*delete inappropriate item*)

NATURE OF WORK TO BE DONE BY YOUNG PERSON   ....

....

....

Please give the following information so far as it is known:—

SURNAME OF YOUNG PERSON (*capitals*)   ....

CHRISTIAN NAME (*OR FORENAME*)   ....

ADDRESS   ....

....

....

DATE OF BIRTH   ....

NAME AND ADDRESS OF LAST SCHOOL ATTENDED   ....

....

....

Signature   .... Date   ....

Position in firm   ....

Fig. 6.5.5

Health and Safety Executive
Health and Safety at Work etc Act 1974
Reporting of Injuries, Diseases and Dangerous Occurrences Regulations 1985

Spaces below
are for office
use only

Report of an injury or dangerous occurrence

● Full notes to help you complete this form are attached.
● This form is to be used to make a report to the enforcing authority under the requirements of Regulations 3 or 6.
● Completing and signing this form does not constitute an admission of liability of any kind, either by the person making the report or any other person.
● If more than one person was injured as a result of an accident, please complete a separate form for each person.

A   Subject of report *(tick appropriate box or boxes)* — *see note 2*

| Fatality ☐ 1 | Specified major injury or condition ☐ 2 | "Over three day" injury ☐ 3 | Dangerous occurrence ☐ 4 | Flammable gas incident (fatality or major injury or condition) ☐ 5 | Dangerous gas fitting ☐ 6 |

B   Person or organisation making report (ie person obliged to report under the Regulations) — *see note 3*

Name and address —

Post code —

Name and telephone no. of person to contact —

Nature of trade, business or undertaking —

If in construction industry, state the total number of your employees —

and indicate the role of your company on site *(tick box)* —

| Main site contractor ☐ 7 | Sub contractor ☐ 8 | Other ☐ 9 |

If in farming, are you reporting an injury to a member of your family? *(tick box)*   ☐ Yes ☐ No

---

C   Date, time and place of accident, dangerous occurrence or flammable gas incident — *see note 4*

Date ☐ ☐ 19 ☐
day month year

Time —

Give the name and address if different from above —

Where on the premises or site —
and
Normal activity carried on there

ENV

Complete the following sections D, E, F & H if you have ticked boxes, 1, 2, 3 or 5 in Section A. Otherwise go straight to Sections G and H.

D   The injured person — *see note 5*

Full name and address —

Age ☐   Sex ☐ (M or F)

Status *(tick box)* —

| Employee ☐ 10 | Self employed ☐ 11 | Trainee (YTS) ☐ 12 |
| Trainee (other) ☐ 13 | | Any other person ☐ 14 |

Trade, occupation or job title —

Nature of injury or condition and the part of the body affected —

F2508 (rev 1/86)   **Fig. 6.5.6**

114

Health and Safety Executive

Health and Safety at Work etc Act 1974
Reporting of Injuries, Diseases and Dangerous Occurrences Regulations 1985

For HSE use

Report of a case of disease

- This form is to be used to make a report to the enforcing authority under the requirements of Regulation 5.
- Completing and signing this form does not constitute an admission of liability of any kind, either by the person making the report or any other person.

A Person or organisation making report
(ie person obliged to report under the Regulations)

Name and address

Post code

Name of person to contact for further inquiry

Tel. No.

Nature of trade, business or undertaking

B Details of the person affected

Surname _____ Forenames _____

Date of birth ☐☐ ☐☐ 1 9 ☐☐   Sex (M or F) ☐
day   month   year

Occupation _____

Please indicate whether   Employee ☐
(tick box)
Other person ☐

If not an employee, what is the ill person's status?
(eg self-employed or trainee)

F2508A (1/86)

Fig. 6.5.7

Chapter 7
Supervisor's responsibility relating to the use of materials, plant and equipment

7.1 Calculating quantities of materials, plant and equipment

Too often site supervisors are expected, in addition to ensuring satsifactory workflow levels and sound man management, to schedule and order their own material, plant and equipment. This is time consuming. Nonetheless, it can be advantageous because the materials when ordered from site tend to be the exact requirements of the site supervisor, and that any mistakes fall squarely with the supervisor or his subordinates. Also, the materials will generally be received exactly when required, and not at the discretion of the central purchasing department, which sometimes overlooks the limited storage facilities on-site and the level of labour needed to unload certain consignments.

The site supervisor, apart from being proficient at reading and understanding the details on working drawings, should be capable at taking-off requirements. Suitably designed scheduling and, if necessary, requisitioning sheets should be made available for site use by head office, and if they are not forthcoming the layout suggested in Fig. 7.1.3 could be used.

When calculating the quantities of materials required while scheduling a number of important details must be observed and which, if studied, may save the site supervisor many hours of wasted time and energy. Take, for instance, the following examples dealing with the different areas of work:

Excavations

In calculating volumes of earthwork to be removed from foundation trenches the following is considered:

(a) length of trench (calculated as the centreline of the trench);
(b) width of trench;
(c) depth of trench (remembering that the surface strip may have already been removed).

Figure 7.1.1 represents the plan layout of a simple ordinary strip foundation showing the distances A, B, C, D, E and F which are added together to determine the centreline of the trench excavation. Trench excavation for internal walls are measured from the edge of the perimeter excavation, as in G (Fig. 7.1.1), and from the inside edge of the perimeter brickwork/blockwork for length of walling.

Volume of trench excavations = length of centreline × width × depth for perimeter wall, plus volume of internal wall excavations.

Note: This subsoil should be removed from site, less any amounts required for back filling.

Concrete in foundations

The concrete volume required for foundations depends on how accurate excavation work has been carried out. More concrete may have to be used if foundation trenches are overdug in depth or width.

Once the centreline length of the trench has been established it can be used to calculate the volume of concrete required for the foundations, this is achieved as follows:

Volume of concrete = the centreline length of the trench × width × foundation thickness.

One must remember that this calculation is for the perimeter foundation concrete. The crosswalls/interior loadbearing walls foundation concrete should be added to give the grand total of foundation concrete required.

Ready-mix concrete
If ready-mix concrete is to be used the volume of concrete measured from drawings or extracted from the bills would be ordered after a waste allowance of $2\frac{1}{2}$ per cent is added.

Mixed on-site concrete
A percentage is added to both the aggregates and cement to allow for a decrease in volume when mixed with water, and for waste due to losses in storage and while transporting the mixed materials to the place of deposit. Allowances of 40 per cent and 10 per cent are made for decrease in volume and waste respectively, with 40 per cent and $2\frac{1}{2}$ per cent being allowed for cements.

Hence, consider the concrete requirements for, say, foundation work to be 140 m^3 ($1:2:4$ mix). Remember the mix is usually given by weight and not volume, but in this example the weights per m^3 of each material is approximately the same. As a precaution proper weigh-batching calculations should be used for very large

FOUNDATION EXCAVATIONS

Calculation of foundation excavation volumes

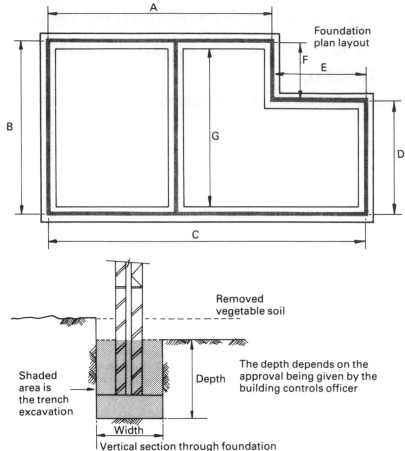

The volume of excavation = lengths A + B + C + D + E + F (centreline) × width of foundation × depth of excavation. This is for the external wall foundations excavations. Now calculate the volume of excavation for internal wall foundations and add to the previous volume to give total volume of excavations.

Always double check that internal walls are in fact, built on foundations and not on the sub-concrete floor. If the latter is the case generally they will require no excavation.

Fig. 7.1.1 Foundation excavations

quantities although it is unusual to order all materials at once, so as the first deliveries are used up one can then assess future additional needs, with deliveries being made when required.

The ratio is 1 : 2 : 4 = 7 parts.
Then the amounts required for the concreting materials are worked out by first valuing 1 part, i.e.: $\dfrac{140\,m^3}{7\ parts} = 20\,m^3$

(weights of materials per m^3 = 1500 kg approx.)

Therefore, cement is 1 part of $20\,m^3$ $=\ \ 20\,m^3$
and the fine aggregate is 2 parts of $20\,m^3$ $=\ \ 40\,m^3$
and the course aggregate is 4 parts of $20\,m^3 =\ \ \underline{80\,m^3}$
$$\text{check}\ \underline{140\,m^3}$$

(*a*) Cement quantity
As cement is approximately
$1500\ kg/m^3$
Hence, $1500\ kg \times 20\,m^3$ = 30 000 kg of cement

Add for decrease in volume
and waste $42\tfrac{1}{2}\%$ = 12 750
Total $\underline{42\,750\ kg}$ cement

As bags of cement
weigh   50 kg,
then $\dfrac{4\,2750\ kg}{50\ kg} = \dfrac{855\ bags\ cement}{are\ required}$

(*b*) Fine aggregate

Fine aggregate
required is   40 m³

Add 50% for decrease
in volume and waste   20 m³
                         ────────
                         60 m³ required (or 60 ×
                                   1500 kg = 90 000 =
                                   90 tonnes).

(*c*) Course aggregate

Course aggregate
required is   80 m³

Add 50% for decrease
in volume and waste   40 m³
                         ────────
                         120 m³ required (or 120 ×
                                   1500 kg =
                                   180 000 =
                                   180 tonnes).

Any form of concrete required on site can be calculated in a similar manner.

Brickwork

Facings

The centreline of the outer facing skin of brickwork can be calculated from the drawings, or details can be extracted from the bills. Using the centreline as the length of the brickwork, and the height of the wall which contains the facing bricks, the area, in square metres, can be ascertained. One must remember that there are normally two courses of facing bricks below the damp proof course level which need to be included in the total height for facing brickwork.

For every square metre there are approximately 60 bricks per half-brick wall in stretcher bond, and if the wall is one brick thick it will contain 120 bricks per metre square. Where special bonding is used the number of bricks per square metre varies as follows:

Half-brick wall in English Bond =   90 bricks/m^2.
Half-brick wall in Flemish Bond = 80 bricks/m^2.
Half-brick wall in English Garden Wall Bond =
                                   75 bricks/m^2.

An allowance of 5 per cent for waste is normally made for facing bricks, and 2½ per cent for common bricks.

Mortar materials are assessed in the same way as for concrete using the ratios as before. However, one can assess the amount of mortar per square metre of brickwork per half brick wall in stretcher bond to be 0.03 m^3, and 0.07 m^3 (a little more than double) for one brick walls. Allow up to 10 per cent for waste on mortar.

Note: As a rough rule-of-thumb when ordering sand or ready mix mortar for facing brickwork allow 1 m^3 for every 1000 bricks. This allows for the internal blockwork, decrease in volume, and waste. Also allow 5 bags of cement (50 kg bags).

Blockwork

Allow 10 blocks per square metre of walling.
Allow 0.005 cubic metres of mortar per square metre of blockwork up to 100 mm thick blocks. 10 per cent is added for waste.

Damp proof course

Find the length of the wall and then divide it by the length of a roll of damp proof course which is 8 metres. This gives the number of rolls which should be ordered, but always allow an additional 5 per cent for laps and waste before doing so.

Hardcore

To find the volume of hardcore for ordering purposes multiply the area of hardcore required by the thickness of the bed (in metres – a 150 mm thick bed is 0.15 metres). Allow 20 per cent extra for consolidation.

The sand or other blinding is calculated in the same way, but allow 50 per cent extra for waste and consolidation.

Timber

(*a*) *Flooring*

The following shows the length of tongued and grooved flooring board required for 1 metre square of floor. One should remember that the widths are not finished widths but sawn widths:

74 mm wide board = 15.4 metres approximately
100 mm wide board = 11.11 metres approximately
125 mm wide board = 9   metres approximately
150 mm wide board = 7.1   metres approximately

Add 5 per cent for waste.

(*b*) *Joists*

The lengths should be sufficient to allow the supporting ends to rest on the walls by half a brick on cavity construction, or by 225 mm on 1 brick walls. Once again allow 5 per cent for waste, or 1 extra joist per house.

When ordering timber for structural members the lengths should be stated in metric standard lengths, i.e. in stages of 300 mm.

Rafters, etc.

The lengths for rafters should allow for the foot and ridge cuts with a small addition of 150 mm to allow for discrepancies in construction. The same applies for ceiling joists. Add 5 per cent for waste.

Slates for roofs

Use the following formulae for calculating the gauge. The gauge is the centre at which the slaters'/tilers' battens are fixed.

(*a*) Slates (header nailed):

$$\text{Gauge} = \frac{(\text{length of slate} - 25\,\text{mm}) - \text{lap}}{2}$$

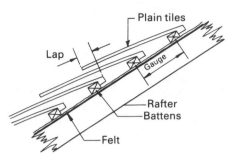

Fig. 7.1.2

(b) Slates (centre nailed):

$$\text{Gauge} = \frac{\text{length of slate} - \text{lap}}{2}$$

(c) Plain tiles:

$$\text{Gauge} = \frac{\text{length of tile} - \text{lap}}{2}$$

(d) Interlocking tiles:

$$\text{Gauge} = \text{length of tile} - \text{lap}$$

(The gauge and lap are illustrated in Fig. 7.1.2.)

To calculate the numbers of tiles/slates use the following formula:

number of tiles/slates per metre square =

$$\frac{1 \text{ m} \times 1 \text{ m}}{\text{width of tile/slate} \times \text{gauge.}}$$

The interlocking tile numbers per metre square =

$$\frac{1 \text{ m} \times 1 \text{ m}}{\text{width of tile less the side lap} \times \text{gauge.}}$$

Example:

number of Plain tiles

$$(260 \text{ mm} \times 160 \text{ mm}) \text{ per m}^2 = \frac{1 \text{ m} \times 1 \text{ m}}{\text{width of tile} \times \text{gauge}};$$

Change the metres into millimetres.

Therefore, number of tiles =

$$\frac{1000 \text{ mm} \times 1000 \text{ mm}}{160 \text{ mm} \times 100 \text{ mm (usual gauge)}}$$

then number of tiles =

$$\frac{1000}{16}$$

Number of tiles = 63.

Plastering

Plaster will spread at the approximate rate per 50 kg bag, which allows for compression and waste, as follows:

(a) Premixed Carlite browning plaster 10 mm thick = 7 m² (140 m²/tonne).
(b) Premixed Carlite finish 3 mm thick = 18 m² (350 m²/tonne).
(c) Gypsum bonding plaster 10 mm thick = 12 m² (225 m²/tonne).
(d) Gypsum finish plaster 3 mm thick = 15 m²/ 300 m²/tonne).

Painting

Paints can be used at the following spreading capacities:

(a) Woodprimer/undercoat or gloss, 13 m² per litre, or 0.075 litres per m².
(b) Emulsion 10 m² per litre, or 0.1 litres per m².

These figures relate to large flat areas. Where narrow or small areas are being painted, e.g. window frames and skirting boards, etc., it is usual to allow for the use of more paint because it is difficult to apply thin coats to small areas.

The figures previously given are a general guide to the quantities required, and for other materials readers are advised to seek out manufacturers' or suppliers' recommendations.

One can be too generous when scheduling for material, the result being that the site supervisor is criticised when too many materials are left unused at the end of an operation or contract. On the other hand, if too few materials are scheduled/ordered, resulting in small subsequent extra orders having to be made later, again there is criticism because for small orders the same generous discounts as for larger orders may not apply, so resulting in additional expenditure by the firm. Careful attention should be paid on calculating quantities so that these two problems may be overcome.

Plant and equipment

The more plant that can be used on-site the less demand there is for labour. Many jobs are simplified by the introduction of machines and ancillary equipment. The strain

BUILDING CONTRACTORS LTD., MERROW, GUILDFORD 2657 MATERIALS SCHEDULE/REQUISITION				
Contract: Job no: Prepared by:			No: Date:	
Bill/ programme ref.	Description	Quantities	Delivery date/s	Remarks/diagrams

Fig. 7.1.3

is taken out of the work which is expected these days by the operatives. Also, it is sometimes difficult to obtain suitable levels of labour so by the introduction of plant operations are speeded up.

It is normal practice for the planning officer within a company, in conjunction with the estimator, contracts manager and plant manager, to fix the levels and types of plant needed for key operations. The site supervisor has, therefore, no alternative but to follow instructions regarding types of plant to be used.

There are times when the site supervisor is expected to use his/her own initiative and to choose, order and use plant, the suitability of which would depend entirely on the supervisor's past experience. If the supervisor has had a wide range of experience his/her judgement is perhaps as good, if not better, than the planner's. On the other hand, some guidelines should generally be offered by head office; this is prudent management.

Plant and equipment falls into three main catagories:

1. Small plant (non-mechanical or mechanical).
2. Large plant (non-mechanical or mechanical).
3. Administrative and sundry plant.

Small plant
One normally refers to the following where small plant is concerned:

1. Non-mechanical:
    (*a*) barrows, buckets, drums;
    (*b*) spades, picks, trowels, sieves, rakes, hods;
    (*c*) ropes, chains, pullies, hoses;
    (*d*) tampers, hammers, chisels, crowbars cutters, screwdrivers;
    (*e*) spanners, jacks, drain plugs and rods;
    (*f*) saws, axes, brushes tarpaulins/dust sheets;
    (*g*) ladders, steps, trestles.
2. Mechanical
    (*a*) electric rotary drills, sanders, screwdrivers, grinders.
    (*b*) hand power saws, routers.

Large plant
1. Non-mechanical
    (*a*) scaffolding-tubular or 'H' scaffold;
    (*b*) formwork – proprietary types;
    (*c*) shoring, hoarding, gantries;
    (*d*) trench timbering, sheeting and jacks.
2. Mechanical
    (*a*) concrete mixers;
    (*b*) transportation – lorries, dumpers, fork-lift trucks;
    (*c*) cranes – tower (static or railed);
    (*d*) excavators – back-acters, face shovels, scrapers;
    (*e*) dozers, graders;

(*f*) pumps;
(*g*) loaders – crawler or wheeled;
(*h*) compressors;
(*i*) demolition plant – breakers, crushers, pneumatic drills;
(*j*) rollers – vibrator/non-vibrator;
(*k*) bench tools – rip and cross cut saws, planers;
(*l*) pile drivers, etc.

Administrative and sundry plant
1. Site hutting (toilets, offices, etc.).
2. Tables, chairs, cabinets.
3. Lighting equipment and heating.
4. Surveying equipment.
5. Compounds, fencing.
6. Helmets, gloves, goggles, boots, protective jackets, etc.

When there is a need for plant on-site it may be obtained by one of the following methods:

1. Hiring from a plant hire company. This required no expensive plant department set-up.
2. Hiring from own company's plant department at a hiring rate per hour/day/week.
3. Buying plant when required with the full cost being charged to the contract. The contract is then credited later when the plant is revalued after it is sent back to the yard, or is sold.

When hiring plant there can be two rates; one which includes the accompanying driver; the other which does not include a driver.

Supervisors should first obtain approval from head office or the contracts manager to hire plant and equipment from outside the company's resources. If permission is granted an assessment by the site supervisor may have to be made for the correct types of plant to be used, and the quantity, to give the optimum of output.

Balancing plant

Take, for example, the proposed excavation for mass, reduced level dig in medium ground, and the excavated material is to be removed immediately by lorry to a tip five miles away.

Consider the type of excavator which would be most suitable depending on the time of the year, and whether the site is accessible to lorries once excavation operations begin. Assuming a tracked excavator with bucket size of 0.25 m³ would be satisfactory to give an output of, say, 20 m³ per hour provided that no delays occur through the excavator standing idle. Idleness of the excavator results when the lorry or lorries are delayed in traffic jams, or the excavator/lorry balance is inadequate. There should be a sufficient number of lorries to meet the requirements of the output of the excavator. Therefore, a little calculation should be made by the supervisor as follows:

Output of excavator per hour (0.25 m³ bucket) in medium ground stated to be about 20 m³ by the manufacturer of the excavator, or from past records from head office, always considering:

1. Winter or summer working.
2. Good or inexperienced operator.
3. Distance that excavator may have to travel to the lorry (lorry should be close to excavator).
4. Depth and arrangement of existing services below ground level.
5. Mass or surface excavation.

Assume the capacity of dug material after 10% bulking to be 20 m³ per hour $\times \frac{110}{100}$ per cent = 22 m³ per hour.

Now assuming the lorry capacity is 10 m³ and the tip is 5 miles (8 km) away.
One round trip by lorry is:
(*a*) travelling to tip assessed at 25 minutes
(*b*) tipping time, 10 minutes
(*c*) return trip, 25 minutes
Total 60 minutes.

The loading of 10 m³ lorry if output per hour of excavator is 22 m³

$= \frac{10}{22} \times 1$ hour (change to minutes)

$= \frac{10}{22} \times 60$ minutes $= 27.3$ minutes.

= say, 30 minutes.

∴ Number of lorries travelling to and from site

$= \frac{60 \text{ mins}}{30 \text{ mins}}$

= 2 lorries, plus 1 being loaded
= 3 lorries.

Finally, if the excavation works needs to be speeded up an introduction of more and bigger excavators would achieve this provided that the site was not congested.

The site supervisor may have other special plant to consider, the work capacity/output depending on the contract programme and the amount of work to be done.

Concreting work

Where small quantities of concreting work are required at regular intervals and the site is not congested, mixing on site may be the answer. Where large quantities are required periodically delivered ready-mix concrete could be arranged. Some form of calculation or assessment of requirements should be made, however.

Calculation of output
Output of a mixer governs the amount of concrete (if mixed on-site) which can be placed per hour or day. Assuming that the only mixer which is available is a 14/10 (output being 0.37 m³ per mix).
Time taken for one output is:
(*a*) filling the skip with aggregate and cement and emptying into mixer drum and adding water = 5 minutes
(*b*) mixing time = 4 minutes
(*c*) emptying into a dumper = 1 minutes
Total 10 minutes

Output in one hour: $\frac{60 \text{ mins}}{10 \text{ mins}} = 6$ mixes.

Therefore, $6 \times 0.37 \text{ m}^3 = 2.22 \text{ m}^3$ per hour.
Say the mixer is worked $7\frac{1}{2}$ actual hours,
Then, $7\frac{1}{2} \times 2.22 \text{ m}^3 = \underline{16.65 \text{ m}^3}$ per day

This amount of output may be satisfactory for small jobs but for mass concreting ready mix could supplement the mixer. On the other hand, a larger mixer could be hired if one is available.

The site supervisor may have to schedule and requisition his/her own needs in the way of plant and equipment, so as a guide the form in Fig. 7.1.4 could be adapted.

7.2 Maintenance of plant and equipment

Plant, whether mechanical or non-mechanical, should be maintained in a sound condition. Invariably this is the responsibility of the plant department, but the site supervisor cannot stand idly by if plant while being used on his/her site breaks down; is being operated in a dangerous state; or the efficiency is below the standard expected according to the contract programme and progress chart. Not only will the site supervisor be accountable for the progress made on-site, but also would be answerable in law to the Factory Inspectorate through the Health and Safety at Work Act, and the police – particularly where dangerous, unroadworthy vehicles are being used.

Maintenance of all plant is essential for the following reasons:

1. It contributes towards the safety of employees and third parties. Any accidents, no matter how they are caused, are expensive and unnecessary.

2. It ensures efficiency of the mechanical plant because there are less breakdowns which can reduce output.

3. It prevents frustrations in operatives, particularly if their earnings are affected by plant breakdowns and poor output.

4. Site supervisors are kept happy where the plant department offers efficient servicing of mechanical plant. Plant with a poor performance record should not be issued but should be got rid of by the plant department.

5. It enhances the value of plant and deterioration, and hence, depreciation, is slowed down.

6. It is generally a guarantee that operatives take more care when using the plant. Dilapidated plant is more misused and abused because operatives treat it with contempt.

7. Well-kept plant is a good public relations exercise. Those from outside tend to recognise the firm as an efficient, well-managed unit where the vehicles, plant and equipment are clean and well looked after.

8. Major breakdowns of mechanical plant are reduced if a sound maintenance scheme is in operation.

9. The work, and responsibility of the firm's Safety Officer is simplified, and the Factory Inspector is pleased.

10. Little or no trouble is experienced with the police as far as the firm's vehicles are concerned.

BUILDING CONTRACTORS LTD., MERROW, GUILDFORD 2657				
PLANT AND EQUIPMENT SCHEDULE/REQUISITION				
Contract: Job no: Prepared by:			No: Date:	
Plant type and equipment	Dates required		Plant hire firm/ supplier	Remarks
	Date on-site	Date off-site		

Fig. 7.1.4

If maintenance is to be effective the supervisor should understand that most plant managers insist on the following divisions as far as mechanical plant is concerned:

1. Servicing and cleaning.
2. Preventative maintenance.
3. Planned maintenance.

1. Servicing and cleaning

The responsibility for servicing is the site supervisor's who then generally delegates the responsibility to the chargehands, foremen or gangers who have plant within their control. Each would ensure that the machine operators regularly clean and lightly oil the bodywork of their machines which assists in making subsequent cleaning easier. Additionally, the following operations fall into the servicing category:

(*a*) topping up engine's, and other working parts, oil levels;
(*b*) greasing the points recommended by the manufacturer's catalogue;
(*c*) filling fuel tanks;
(*d*) providing water for the cooling systems;
(*e*) maintaining anti-freeze levels in winter, or draining off water at the end of a working day;
(*f*) inspecting for fuel, oil or water leaks;
(*g*) tightening loose screws and bolts where the services of a trained mechanic/fitter is not required. Operators/drivers of machines, etc., should be allowed some time at the end of each working day or when the machine is no longer used on the completion of an operation to clean down, check and secure against vandalism or theft.

2. Preventative maintenance

This responsibility falls squarely on the shoulders of the plant manager or plant hire firm. Naturally, the site-based or travelling maintenance engineer/fitter/mechanic acts on behalf of the plant department or plant hire firm, ensuring breakdowns are kept to a minimum.

A regular servicing arrangement is provided to prevent deterioration and breakdown of plant, and the typical areas in which maintenance is carried out relate to the following.

(*a*) engine oil change;
(*b*) flushing and cleaning the coolant system;
(*c*) replacing oil and air filters;
(*d*) spark plugs adjusted or changed;
(*e*) thoroughly checking the bodywork and outer engine bolts and screws, and either tighten or provide new ones if necessary;
(*f*) exchanging other parts of the plant which are worn or have broken;
(*g*) carrying out tests to ensure machines, etc., are working correctly;
(*h*) adjusting mechanical parts which have worn, such as brake shoes or contact points in distributors;
(*i*) generally carry out the servicing instructions issued by the manufacturers of the plant or equipment.

3. Planned maintenance

It is obvious that sound servicing and preventative maintenance reduces breakdowns and excessive wear and provides safe machinery, etc., but there comes a point in time where mechanical plant ceases to work efficiently and it is, therefore, the plant manager's responsibility to assess from past experience and records at what time vital parts require to be adjusted or changed, thereby preventing expensive and inconvenient breakdowns occurring which could seriously affect output on-site and lead to further additional damage of the plant.

Records are kept by the plant department on each piece of mechanical plant (vehicles, excavators, etc., electrical equipment) so that when the time is right for temporary withdrawal of plant for planned maintenance, it can be conducted with the minimum of disruption. Withdrawal from circulation would take place when one site has returned the plant to the yard; the planned maintenance would then take place before redistributing it to another site.

The only alternative to planned maintenance would be to withdraw from use any plant which is beginning to break down frequently or is giving other regular trouble. This causes inconvenience to everyone, particularly the site supervisor, who would be aggrieved because of the constant breakdowns of the troublesome machine, etc., but further aggravation would result while waiting for a replacement machine, especially if the operation being executed is a critical one on the contract programme and progress chart.

Plant and equipment records

The plant manager opens up a file on each piece of plant purchased, and all relevant details are recorded regarding the plant history, e.g. manufacturer's handbook; date of purchase and cost; supplier's name and address; the code number, modifications; cost of repairs; running costs; internal hire and external hire charges; fuel; maintenance; damages and records of repairs; deficiencies; expected life and residual value; depreciations; tax and insurance; record of use (worked hours, idle time, breakdown hours, delivery cost, and erection, dismantling and removal costs – if any).

Some form of 'charging-out' of plant, etc., must be devised by the plant manager, and a plant movement schedule/chart may be operated by one of the plant manager's subordinates as a visual record of where pieces of plant are at any time.

Some forms associated with the use of plant are:

1. Plant Utilisation Sheets.
2. Plant Transfer Sheets.
3. Plant Out and In Chart.
4. Plant-Hire Charges.
5. Inspection Records and Maintenance Sheets.
6. Delivery Tickets, etc.
7. Lorry Driver's Log Book.
8. Inspection Records and Testing Result's Forms.
9. Factories Act (construction and other Regulations).

Site managers are advised, when requisitioning or ordering plant, not to rely entirely on a telephone call but to verify the order in writing to ensure that the correct type of equipment, etc., and size are received: less disputes occur between plant departments and site supervisors when this method is adopted. Also, charging out of plant is more accurately predicted as the date of requirement is shown on the requisition/order.

Receipt of plant, etc., on-site

The site manager, or persons delegated with the responsibility to take receipt of plant deliveries, should ensure that the plant is in good working order, and that any deficiences, damages, etc., are noted and stated on the delivery or transfer ticket to prevent any additional charges being made to the contract which were not justly incurred. A refusal of the plant delivery would be in order if it is the wrong type and size, also, if it was faulty and dangerous.

At the same time as taking receipt of plant, etc., special servicing equipment should also be made available from the plant department, such as:

1. Simple maintenance tools.
2. Cleaning gear.
3. Fuel, if not already available on-site.
4. Oil, grease, anti-freeze.
5. Recording dials, gauges, etc., for tyre pressures.
6. Colour code paints if plant is required for long use.
7. Instructions on starting and securing where driver/operator is not provided.
8. Chains and ropes, if needed, and other spares.

Finally, someone should be entrusted with the responsibility for each machine, and its servicing.

7.3 Minimising waste on-site

It is recognised that wastage of materials will result on a construction site due to many reasons, therefore estimators, when pricing for work, allow within their 'unit-rates' (cost to do a unit of work, i.e. cost per square metre of brickwork, cubic metre of concrete) a wastage percentage ranging from as little as 1 per cent to a disproportionate figure of 15 per cent which may or may not be necessary, but which is neither an advantage to the contractor nor the client, and which unfortunately increases prices generally for both parties.

Contractors who are able to reduce wastage through being well organised on-site can estimate more competitively than others (who are not conscious to the benefits of sound supervision regarding materials handling), and increase their chances of bidding successfully for work, and create more wealth for shareholders and employees alike. With the maintenance of continuous employment and perhaps high earnings, stability of labour levels is created to the advantage of all concerned.

Wastage of materials can result from either the head office or site staff's inability to schedule materials accurately, with delivery dates inconsistent with requirements leading to possible storage difficulties because of inadequate space on confined sites, and the inability of site supervisors to control the whole aspect of materials receipt, storage, distribution and correct usage.

Naturally, if a contract is large enough and costs allow it, suitable handling facilities, access and storage space or accommodation will generally result in an overall improvement regarding reduction of material breakages and its misuse.

The most common problems relating to material supply, etc., revolves around the following:

1. Taking-off and scheduling.
2. Requisitioning and ordering.
3. Receipt and checking of deliveries from supplier's or contractor's own yard.
4. Off-loading and handling.
5. Storing and protecting.
6. Issuing and distributing.
7. Use of materials.
8. Quality control and supervision.
9. Security.

1. Taking-off and scheduling

The buyer/purchasing officer can have the responsibility to take-off and schedule materials from the bills of quantities or from the production/working drawings. Where the site supervisor is given this role he/she must be careful to ensure that correct quantities and details of quality of materials are extracted from the drawings and specifications. It is emphasised that it is necessary to make allowances for material wastage, and adherence to a realistic percentage from correctly recorded estimator's or purchaser's figures obtained through previous experience must be made.

Taking-off material measurements from drawings should be meticulouly made without the need for allowing excessive extra lengths or quantities on, say, joists to give a suitable margin of safety. Too many examples can be witnessed on-site of long off-cuts from rafters, joists and other structural members due to generous allowances having been made at the taking-off and scheduling stages. Also, extra members can be seen lying around the site unused.

2. Requisitioning and ordering

The purchasing officer has a number of duties to perform before orders are placed for materials. When schedules have been prepared, or requisitions have been received from various departments or site supervisors, enquiries are made to suppliers to enable the most competitive prices to be accepted. Purchase orders then follow the enquiry's stage to the successful suppliers whose quotations come up to expectations.

Site supervisors, if allowed to make orders themselves, would find it advisable to enquire from head office with which suppliers or manufacturers orders should be placed for the various materials.

Where bulk orders have been previously placed with the suppliers by the purchasing department in a centralised purchasing set-up and site supervisors are given the responsibility to call forward these materials when required, care must be exercised in calling forward sufficient for the storage areas available, and sufficient to maintain the levels of production without the need to hoard. Where materials arrive too early on-site before they are required there is more chance of damages and loss arising due to vehicle and other mechanical damage, damages due to the weather, pilfering and theft.

3. Receipt and checking of deliveries from suppliers, or contractor's own yard

Site supervisors should expect some notification from the supplier or sender of materials of impending deliveries. This can be by a posted advice note or by a telephone call, unless, when an order was made originally, phased delivery dates had been agreed between the parties. If forewarned of imminent deliveries the site supervisor can make arrangements for safe storage, and labour and plant can be provided to facilitate off-loading and other essential handling procedures to prevent undue damages. It may also be necessary to arrange for the testing of materials if thought necessary.

Delivery notes should accompany delivered goods and must be presented by the delivery drivers as proof of the consignments' quantity and quality. Although it is more easily said than done, meticulous checks should be made of every consignment by the person delegated to receive the goods. Delivery notes should not be signed until the goods are checked, and if damages have arisen during transit, or there are discrepancies between the delivery notes and deliveries, the amounts should be indicated on both copies of the delivery notes and records should be made for future reference (see Fig. 7.3.1).

Material transfers from other sites should be accompanied by a copy of a Materials Transfer Sheet. A check should again be made to ensure that the materials agree with the Transfer Sheet (see Fig. 7.3.2) before a signature is added.

In all cases when there is a receipt of materials/goods an entry should be made on a Material/Goods Received Form for the eventual despatch to the head office, or an entry could be made in the Daily/Weekly Report.

4. Off-loading and handling

Careless off-loading and handling of material adds to wastage on-site and each supervisor is advised to pay sufficient attention to the methods used depending on the materials being dealt with. Facing bricks may be delivered on pallets which, unless the delivery lorry is self-unloading, fork-lift trucks should be on hand. The alternative is to unload manually, which is time consuming and, hence, wasteful. Prior discussions with the supplier

BUILDING CONTRACTORS LTD., MERROW, GUILDFORD. 2657. MATERIALS/GOODS RECEIVED					
Contract: Contract no: Prepared by:					No: 12.
Date	Delivery Note No.	Supplier	Materials/goods and quantity	Rate	Total value

A line is filled in by the site supervisor or assistant for each quantity of materials received. The rate and value is filled in by head office.

Fig. 7.3.1

```
┌─────────────────────────────────────────────────────────────┐
│      BUILDING CONTRACTORS LTD., MERROW, GUILDFORD. 2657.      │
│                 TRANSFER OF MATERIALS                         │
│                                              No:              │
│                                              Date:            │
├───────────────────────────┬─────────────────────────────────┤
│ From site:                │ To site:                        │
├───────────────────────────┴──────────────────┬──────────────┤
│ Description:                                  │ Quantity     │
│                                               │              │
│                                               │              │
│                                               │              │
│                                               │              │
│                                               │              │
│                                               │              │
│                                               │              │
│                                               │              │
│                                               │              │
│                                               │              │
├──────────────────┬────────────────┬──────────┴──────────────┤
│ Issued by:       │ Received by:   │ Driver's signature:     │
└──────────────────┴────────────────┴─────────────────────────┘
```

This document is made out in triplicate: top copy sent to head office second to site receiving materials third copy retained.

Fig. 7.3.2

with, perhaps, an agreement on methods of delivery and off-loading could save the site supervisor time and effort in dealing with each consignment.

Tipper lorries for the delivery of certain common bricks may be acceptable on some contracts but, generally, palleted or banded bricks have a few cost-saving features, e.g. ease of handling with correct plant; ease of storing in neat stacks; and less breakages and waste because the bricks are manhandled very little. The banding keeps the bricks together until the work for which they were ordered commences. Tower cranes can handle heavy loads from the backs of lorries lifting material immediately into their required place, whether it is at ground level at differing positions around a congested site, or at first or any floor level.

The lifting of small loose materials are faciliated better if the suppliers pack their goods in boxes, cases, bags, drums, tins, baskets; this helps to prevent breakages and reduces waste.

5. Storing and protecting

This is a more important element where material wastage can increase substantially if care is not exercised in the selection of suitable storage points. Pre-planning is essential, and the site supervisor who fails in this and decides where to store materials when they have already arrived on-site, is not operating in the way for which he/she was employed.

At commencement of a contract a Site Layout Plan should be drawn up to outline the construction working areas, and to show all the site facilities (huts etc.) and material storage areas which are codified for the various key bulk or valuable materials (see Fig. 7.3.3). A storage compound layout would also serve to highlight to everyone concerned where the various delivered materials are to be placed for safe keeping, so that if the site supervisor is occupied when there is a delivery, other supervisors, storemen, etc., may be able to direct the delivery drivers to appropriate unloading points by referring to the Site Layout Plan retained in the main site office.

The storage areas would have to be clearly marked with notices so that proposed drain runs and other service positions are not encumbered (see Fig. 7.3.3).

The following are a few pointers regarding storage:
1. Hardstands for heavy material should be provided.
2. Provide concrete blinded areas for loose materials such as sands and gravels, with separate bays for different grades.
3. Show clear marks and laid-out areas for different materials, such as reinforcement bars and mesh, concrete products, blocks, tiles, timber, etc.
4. Provide huts for bags of cement, plaster and lime, taking care to stack materials so that the earliest deliveries are used first.
5. Ensure there are level beds for bricks, etc., with covering provided (polythene or tarpaulin) until

126

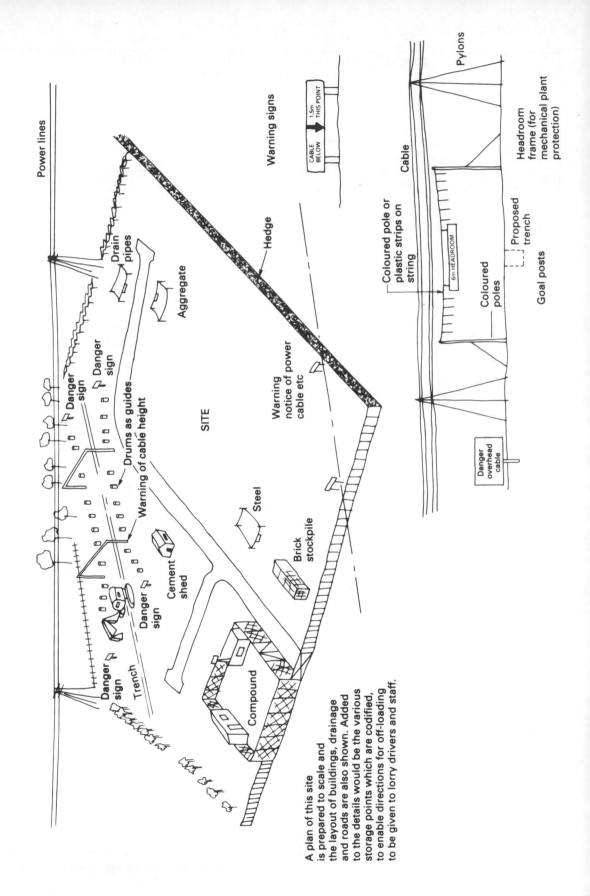

Power lines

Danger sign

Danger sign

Drain pipes

Aggregate

Danger sign

Drums as guides

Warning of cable height

SITE

Hedge

Warning signs

Danger sign

Trench

Danger sign

Cement shed

Steel

Brick stockpile

Warning notice of power cable etc

Compound

A plan of this site is prepared to scale and the layout of buildings, drainage and roads are also shown. Added to the details would be the various storage points which are codified, to enable directions for off-loading to be given to lorry drivers and staff.

Pylons

CABLE BELOW → 1.5m THIS POINT

Cable

Headroom frame (for mechanical plant protection)

Coloured pole or plastic strips on string

6m HEADROOM

Coloured poles

Proposed trench

Goal posts

Danger overhead cable

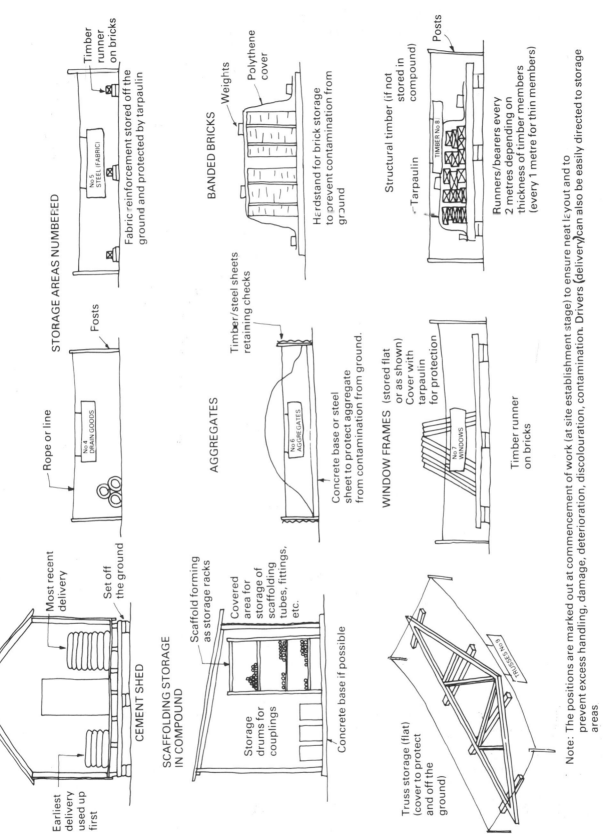

Fig. 7.3.3 Storage and signs

Note: The positions are marked out at commencement of work (at site establishment stage) to ensure neat layout and to prevent excess handling, damage, deterioration, discolouration, contamination. Drivers (delivery) can also be easily directed to storage areas materials are required. If pallets (or polythene on the base) are used they prevent sulphates and other contaminants from the ground affecting the materials. Stacks should be to suitable heights relating to their width and length for stability (height = four times the stack depth).

6. Be sure that stacked materials should not have to be moved to allow operations to be completed, such as: road, services and building works.

7. Lay drainage pipes on their sides in neat stacks, using wooden wedges to prevent movement.

8. Stack steelwork, reinforcement bars and fabric flat on timber runners off the ground, with polythene covers to prevent undue rusting and to keep them clean – especially if not required for some time.

9. Move the residue of any discontinued pile of material, especially bricks, along to an existing or new work area to prevent burials, breakages or their misuse.

10. Store windows and door frames in a covered compound area, or stack them perfectly flat off the ground on sleepers/runners and covered to protect. If standing vertically care should be taken to prevent twisting.

11. Put valuable small items in the charge of a storekeeper or supervisor. Individuals should be made to sign for them when required. Only the quantity for immediate usage should be issued. Items such as: nails, screws, locks, tiles, electric switches, plumber's fittings, paint, etc. fall into this category. A sound method of distribution is important on very large jobs to prevent waste and pilfering, and where major materials are involved the various trade supervisors/foremen should be made responsible for the distribution and use of their own trade's materials.

12. Store scaffolding fittings in strong bins and not sacks, although the latter facilitates ease of handling at the workplace. Scaffold poles when not in use should be stored on racks in the compound area.

The above recommended methods for storing and protecting material is expensive, but is necessary and worthwhile in the long run and prevents a greater loss to the firm/company due to the following:

(a) deterioration due to weather damage;
(b) mechanical damage;
(c) misuse;
(d) pilfering;
(e) theft.

6. Issuing and distributing

An efficient storekeeping system needs to be set up onsite as soon as possible to ensure the operatives are issued, when required, with the correct quantity and quality of materials. On smaller jobs the supervisors/foremen could take on this responsibility.

The store is best situated adjacent to any other lock-ups and storage compound. For ease of vehicular access it may be necessary to have two access gates to a compound, especially if space is limited, to allow manoeuvrability of

vehicles; one gate being kept locked until its use is essential. The main access to the compound must be close to the storeman's or supervisor's office so that it is overlooked for security. A booking-out system of special and valuable materials should be operated so that carelessness by operatives in using the materials issued can be pinpointed, which, although damages or loss cannot be charged to operatives, puts the issued operatives in an awkward position in which they would not normally wish to find themselves, making them more careful in the future and, further, contributes to reducing material losses. (See Fig. 7.4.2.)

7. Use of materials

Most operatives appreciate the value of the materials with which they work and act responsibly when using them. One can soon check on, say, a bricklayer, by observing how many half bats or broken bricks are lying unused at the workplace. Responsible bricklayers use up all their materials, leaving little waste. A good concretor mixes only sufficient material for his/her immediate use, and even if small amounts are left surplus to requirements they attempt to dispose of the materials usefully within the structure. Short ends of timber are retained to be used elsewhere by cost-conscious carpenters and joiners.

There will always be the careless and irresponsible operatives who have little or no regard for others' property, particularly the contractor's who can be seen operating amongst what they refer to as waste, when in fact the so-called waste could be applied usefully within the structure. These individuals need controlling and where necessary approached to highlight the materials value.

Some of the misuses of materials are shown as follows:

1. Insufficient care in the use of off-cuts.
2. Operative's inability to measure lengths of wood to obtain the most economical cuts which result in waste.
3. Facing bricks being used where common bricks are not readily at hand.
4. Longer lintols being used where shorter ones are available.
5. Additional joists being incorporated within a floor compared to the numbers shown in the design.
6. Small cuttings off full plywood sheets when there are off-cuts already in existence from previously cut sheets.
7. Plasterers allowing too much plaster to fall to waste on to the floor when applying the wet material to the walls, with the labourer having to work harder to produce more as a replacement.
8. Centres of first floor joists inadequately arranged so that the plasterboard on the underside has to be uneconomically cut to fit.
9. Surplus plumbers' expensive capillary or compression joint fittings left to be swept away as rubbish, including short lengths of copper tubing.
10. Valuable electric fittings and wire not being transferred to the next work position but being left to be

casually removed by other operatives for little jobs at home.

8. Quality control and supervision

Materials used should conform to the types and standards laid down in the specifications. It would improve matters if all delivered materials were checked by those taking receipt of them. If unspecified materials were used and were rejected later by the clerk of works, it would be expensive to cut out and provide with the proper type: stricter control by the supervisors would prevent this kind of inconvenience and expensive rejection occurring.

Where architect's design and supervise the works he/she expects specimen materials to be submitted by the main contractor for his/her approval before use. Also, specimen samples of work may have to be provided, on-site, at the commencement of a contract so that they can be used as a guide to everyone, particularly the architect and his representative, the clerk of works; regarding the quality expected. If future work falls below these standard specimens they may be rejected. A typical sample usually takes the form of a brick panel being built as a display to show the bonding, mortar thicknesses and the standard of pointing and straightness of the courses.

The testing of some materials on-site is normal practice, and results would have to be made available to the architect and, once again, if the results fall below the British Standard (BS) and the specifications, the materials would be rejected or the part of the building under construction affected by the substandard materials would have to be removed.

Supervisors, site engineers or other designated employees may have the responsibility to ensure that tests on materials received, or about to be used, are checked, and they should have some knowledge of the tests and check to be made to each type, e.g.

(a) Aggregates.
  (i) silt tests on fine aggregates (sand);
  (ii) bulking tests on fine aggregates;
  (iii) sieve tests;
  (iv) moisture content tests on course aggregate, etc.
(b) Concrete.
  (i) slump tests on wet concrete;
  (ii) compression tests on cured concrete (cube tests);
  (iii) compaction factor tests on wet concrete, etc.
(c) Bricks.
  (i) check dimensions according to the relevant BS, if necessary;
  (ii) check for good arises, regular colour, and see the bricks are hard and well burnt with no cracks, etc.

A visual inspection of deliveries can highlight defective materials, and some measure of protection should be evident to show that care had been taken by the supplier to safeguard against damage.

One should never allow materials to be unloaded if there is doubt about their standard.

Random statistical sampling of, say, between 100 to 200 bricks from a load will highlight the percentage of defects. if five bricks from a sample of 100 are defective, this is taken as 5 per cent defects.

(d) Timber.
  (i) check for woodworm or other infestations or diseases;
  (ii) check for dead knots, etc., depending on where the timber is to be used;
  (iii) signs of twisting, cupping, splitting and bowing may mean rejection. Better to reject before unloading than try to get replacement load later.

9. Security

On most sites security of materials poses the biggest problem to site supervisors. Theft, pilfering, vandalism and other losses adds to the value of materials to be written off as a loss or waste. The problem of security is dealt with in section 7.4.

In constantly being vigilant to the serious problem of waste, a suitable system of materials control should be adopted using specially designed forms to show: present stock, deliveries, amounts used and percentage waste. This monitors material used and, hence, material wastage.

Site cleanliness and layout also gives some indication of how well a contract is being run. On some mismanaged sites one can barely take a step without treading on discarded but otherwise sound materials.

7.4 Site security

To make a site as secure as is possible requires a financial outlay which many contractors feel is unjustified considering the comparatively small actual financial losses they normally sustain regarding materials, plant and equipment, etc. There are, it seems, always losses that occur which are more an exception than the rule, however, and one should always be alert to the possibility of a high loss due to the following reasons:

Types of loss

1. Ordinary theft – by the amateur.
2. Criminal activity – losses caused by organised criminals of payrolls, vehicles, plant and equipment, and materials.
3. Pilfering – by site operatives, subcontractors and administrative staff.
4. Vandalism – caused by children and adolescents.
5. Short deliveries – by suppliers of materials or their delivery drivers (provide a weighbridge for checking loose material).
6. Fraud – from within the company or from outside. Usually associated with the falsification of documents.

Prevention of loss

The method of prevention of loss depends generally on where the site is situated and the type of siteworks to be undertaken. Large housing estates work tends to be more difficult to secure, while compact sites in the centre of towns require the minimum of effort and cost to keep out intending trespassers, as the general public act responsibly when intruders are seen at work.

Sites can be made secure to a certain degree, but where determined criminals are concerned more drastic methods may have to be employed as a counter-measure.

The services of Construction Security of the Building Employers Confederation (Consec) may be enlisted and, as consultants on all security measures, are a valuable organisation to the construction industry. The following are the more general methods used to prevent losses occurring on construction sites.

Fencing off the site

Where a site is small the entire area may be encircled by means of the following fences:

1. Low chestnut paling or wire fence to define the site boundary, and which highlights to all concerned the limits of pedestrian and vehicular rights of way, and prevents everything and everyone from straying on to the site by accident – it also has the added advantage of ensuring workpeople remain on site. It keeps unwanted, inquisitive individuals away from the work area, which is good from a safety point of view.
2. High wire (chain link) or close boarded fence with barbed wire fixed at the top which not only defines the boundary of the site, but prevents anyone from gaining access without some vigorous effort. Persons found trespassing after such measures have been taken would have a difficult time explaining the reason why. A good viewing platform should be provided where possible, which may be used for viewing the site by security officers and others, and which would make it difficult for the thief or criminal to go about his/her deeds unobserved.

    Close boarding has the advantage of minimising the nuisances of dust and noise to the surrounding building occupiers.

    Doors/gates and locks should be commensurate in strength with the boundary fences or hoardings in which they are used.
3. Open site access roads should have a gate erected across them (see Fig. 7.4.1), or at least be sealed off with planks or drums filled with sand at night or at weekends to frustrate organised criminals, and to make it obvious to anyone that beyond these points they would be trespassing.

Compounds

These are generally associated with the larger open sites and are erected to protect and safeguard valuable plant and equipment, fuel and materials.

The type of fencing used depends on that which is available to the contractor, and may be erected with careful consideration to the following points:

1. Chain link fences 2–3 metres high should be fixed to concrete, timber or steel posts which are concreted into the ground.
2. The top of the chain link fence should be finished off with two strands of barbed wire.
3. The base of the wire would be sunk into the ground about 250 mm to prevent anyone from gaining easy access by tunnelling or undermining.
4. Care should be exercised to provide a gate (or gates) and padlocks of similar strength and security as the compound fencing. The gates should be hung carefully to limit the space at the bottom with, perhaps, a bed of concrete across the threshold to prevent thieves from gaining access by undermining.
5. Chains should not be used to secure gates, and the padlock should be of the extra security type which makes it extremely difficult to cut with bolt-croppers.

There are many points which need to be observed as additional security measures relevant to compounds which are as follows:

(a) materials should be stored away from the compound fencing inside and particularly outside, to prevent anyone from using the stacks or piles as access bridges to the inside compound area;
(b) internal huts are best kept away from the compound fencing unless the windows are facing inwards. Where the windows face outwards anti-burglar bars should be fitted on the internal faces and, perhaps, security wire could be fixed on the outside as an anti-vandalism measure;
(c) ladders, picks, shovels and bolt-croppers should be locked away (inside the compound area) to prevent thieves and criminals using them as access implements;
(d) machines should be disengaged and locked to prevent their use as battering rams, but, above all to stop them from being driven away or removed from site by vandals or criminals;
(e) internal floodlighting could be provided at night to frustrate thieves and make them visibly vulnerable to the passing public, police or security officers while gaining unlawful access;
(f) burglar alarms (there are numerous types) could be installed which activate one of the following: siren, bell, buzzer or wailer, and which also could operate a flashing light set at the top of a tower or high hollow tube cast into the ground to alert or direct the police to the exact spot of the break-in. The flashing light arrangement is particularly important on large open sites where there may be two or more compound or storage areas, the direction from which buzzers and bells are sounding being sometimes difficult to assess (see Fig. 7.4.1).

Hutting

Huts where valuable items of small plant, equipment and materials are stored should be well constructed and should have strong doors, windows and locks. Where possible they should be situated inside the compound as an added security measure, or should be grouped together for ease of patrolling by the security officer or police when the site is closed.

Administration offices

There are many valuable items of stationery and equipment which are retained for the proper processing of information within offices; conducting of administrative or management duties; and which, if obtained by the organised criminals would be very rewarding for the minimum of effort on their part. The items in question are:

1. Filing cabinets.
2. Duplicating or printing machines.
3. Plan chests, desks and chairs.
4. Drawing boards or easles.
5. General stationery.
6. The contents of the safe if not secured, i.e. postage stamps, holiday stamps, petty cash, unclaimed wages.
7. Priced bills of quantities (sell to competitors – trade secret).
8. Protective clothing.
9. Testing equipment.
10. Setting out instruments and equipment, i.e. surveyors' levels, theodolites, lasers, electronic distance finders, autoplumbs, tapes, etc.
11. Cameras, communication radio equipment and many more items depending on the complexity and extent of the works.

Therefore, the extra measures expected to be taken on behalf of the company by the site supervisor to protect the administrative area from intruders both during working hours and after are:

(a) Operate a sound control of security keys to prevent the wrong persons from obtaining the original or from copying them.
(b) A night light could be left on in the main office if the the size of the administrative area warrants it.
(c) Important office doors could be locked when not occupied during the day; but all doors should be locked at night to prevent thieves who have gained access from passing from one office to other offices easily.
(d) Notices should be strategically placed around administrative areas and access roads warning every visitor to report to the main or other offices before proceeding further.
(e) Other warning notices such as 'Trespassers Will Be Prosecuted' may deter the less determined opportunist thief.
(f) A more detailed notice could be displayed for the general public's benefit. stating, as an example, 'Persons observed on site after 5.30 p.m. until

7.30 a.m. and at weekends are trespassers, and that a reward of £50 (or more) is offered to those who call the police and which leads to the trespassers being prosecuted.'

Wages and other cash

On small sites wages and salaries can still amount to a considerable sum and therefore is a temptation to the determined thief/criminal if security precautions are minimal. In such cases, and especially where there are large concentrations of labour, extreme care must be exercised by management and the site staff to minimise the ease in which the payroll can be snatched.

There should always by a company policy dealing with payroll handling, but where there is insufficient attention given to such an important issue, the site supervisor should give some consideration to the following points:

1. The office safe should be secreted behind a panel or cabinet and built into a brick or concrete surround. An added precaution would be to strap and bolt the safe into a concrete base.
2. The wages office on-site should be secured further by fixing mild steel bars across the glass from the inside, or by providing bullet-proof glass where necessary. Doors should be extra strong and should be closed and locked while the processes of receiving money, bagging-up and distribution of wages is under way. The provision of warning buzzer or bell would add to the morale of wages personnel. One should always be vigilant, and implement ideas which will frustrate criminals' activities.

Cash in transit

Most payroll robberies occur while the money is in transit from the bank to the site. Therefore, firms are advised to consult with the police if they are unsure of a suitable procedure for dealing with the payroll for major projects.

If a security firm (e.g. Securicor) is not used for the collection and delivery of the payroll the following advice is recommended:

1. Use own trusted employees (males) to collect the money from the bank, with specimen signatures issued to the bank beforehand of those authorised to make the collection.
2. Use a suitably constructed vehicle, such as a van, without making it obvious to whom the van belongs.
3. Vary, each week, the time for collection and delivery, and the route to be taken. Do not use narrow side roads.
4. Carry cash in security cases which, if snatched, give off an alarm, smoke or dye.

There are other points to consider regarding the handling of wages, such as:

(a) Correct value of cash to be collected from the bank is first calculated from the wages sheet.
(b) Trusted individuals used to place money into wages packets.

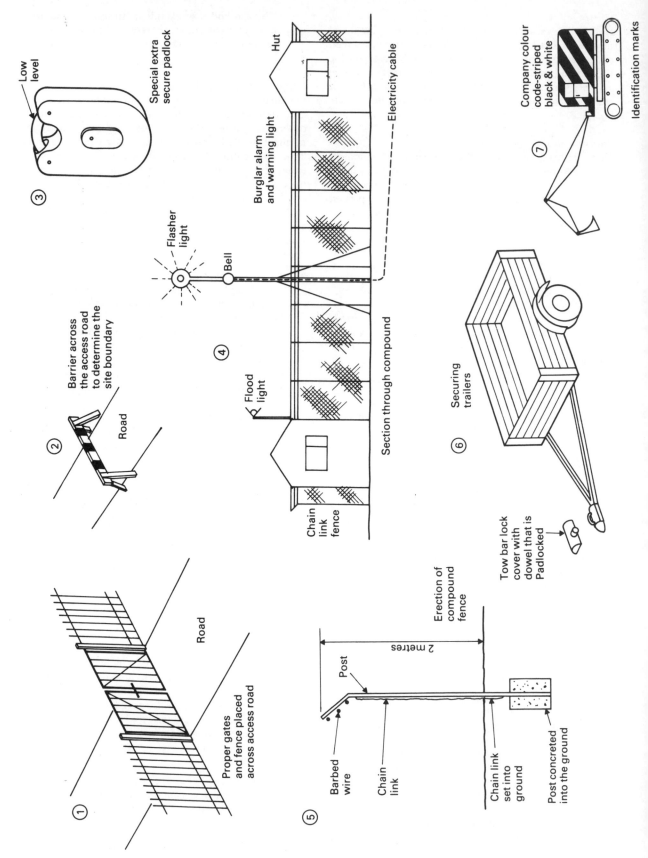

③ Special extra secure padlock

Low level

② Barrier across the access road to determine the site boundary

Road

① Proper gates and fence placed across access road

Road

④ Section through compound

Flasher light

Bell

Flood light

Burglar alarm and warning light

Chain link fence

Electricity cable

Hut

⑤ Erection of compound fence

Post

2 metres

Barbed wire

Chain link

Chain link set into ground

Post concreted into the ground

⑥ Securing trailers

Tow bar lock cover with dowel that is Padlocked

⑦ Company colour code-striped black & white

Identification marks

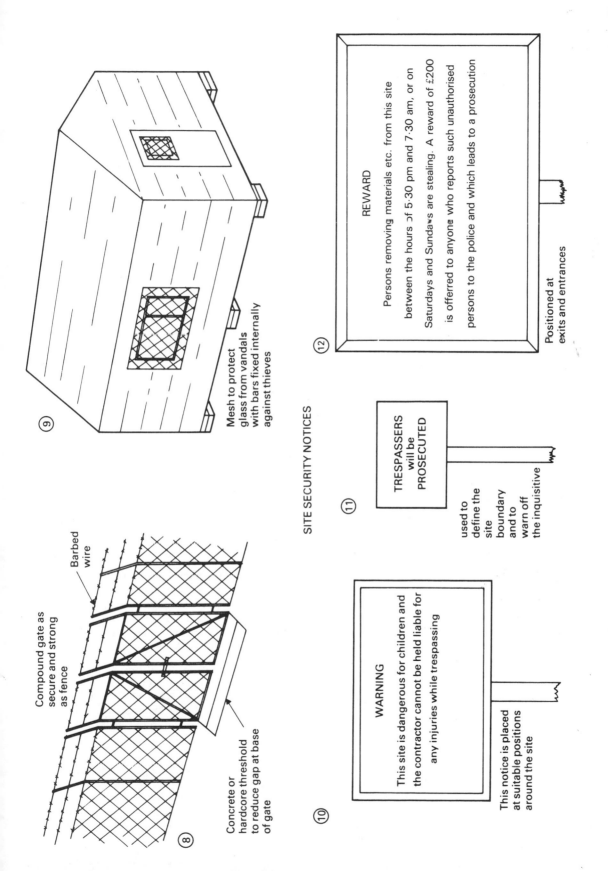

Mesh to protect glass from vandals with bars fixed internally against thieves

⑨

Barbed wire

Compound gate as secure and strong as fence

Concrete or hardcore threshold to reduce gap at base of gate

⑧

SITE SECURITY NOTICES

REWARD

Persons removing materials etc. from this site between the hours of 5·30 pm and 7·30 am, or on Saturdays and Sundays are stealing. A reward of £200 is offerred to anyone who reports such unauthorised persons to the police and which leads to a prosecution

⑫

Positioned at exits and entrances

TRESPASSERS will be PROSECUTED

⑪

used to define the site boundary and to warn off the inquisitive

WARNING

This site is dangerous for children and the contractor cannot be held liable for any injuries while trespassing

⑩

This notice is placed at suitable positions around the site

Fig. 7.4.1 Security

(c) It is better to get approval from employees to accept cheques, this minimises the need to handle cash.

(d) Ensure a sound system of issuing and recording of wages is maintained to prevent the wrong individuals from being given someone else's pay packet (some means of identity, or a signature being obtained).

(e) Absentees' wages should be sent back to head office if not claimed, unless the employees have made other special arrangements.

Plant and machinery

The most suitable way to make the stealing of plant, etc., more difficult is to remove those pieces no longer required from the site, or to place them within a secure compound. Other mechanical plant may be secured temporarily by one or more of the following:

1. Disengage by removal of starter devices.
2. Cabins locked up.
3. Simple but valuable parts removed, e.g. spark plugs, etc.
4. Provision of wheel locks – prevents wheels being removed.
5. Tow bar locks – to prevent the easy coupling-up and towing away of generators, trailers, compressors, and even caravans.

To leave any type of plant or equipment unattended close to main highways is unwise, particularly as there appear to be many organised criminals operating throughout the country. Also, it is a temptation to other weak-willed opportunists who are always on the look-out to make easy money particularly if the item of plant is small and can be simply concealed in the boot of a car or van. As a precaution against petty thieving hand tools and small plant can be locked away in a security chest (made of steel) with a good lock which can be installed close to a work area; and because of the special way patent security chests are made and weighted, the amateur thief would have great difficulty breaking them open or removing the chests from the site.

Colour coding:

Colour coding: Most firms believe that every effort should be made to frustrate criminals, thieves and pilferers, and insist that their equipment, etc., is painted or marked in some way to make identification easier. Any items of equipment stolen and which may be recovered by the police would very soon be claimed because of the firm's special markings or coding. Colour coding or marking is deemed essential by members of the BEC. So, one firm may use yellow markings on all its equipment, plant and vehicles while another may use red, and so on. If equipment is stolen the thief would first have to obliterate the special markings or remove the paint, making it more difficult to pass on the equipment for sale to others.

The site manager should ensure that where plant, etc., is used on-site for a considerable time the identifications markings should be maintained.

Finally, it is reassuring to know that the police make periodic spot checks on motorways and other roads, particularly at night or weekends, on transportation vehicles, and are often successful in recovering valuable plant and equipment. Unfortunately, on the other hand, some of the plant recovered remains unclaimed in the police impound areas because they cannot trace the owners due to lack of company identification marks on the plant, etc.

Pilfering

This occurs due to the contractor's own staff's dishonest activities. In most cases it is on a small scale, but if the entire proceeds from pilfering is added together within the construction industry large amounts of money are involved. A pocketful of nails taken by an employee for a private job at home is a not too infrequent occurrence; short lengths of timber or a plastic bag filled with cement removed via an employee's car-boot adds to the yearly losses for a contractor.

The type of pilfering mentioned is most difficult to detect, but to prevent bigger losses being sustained a proper control of materials and small plant should be maintained by suitable storekeeping procedures. Booking out and booking in of the aforementioned minimises the temptation for employees to steal from the employer. The following are ways in which a site supervisor can help to prevent serious losses occurring on-site:

1. Carry out spot checks on site employees and their vehicles (usually carried out without their knowledge by simple, discreet observation).
2. Ensure employees' car parking areas are situated away from the work area and preferably near to the site offices to prevent the transference of pilfered goods to their vehicles.
3. Rubbish skips should be checked by observing collection drivers unloading at the tip. Valuable items are sometimes smuggled off site in the bottom of the skip. The depositor of stolen materials is usually in league with the skip driver.
4. A 'materials/small plant issued and returned' book should be kept by the storeman or supervisor which checks on the persons who have been issued with the goods, and there should be a strict ruling that when not in use or at the end of the working day operatives should return issues back to the store (see Fig. 7.4.2).

Operatives tend to be more responsible for small plant and valuable materials if they have had to sign for them. Unfortunately, items such as electric drills are lost more frequently than other handtools and therefore should not be left unattended. Special care is needed here.

Vandalism

Although no one likes to be unpopular, a site supervisor cannot afford not to be where children or adolescents are concerned. The youngsters should never be encouraged near or on the site. In keeping them away from the site little vandalism will occur during working hours, and injury to themselves is prevented.

B. BUILDER, LTD.						
MATERIALS BIN CARD						
Type of materials: Re-order quantity:				Maximum quantity: Minimum quantity:		
Date	Supplier ref:	Requisition reference	Signature	Quantity inwards	Quantity outwards	Balance

NOTE: Bin cards are hung beside the relevant materials as a control sheet of materials in and out of the stores. The balance column is useful for continuous stock-taking and as a guide for reordering levels.

Fig. 7.4.2

A notice could be displayed stating that parts of the site are dangerous and that children are warned not play there. This appears to be observed more than the sign 'Trespassers will be prosecuted' as far as children are concerned, but also the parents stop their children from trespassing if there is a warning of danger. Where a child is injured on site after being chased or warned off a few times there would be little chance of the parents successfully claiming damages against the firm (see Fig. 7.4.1).

Chapter 8
Supervisor's appreciation of effective working methods and accurate measured performance

8.1 Method study to find the most effective working method

In order to achieve better productivity organisations have for years placed a great deal of faith in their work study officers/engineers.

Work study was evolved and used quite frequently more then 100 years ago by a few scientific managers in order to increase output, to simplify production techniques and as a basis for the eventual introduction of piecework rates and bonus payments – although bonus payment through measured incentives were not introduced significantly until firms like John Laings and Sons saw the need during the Second World War. Since then the use of bonus systems has increased, with the result that most construction companies now operate such systems, mainly with the approval of the construction trade unions, as agreed through the NJCBI and the CECCB.

To operate such a scheme careful consideration must be given to fixing the correct rate for each job or operation. Firms must be better organised than previously, and with this in mind have seen the necessity for maintaining a work study section or department, or, if this has proven uneconomical, have approached management consultancy firms who offer such a service. Only the large firms, unfortunately, tend to be in a position to employ work study officers on a full-time basis. Nonetheless, rates of work are built up which are then synthesised and these records are kept for future use.

Work Study is divided into two main areas:

1. Method Study – to improve the way work is done.
2. Work Measurement – mainly to fix a standard time for doing various operations (see Fig. 8.1.1).

Method study

A process or job would be chosen which is usually repetitive in nature and which may appear at the time to be inefficiently carried out, or the workplace seems to be laid out wrongly. Also, it is good policy to analyse a proposed method before approval is given to install. As in all management problems, analysis in a systematic manner allows for the best decisions to be taken on the way work is to be executed (see Fig. 8.1.2).

Bearing in mind the details shown in Fig. 8.1.2 on method study after selection of a process which is a cause

for concern by management or the site manager, there are various ways of making recordings depending on the type of process being undertaken. There are many more techniques used by work study officers, but the site manager would do well to study and understand the more common ones which can be applied to processes associated with site work and layout of sites, as listed in Fig. 8.1.3.

In work study symbols are used to denote certain activities, particularly where process charts and flow diagrams are concerned, and these help to reduce the written work considerably. The symbols also help to clarify the important from the unimportant operations movements, etc., of labour, plant or materials (see Fig. 8.1.4).

Systems of recording in method study

Outline process charts
These are useful charts to record the order in which processes are carried out before deciding to undertake a more detailed study. Only the main operations and inspection symbols are used but with a brief description on each. Fig. 8.1.5 shows the outline process of gauging and mixing composition mortar.

Flow process charts
These, generally, are prepared from standard printed sheets with symbols already included. The work study officer would then use the sheets to record a process by sequence, but could use the information so recorded to scrutinise an outline process chart, or, in fact, a flow diagram.

If a flow diagram is to be used in a study the existing sequence of a process may be recorded first by this method. As a back-up a flow process chart is also used to record the same information but travel distances and times for operations may be included. By examining the original system it is hoped that an alternative, improved one can be discovered (flow diagrams redrawn). Additionally, a second flow process chart of the improved method is prepared and a final comparison is made between the original and improved total times, distances, operations, delays, etc. (See Fig. 8.1.6 and 8.1.7, which are recordings of the same hypothetical process.)

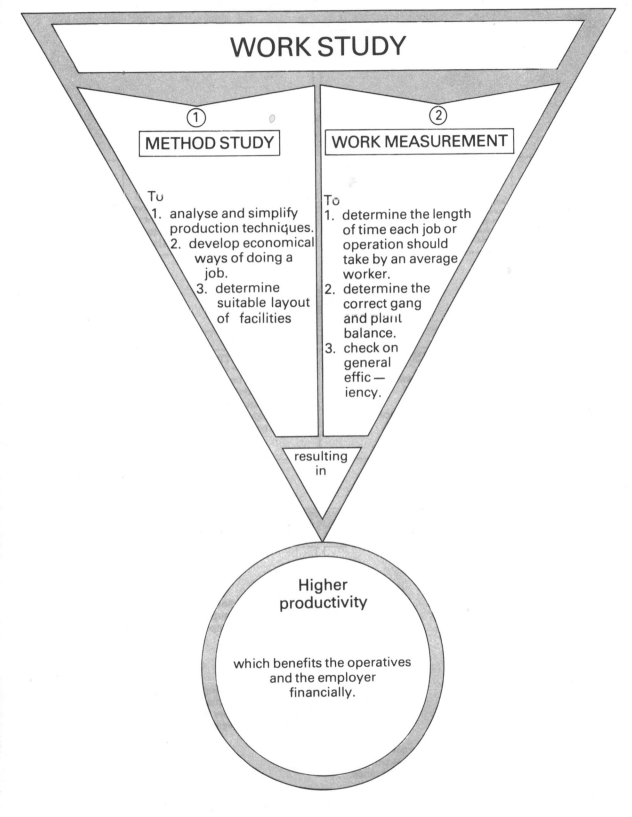

WORK STUDY

① METHOD STUDY

To
1. analyse and simplify production techniques.
2. develop economical ways of doing a job.
3. determine suitable layout of facilities

② WORK MEASUREMENT

To
1. determine the length of time each job or operation should take by an average worker.
2. determine the correct gang and plant balance.
3. check on general effic — iency.

resulting in

Higher productivity

which benefits the operatives and the employer financially.

Fig. 8.1.1

138

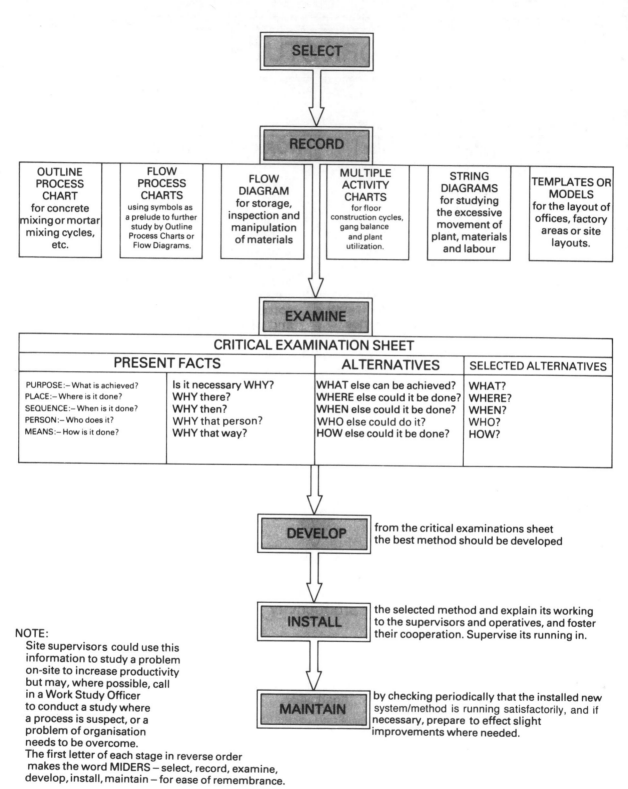

Fig. 8.1.2 Method study (stages in its application)

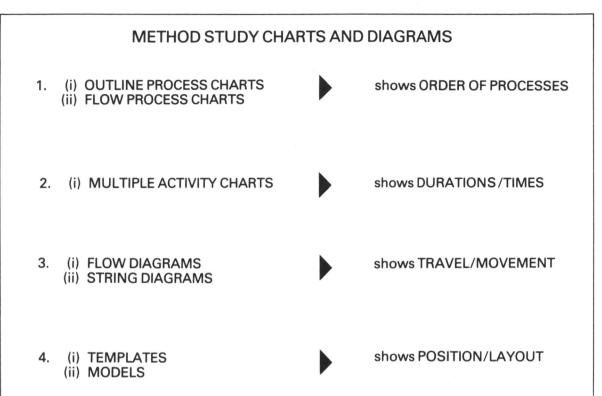

METHOD STUDY CHARTS AND DIAGRAMS

1. (i) OUTLINE PROCESS CHARTS   ▶   shows ORDER OF PROCESSES
    (ii) FLOW PROCESS CHARTS

2. (i) MULTIPLE ACTIVITY CHARTS   ▶   shows DURATIONS / TIMES

3. (i) FLOW DIAGRAMS   ▶   shows TRAVEL/MOVEMENT
    (ii) STRING DIAGRAMS

4. (i) TEMPLATES   ▶   shows POSITION/LAYOUT
    (ii) MODELS

Fig. 8.1.3

PROCESS CHARTS AND FLOW DIAGRAM SYMBOLS AND MEANINGS

SYMBOL	ACTIVITY	MEANING
◯	OPERATION	The actual production of work which leads to output i.e. laying bricks, fixing cupboards, spreading concrete.
▢	INSPECTION	Checking, weighing, comparing for quality or quantity.
⇨	TRANSPORT	Movement of materials, plant or operatives from one place to another.
◗	TEMPORARY STORAGE OR DELAY	The unnecessary storage or laying aside of material/work due to bottlenecks in production.
▽	PERMANENT STORAGE	Safe keeping of materials/products until required.
◙	COMBINED OPERATION AND INSPECTION	This shows an operation which at the same time has to be checked, weighed, etc. such as filling and checking quantity at the same time (filling and checking weigh-batching machine with aggregate.)

Fig. 8.1.4

140

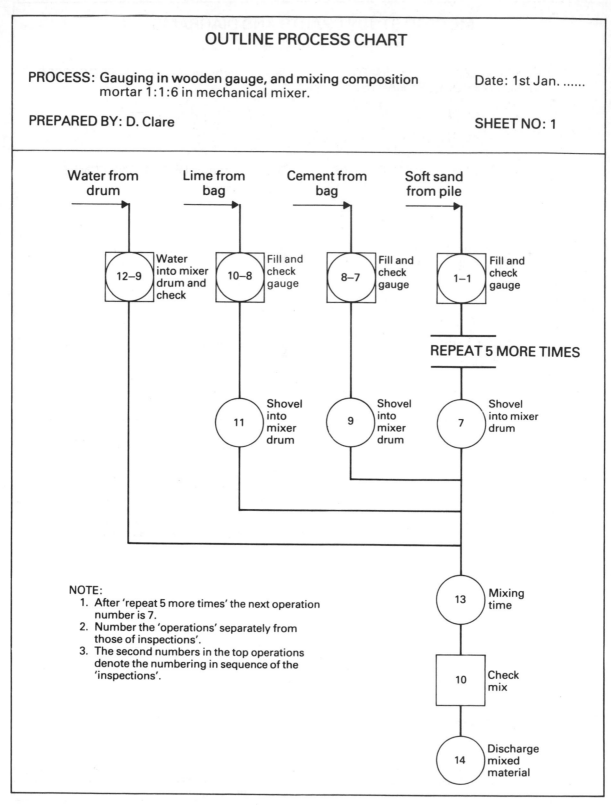

OUTLINE PROCESS CHART

PROCESS: Gauging in wooden gauge, and mixing composition
mortar 1:1:6 in mechanical mixer.

Date: 1st Jan. ......

PREPARED BY: D. Clare

SHEET NO: 1

Water from
drum

Lime from
bag

Cement from
bag

Soft sand
from pile

12–9 — Water into mixer drum and check

10–8 — Fill and check gauge

8–7 — Fill and check gauge

1–1 — Fill and check gauge

REPEAT 5 MORE TIMES

11 — Shovel into mixer drum

9 — Shovel into mixer drum

7 — Shovel into mixer drum

13 — Mixing time

10 — Check mix

14 — Discharge mixed material

NOTE:
1. After 'repeat 5 more times' the next operation
   number is 7.
2. Number the 'operations' separately from
   those of inspections'.
3. The second numbers in the top operations
   denote the numbering in sequence of the
   'inspections'.

Fig. 8.1.5

FLOW PROCESS CHART

SUBJECT: *Unloading, Storing + Preparation of timber for structural work.*

PREPARED BY: *D. Clare.*

TRADE/OPERATOR: *C + J*

DATE: *1st January 19........*

SHEET NO: *1*

NO	OPERATION/DESCRIPTION	DIST.	TIME	○	⇨	D	□	▽	
1	Transport by suppliers lorry of timber.				●				
2	Unload timber by hand – 2 labourers			●					2 labourers
3	Inspect quality						●		1 carpenter + Joiner
4	Transport manually to storage area.	9 m			●				2 labourers.
5	Stored awaiting sawyer							●	
6	Move timber to cross-cut saw	8 m			●				2 Carpenters + Joiners
7	Cross-cut timber into predetermined lengths			●					2 Carpenters + Joiners
8	Move onto ripsaw	10 m			●				2 Carpenters + Joiners
9	Rip-timber into required strips or widths			●					2 Carpenters + Joiners
10	Move ripped timber to C+J workshop.	13 m			●				2 labourers
11	Cut timber awaiting notching and housing					●			
12	C+J houses/notches the timber			●					1 Carpenter + Joiner
13	Final inspection for correct size and detailing						●		1 foreman C+J
14	Transport manually to C+J store	15 m			●				2 labourers
15	Store prepared timber until required on site							●	
16	Transport, by tractor and trailer the prepared timber to site.				●				2 labourers + tractor.

NOTE:
○ OPERATION   □ INSPECTION
⇨ TRANSPORT   ▽ STORAGE
D DELAY

TOTALS	55 m		4	7	1	2	2

Fig. 8.1.6

142

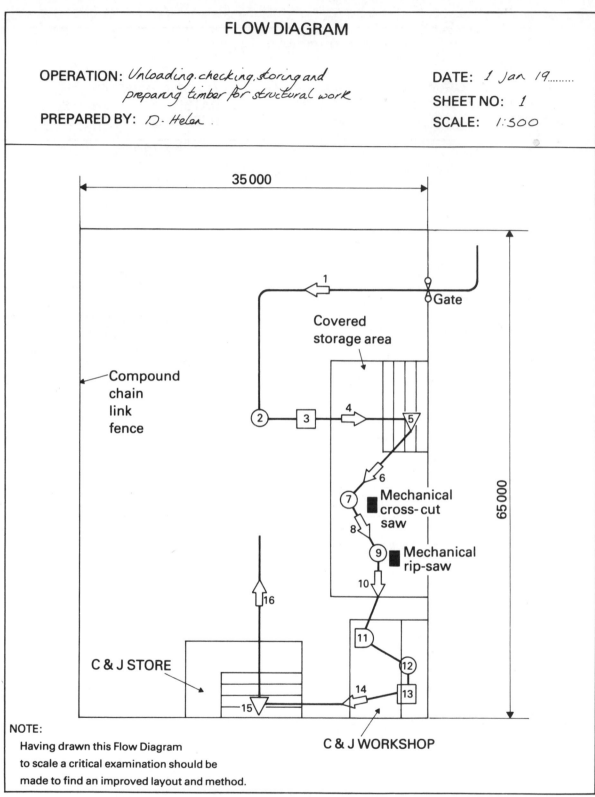

FLOW DIAGRAM

OPERATION: *Unloading, checking, storing and preparing timber for structural work*

PREPARED BY: *D. Helen .*

DATE: *1 Jan 19........*

SHEET NO: *1*

SCALE: *1:500*

35 000

1

Gate

Covered
storage area

Compound
chain
link
fence

2   3   4   5

65 000

6

7   Mechanical
cross-cut
saw

8

9   Mechanical
rip-saw

10

11

16

12

C & J STORE

14   13

15

C & J WORKSHOP

NOTE:

Having drawn this Flow Diagram
to scale a critical examination should be
made to find an improved layout and method.

Fig. 8.1.7

Flow diagrams

These diagrams are most useful to highlight the travel routes and distances where a process is being undertaken; they are more detailed in presentation than outline process charts and give a better picture of achievement because they are drawn to scale. A flow process chart, as an additional aid, may be prepared so that by careful study of the data contained therein improvements can be determined to the process. A flow diagram of the improved method can be drawn, and a new flow process chart could be used to check the savings in travel distance, time and operations, etc. The usual symbols are incorporated at each stage of a process, as shown in Fig. 8.1.7.

Multiple activity charts

As a useful aid for studying a process which is being undertaken by two or more individuals a multiple activity chart could be used with good effect. Accurate recording is best achieved where the maximum number in a gang of operatives under observation is limited to between four and six.

The usefulness of such a chart is to check on the correct gang balance or gang/plant balance. By careful study the inactivity of individuals within a gang are highlighted, or alternatively, it may pinpoint the need to increase the gang size to effect higher productivity with the minimum period of inactivity.

Similarly, where operatives are required to work with the use of plant or machinery idle or wasted time of either operatives or plant can be kept to a minimum.

The chart is prepared in a series of bars against a time scale after studying and recording the activities of each member of the gang under observation. This is achieved by writing down the various operations being done by each operative during a process, and at the same time recording times for each operation with the use of an ordinary wrist-watch.

Each operative is allocated a bar and both active and inactive times are drawn against a time scale. (See Fig. 8.1.8 for an example of a gang of four labourers erecting timber interwoven fencing, where the original method is shown first, and the improved method shows the savings in time/duration.)

The objects of such a chart, therefore, can be stated as thus:

(a) to balance the gang;
(b) to balance the gang against plant;
(c) to reduce idle/waiting/inactive time;
(d) to arrive at a quicker method of doing the work;
(e) to arrive at a better method of doing the work.

String diagrams

A string diagram is prepared after studying the operations carried out by a mobile crane, lorry, fork-lift truck, etc., or operatives' movements. The initial study is best undertaken on the spot, but, alternatively, details could be extracted from movement sheets, allocation sheets and time sheets to reveal most of the required information.

The survey is usually called for by the contracts manager or production manager as a check on suspected bad site supervision and control of mobile plant, but particularly where there is low productivity of plant.

A scale drawing of the site or factory layout is needed when the work study officer next decides to pictorially display the loading points and travel routes of the plant/operative. Mapping pins are positioned at the loading and unloading points (and any changes in direction of movement), and then string or cotton is threaded around the pins and along the routes taken by the plant, etc., and in the correct sequence.

After the practical study and preparation of the string diagram the officer checks which are the most used routes taken. The string or cotton can be unthreaded and its total length measured (to scale) as a check on the distance travelled. it must be remembered that normally while a piece of plant – say, mobile crane – is travelling it is not producing any work. The shorter the travel distance the better.

See Fig. 8.1.9 for a typical study of a mobile crane used for heavy lifting on a new housing contract: there appears to be unnecessary amount of travelling which will affect productivity. One should ask the question: Can some of the lifting be done manually? or, are the houses built in the correct sequence?

Templates or models

The application of either templates or models is unlimited, but the most usual relates to:

(a) layout of site huts;
(b) layout of storage compounds;
(c) layout of offices/rooms;
(d) layout of plant and machinery in a factory type set-up.

Even the architect/designer uses templates, etc., when arranging an estate of houses on a scale drawn block plan.

The templates and models are prepared to the same scale as the plan of the room, site, compound, etc., to be placed in the most economic or convenient layout position. (See Fig. 8.1.10.)

While carrying out the 'examination process' of the recorded information which is shown on the previously mentioned charts or diagrams one can use the recommended Critical Examination Sheet (example of layout can be seen as part of Fig. 8.1.2). Alternatively, with experience, the sheet is dispensed with but the principle is observed by most work study officers who simply apply the questioning technique mentally to each fact present.

The remaining outline procedures for method study, i.e. develop, install and maintain, are briefly referred to in Fig. 8.1.2.

It must be remembered that method study is only suitable where the process under observation is of a repetitive nature, so that any improvements which are recommended can be implemented to make future savings in time, effort and money.

144

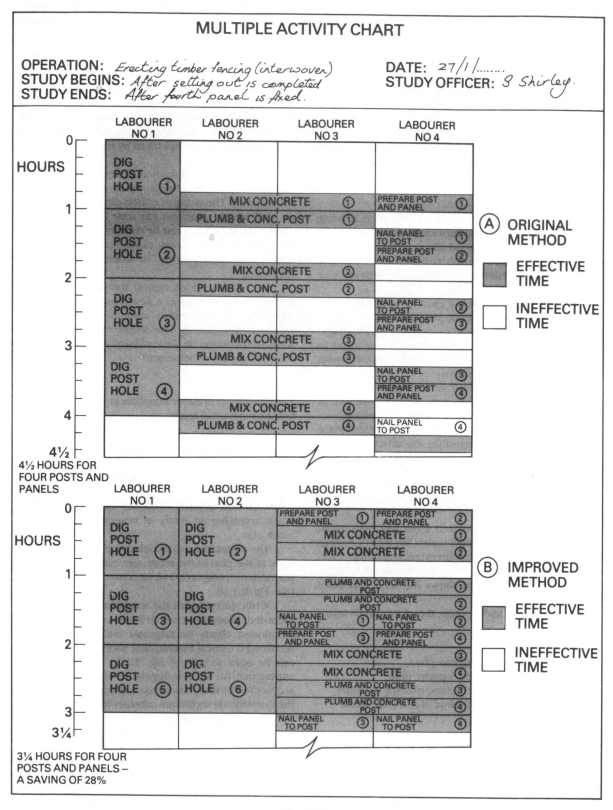

Fig. 8.1.8

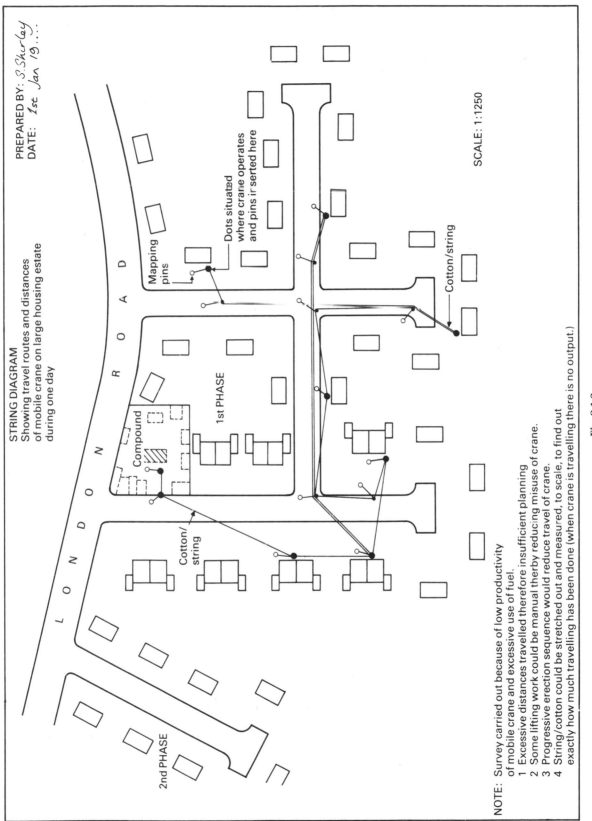

STRING DIAGRAM
Showing travel routes and distances
of mobile crane on large housing estate
during one day

PREPARED BY: S. Shurley
DATE: 1st Jan 19....

Mapping
pins

Dots situated
where crane operates
and pins ir serted here

Cotton/string

Compound

1st PHASE

Cotton/
string

2nd PHASE

SCALE: 1:1250

NOTE: Survey carried out because of low productivity
of mobile crane and excessive use of fuel.
1 Excessive distances travelled therefore insufficient planning
2 Some lifting work could be manual therby reducing misuse of crane.
3 Progressive erection sequence would reduce travel of crane.
4 String/cotton could be stretched out and measured, to scale, to find out
exactly how much travelling has been done (when crane is travelling there is no output.)

Fig. 8.1.9

146

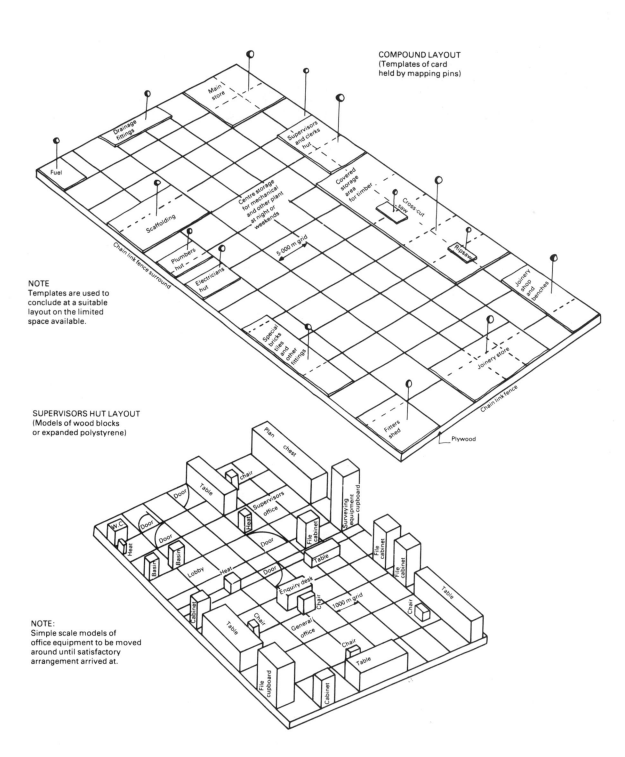

COMPOUND LAYOUT
(Templates of card
held by mapping pins)

Main store

Drainage fittings

Supervisors and clerks hut

Fuel

Covered storage area for timber

Centre storage for mechanical and other plant at night or weekends

Cross-cut saw

Scaffolding

Chain link fence surround

5·000 m grid

Ripsaw

Plumbers hut

Electricians hut

NOTE
Templates are used to
conclude at a suitable
layout on the limited
space available.

Joinery shop and benches

Special bricks tiles and other fittings

Joinery store

Chain link fence

SUPERVISORS HUT LAYOUT
(Models of wood blocks
or expanded polystyrene)

Fitters shed

Plywood

Plan chest

Chair

Surveying equipment cupboard

Table

Door

Supervisors office

Heat

File cabinet

W.C.

Heat

Door

Door

File cabinet

Basin

Basin

File cabinet

Heat

Door

Lobby

Table

Cabinet

Enquiry desk

Chair

1000 m grid

Chair

NOTE:
Simple scale models of
office equipment to be moved
around until satisfactory
arrangement arrived at.

Table

Chair

General office

Chair

Table

File cupboard

Cabinet

Table

Fig. 8.1.10 Templates and models

8.2 Work measurement as a means for accurately measuring performance

Work measurement is the second of the two main parts to work study (see Fig. 8.2.1). This part is divided into the following techniques.

1. Time studies.
2. Activity sampling.
3. Synthesis.
4. Analytical estimating.

Each of the above plays a different role in making results of work study available to management, which can be used in the quest for information and which assists in maintaining a degree of control of an organisation. The uses to which the information can be put revolves around the setting of standards of labour/plant costs and the techniques of controlling them, i.e.

1. Standard costing and budgetary control (pre-determined costs and availability of finance).
2. Estimating – from recorded performance times extracted from synthetic data.
3. Planning (durations of operations extracted from synthetic data).
4. Production control and fixing of bonus/incentive schemes.
5. Supervision (checking on how well jobs are supervised).
6. Gang balance or gang/plant balance.

Most of the methods of work study described in this section are carried out by work study officers, but with a little training the site supervisor could, if time allowed, conduct selected studies if necessary. However, it is normal to have an independent individual conducting the studies, if union stewards request it, when time rates for bonus schemes appear to be inadequate and the site supervisor wishes to remain impartial.

It has been a source of aggravation in the past for operatives to be secretly timed in their work situation, or be studied by management by one or more of the work measurement systems without their prior knowledge, but which eventually they discovered when work study officers were observed hiding behind brick stacks or peeping out of upper windows. Now, the operatives' cooperation is sought, through their union stewards, when a rate for an operation appears to be uneconomical or serious losses are being incurred due to inadequate gang/plant balance. When difficulties are experienced in striking the correct bonus rate for the job and the operatives are aggrieved, work study is used (time study) to arrive at an acceptable standard rate.

It is usually the case that the operatives have the final say before most forms of work study is undertaken, particularly where trade unions dominate, and therefore time studies and other work measurement studies can only be carried out after the usual consultations.

Time studies

Various equipment is used to conduct this type of study e.g.

(a) decimal minute stop watch – fly back type – which is either calibrated for 30 minutes or 60 minutes;
(b) clip board;
(c) study sheets (see Figs. 8.2.2 and 8.2.3);
(d) tape measure;
(e) pencil and rubber;
(f) calculator.

Job selection
Any job or operation could be studied to discover the time taken to complete it, particularly where synthetic data is being compiled for the first time; the most usual jobs, however, in an organisation which has used work study for some period of time are:
(a) newly introduced jobs or operations;
(b) jobs requiring the use of new techniques, new materials or new plant;
(c) those jobs which are showing a worsening trend;
(d) those jobs causing the operatives to be disgruntled.

In the above situations incentive and bonus schemes would be the basis for the studies.

Operatives to be studied
A study is normally taken using the services of what one would refer to as a 'qualified worker' – the type who is an average all-rounder and who has the attributes necessary for the job for which he/she is employed, i.e. degree of intelligence, skill and physical build.

The choice of operatives who are to be studied would normally be by agreement with the trade union steward, and, naturally, individuals would have to consent. Either the site supervisor or the union steward would introduce the individual to the work study officer, who should explain all the facts about how the study is conducted.

How the time study is carried out
It is necessary to begin the study by recording all the existing facts about the job, including the working conditions, degree of skill, effort, etc., required. The work study officer should watch the operative complete one or more cycles of the work to be studied to decide what the 'elements' of the job are, and exactly where the 'break-points' occur.

The following is the order in which the eventual 'standard time' is determined:

1. Choose the job and extent of it (when it starts and finishes).
2. Break the job down into elements and decide break-points, e.g. collect component, measure, cut, fix, etc. (the break-point being the points between the completion of one element and the starting of another).
3. Record the elements, in order of execution, on the Time Study Recording Sheet (see Fig. 8.2.2).

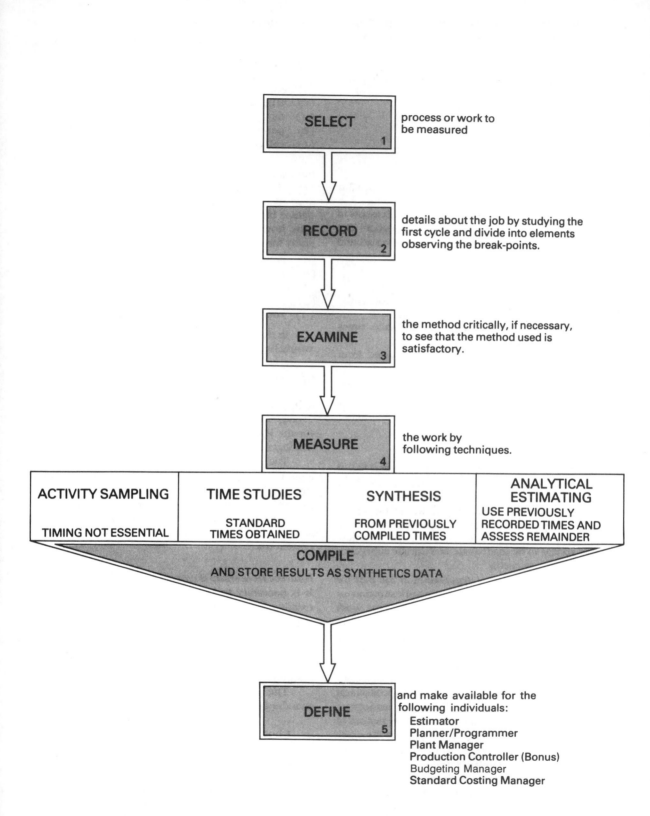

Fig. 8.2.1 Work measurement (stages in its application)

4. Time each element by the 'repetitive' or 'continuous' timing method, and record in the Time Study Recording Sheet O.T. or W.R. column.

5. Simultaneously to timing an element a rating should be recorded in the R column.

6. Calculate the basic times using:

$$\text{B.T.} = \frac{\text{Rating} \times \text{Standard or Observed Time}}{100 \text{ Standard Rate}}$$

7. Calculate the standard time using the Time Study Sheet, as shown in Fig. 8.2.3 (abstract, relaxation allowances and summary).

Rating

Because operatives work at varying degrees of speeds during a work period (depending on whether they have been, or are about to be, refreshed and revitalised with a mid-morning break, etc.) a British Standard Rating Scale is applied by the work study officer to each observed element of work produced during a study. The skill of a work study officer is greatest in applying the rating scale in assessing how fast an operative is working, and above all, how efficiently and effectively the work is being carried out.

The BS Scale ranges from 0–100 where:

(a) 0 represents inactivity;
(b) 50 is very slow and ineffective;
(c) 75 is unhurried, but the operative is meaningful and producing a normal rate of work for a normal rate of pay;
(d) 100 is the standard rate with the operative working at a speed sufficient to earn up to $33\frac{1}{3}$ per cent bonus on top of normal earnings;
(e) 120 means that an operative is very skilful, fast and effective – only a small percentage of workers would be rated thus, especially over a sustained length of time.

The usual rate ranges between 70 and 130.

Abstracting

This is the collection of the basic times for each repeated element which are then added together. The element totals are then divided by how many times they have been repeated to arrive at average basic times (see Fig. 8.2.3).

Relaxation allowances

These allowances are additional percentages which are added to the average basic times to allow for delays caused by an operative having to take care of his/her basic needs, and to allow some recovery time where to complete an operation work has to be undertaken using some effort, concentration, etc., which leads to fatigue. These allowances are:

1. Working position, 0–10 per cent (sitting 0%, standing 2%, stooping 5%, etc).

2. Amount of physical effort, 0–20 per cent (lifting light weights, 0–10%; lifting heavy weights, 11–20%).

3. Conditions of the work area, 0–20 per cent (cold, hot, bad light, dirty, fumes).

4. Degree of attention or concentration, 0–10 per cent (measuring or checking, 0–5%; process which leads to some anxiety, 6–10%).

5. Monotony, 0–5 per cent (degree of repetitiveness)

Because it is expedient to do so, to these relaxation allowances are now added a standard allowance for operatives' standard needs, such as: mid-morning and mid-afternoon tea or rest breaks; attending the toilet; for washing if periodic dirty processes are undertaken.

Normally, for each element undertaken relating to building, the total percentage allowances would not exceed 40 per cent. The relaxation allowances and standard allowances are shown combined under one heading in Fig. 8.2.3, and can be remembered using SPECAM (standard, position, effort, conditions, attention and monotony).

Summary: Both the total average basic time and total relaxation allowances are added together to give total times for each element. These totals are next added to give a grand total (see Fig. 8.2.3). There now remains for only one final additional allowance to be added to the grand total time to arrive at a performance/standard time per job, which is called a contingency allowance.

Contingency allowance

This allowance varies between 0 and 5 per cent, and is added to the grand total time for infrequent periods of delay which the work study officer expects will happen at some time, depending on the operation studied, e.g. infrequent items of work (maintenance of machines or tools); unexpected interruptions caused by other trades.

Activity sampling

Although management may have the foresight to plan ahead and use various control techniques, it may be necessary during the course of construction work to take action on site where difficulties are encountered, e.g. the level of productivity is falling and it is suspected that all is not right with the site supervision, or the gang balance of certain groups of workers is incorrect.

The work study officer could be called in by the contracts manager to carry out some form of activity sampling as a check on actual output or quality of work with that which was planned or estimated.

The two main activity sampling systems known in construction studies are:

(a) Field Activity Counts;
(b) Random Observation Studies.

Field Activity Counts

This is a technique of study sometimes carried out by the contracts manager as a check on the degree of supervision of operatives on-site. It would be undertaken where there appeared to be a labour problem and the productivity is

TIME STUDY RECORDING SHEET

LOCATION: Council Housing Development. Contract Nº4

OPERATION/PROCESS: Hanging softwood panelled doors size 50mm x 2000x750 (no furniture).

PLANT USED: —

TRADE/OPERATIVES: C+J (1 No)

WORK STUDY OFFICER: F Joanne

DATE: 1st January ....

STUDY NO: 6

SHEET NO: 1/2

START TIME: 9.45 am

FINISH TIME:

DIFFERENCE

TOTAL OBSERVED TIME:

% AGE VARIANCE:

Note: This side of the page shows the layout for the "Continuous Timing" method.

1st CYCLE
Check time 9.45am and start stop watch

ELEMENT	R	WR	ST	BT
		0		
1. Collect door, change work position and arrange tools.	100	3.55	3.55	3.55
2. Fit hanging edge of door.	90	9.27	5.72	5.15
3. Fit head of door.	90	13.33	4.06	3.65
4. Mark and plane door to frame width.	110	20.67	7.34	8.09
5. Mark and plane bottom of door.	120	27.38	6.71	8.05
6. Offer up the door and mark hinge positions.	105	29.13	1.75	1.84
7. Cut	100	6.98	6.85	6.85

This side of the page shows the layout suitable for the "Repetitive Time" method - note the 3 columns only. This is the more popular system.

2nd CYCLE

ELEMENT	R	OT	BT
1	35	3.69	3.51
2	120	4.88	5.87
3	90	4.50	4.05
4	100	8.00	8.00
Studies the scenery 5	17	(3.77)	
6	95	7.00	6.65
7	90	1.56	1.40
8	100	6.70	6.70
9	100	6.20	6.20
	80	2.40	1.92

Upper data block (4th CYCLE area and preceding rows)

Element	R	WR	OT
(top, cut)	100	3.20	3.20
2	120	4.80	5.76
3	105	7.41	7.78
4	100	7.02	7.02
5	115	1.82	2.09
6	110	6.38	7.02
7	90	6.34	5.71
8	95	2.34	2.22
9	100	1.85	1.85
10	100	1.85	1.85

4th CYCLE

Element	R	WR	OT
1	100	3.44	3.44
2	90	5.60	5.04
3	80	4.28	3.42
4	80	7.62	6.10
5	30	7.09	6.38
6	100	1.93	1.93
7	105	6.60	6.93
1 (Has a smoke)	IT	(2.65)	
8	120	5.81	6.97
9	120	1.96	2.35
10	115	1.74	2.00

Note: After each element reading the stop watch clicked back to zero; hence, in the O.T. column there is a direct repetitive reading

Left lower block

		R	WR		
9	Adjust door	90	14.22	2.22	2.00
10	Clean up shavings.	95	16.13	1.91	1.82

Note: A decimal stop-watch on a long study will keep completing a full revolution (watch used only 30 minute recording) and causes a little extra calculation procedure. See number 6 element. W.R. as hands of watch completes one revolution.

NOTE: R = RATING; WR = WATCH READING; ST = SUBTRACTED TIME; BT = BASIC TIME; OT = OBSERVED TIME

Fig. 8.2.2

TIME STU

Study No. 17
OPERATION/PROCESS Hanging Softwood Panelled doors Size 50mm x 2000 x 750 (no for

1. ABSTRACT (FROM TIME STUDY RECORDING SHEET)

NO.	ELEMENT	BASIC TIMES (ON EACH REPEATED ELEMENT)						TOTAL
1	Collect door, change work position and arrange tools	3.55	3.51	3.60	3.44		minutes	14.10
2	Fit hanging edge of door.	5.15	5.87	5.20	5.04			21.2
3	Fit head of door.	3.65	4.05	5.76	3.42			16.8
4	Mark and plane door to frame width.	8.09	8.00	7.78	6.10			29.
5	Mark and plane bottom of door.	8.05	6.65	7.02	6.38			28.
6	Offer up the door and mark hinge position	1.84	1.40	2.09	1.93			7.2
7	Cut out for one pair of hinges (75mm)	6.85	6.70	7.02	6.93			27.
8	Fit hinges to door, and door and frame	6.02	6.20	5.71	6.97			24.9
9	Adjust door	2.00	1.92	2.22	2.35			8.4
10	Clean up shavings	1.82	1.67	1.85	2.00			7.3

Fig. 8.2.3

CULATION SHEET

SHEET NO: 2/2
PREPARED BY: F. Joanne.

AVERAGE BASIC TIME (MINS)	2. RELAXATION ALLOWANCES							3. SUMMARY		TOTAL: AVERAGE BASIC TIME PLUS TOTAL % RELAX ALLOWANCES
	Standard (5–10%)	Position (0–10)	Effort (0–20)	Conditions (0–20)	Attention (0–10)	Monotony (0–5)	TOTAL % RELAXATION ALLOWANCES	AVERAGE BASIC TIME	TOTAL % RELAXATION ALLOWANCES	
3·53	10	7	10	5	1	1	29	3·53	29	4·55
5·32	10	3	12	5	2	0	32	5·32	32	7·02
4·22	10	3	12	5	2	0	32	4·22	32	5·57
7·49	10	3	12	5	2	0	32	7·49	32	9·89
7·03	10	3	12	5	2	0	32	7·03	32	9·28
1·82	10	2	5	5	3	0	25	1·82	25	2·28
6·88	10	1	8	5	3	0	27	6·88	27	8·74
6·23	10	2	10	5	2	0	29	6·23	29	8·04
2·12	10	1	8	5	2	0	26	2·12	26	2·67
1·84	10	1	2	5	0	1	19	1·84	19	2·19

GRAND TOTAL TIME:	60·23 mins
CONTINGENCY ALLOWANCES: 1%	0·60 mins
PERFORMANCE/STANDARD TIME: Per Door	60·83 mins
OUTPUT PER HOUR: (if required)	

less than was estimated or experienced in past projects; also, the bonus earnings may have dropped and this could be a cause for concern as output could affect the contractor's position regarding the meeting of target dates. Of course, the work study officer could undertake this form of study but without the knowledge of either the operatives or site manager.

The study is conducted about once each hour to obtain the most accurate results. The contracts manager or work study officer simply walks around the site observing as many of the clocked-on operatives in their work situation as possible. The idea is to record, on small mechanical counters (one held in each hand), those operatives who are actively working, and those who are inactive at the moment of observation. Those not active are recorded, say, on the right-hand counter, while those who are active are recorded on the counter held in the left hand.

It is necessary to avoid making each tour of the site every hour on the hour, and using the same route. It is less conspicuous to make a tour at some period in each hour and by taking a different route each time.

Results are recorded at the conclusion of each tour on record sheets, as shown in Fig. 8.2.4, and it is essential that the numbers of operatives present on the site are obtained from the time-keeping records. This enables an observation percentage to be made by calculations using the total numbers of operatives observed (number of inactive operatives plus those that are active). For example:

(*a*)   assume number of operatives on-site to be 120;
(*b*)   total of operatives observed is 110.

$$\text{observation percentage} = \frac{110}{120} \times 100 = \underline{91.7\%}$$

If the answer is greater than the recommended limit of 90 per cent this study is taken as valid. If the percentage is less than 90 per cent the study is discounted.

This form of study would best be conducted over a whole week period for the best results, unless it was so obvious that the operatives were spending too long on their rest breaks during the day without the site manager objecting, or not enough close supervision is being maintained at the workplace.

The study should be undertaken in the early stages of a contract so that any recommendations can be implemented to improve productivity and to maintain some acceptable control over costs. It may lead to the replacement of the site manager.

The results are plotted/drawn on a chart as suggested in Fig. 8.2.5, with the shaded part (usually in red) to show the percentage inactivity. This is a clearer method of illustrating how efficiently the site is being run.

Random Observation Study
This method of observation is used for checking the activity of a gang of workers, but its application is limited to gangs of up to six individuals because of the difficulty in studying a greater number simultaneously.

The recordings of this form of study are used to analyse the operations undertaken by each gang member. The percentage of each operation can also be compared with the other operations in the process, but particularly against the estimated figure.

An observation of each gang member is taken, say, every 10 minutes or more, and a chart or record sheet (see Fig. 8.2.6) has recorded on it: the gang members code (A–D), the operation being done at the moment of observation and the number of the observation.

Synthesis

This is a work measurement technique which is used to obtain a duration or time for doing a job. The data for working out the duration is obtained from records which are retained by the work study department of previous time studies; this data is known as synthetic data.

One or both of the following two durations or times can be recorded and are referred to as synthetic data:

1.   Standard times – the time it takes to hang a door; fit a lock; lay 1000 bricks (with allowances being added for relaxation times, standard allowances and contingency allowances).
2.   Element times – the time taken to do an element of work, such as: mark a hole on steel; drill a hole in 6 mm thick steel; attach a bolt to steel, etc.

In the first example of recording standard times, e.g. hanging a door, generally speaking for each door that will be hung in the future of the same type, thickness and size there are few conditions which could change and have an affect on the time for hanging a door. Perhaps temperature affects a worker's speed, but not normally while working inside a building. Therefore, it is prudent to record the standard time for hanging doors of different types, thickness and size so that these records or data can be used in similar job situations.

Where a one-off, different sized door is to be fixed a standard time can be calculated using the synthetic data which is now available but perhaps is recorded using the element times. One divides the work associated with hanging a door into elements as in time studies, and similar/identical element times are extracted from the synthetic data perhaps with an adjustment for minor variations, as shown in the example in Fig. 8.2.7. This mode of calculation for standard times is known as 'synthesis', and reduces the need to carry out time studies for every job or process as past records for similar type elements are used.

In the worked example (Fig. 8.2,.7) the data extract from synthetic data is for the hanging of a 50 × 2000 × 750 mm softwood panelled door (shown in Figs. 8.2.2 and 8.2.3), but the new door under study is only 50 × 1500 × 500 mm. Therefore, as an example, in element 2 (fitting hanging edge) while the thickness is the same in both cases, the length to be planed is one quarter less than that for a 50 × 2000 × 750 and an adjustment is allowed, as shown, of minus 1.33 minutes from the original element's

155

FIELD ACTIVITY COUNT RECORD SHEET

CONTRACT *Hospital Extension Nº 25*
OBSERVER *D. Heler*
DAY *Monday*
DATE *1ˢᵗ January 19........*

STUDY NO: 36
SHEET NO: 1

TIME OF STUDY	ACTIVE		INACTIVE		NUMBER OBSER—VED	TOTAL NO PRESENT ON—SITE	OBSER—VED % (MIN. 90%)	NOTES
	No	%	No	%				
8.15 – 8.28	49	70	21	30	70	75	93	Temp 16°C
9.05 – 9.15	54	71	22	29	76	78	97	Sunny 18°C
10.15 – 10.22	62	86	10	14	72	78	92	" 20°C
10.50 – 10.59	53	71	22	29	75	78	96	" 20°C
11.30 – 11.38	59	79	16	21	75	78	96	" 21°C
13.08 – 13.20	39	52	38	48	77	78	99	Long lunch 21°C
14.00 – 14.09	68	92	6	8	74	78	95	Sunny 21°C
15.17 – 15.25	50	68	23	32	73	78	94	" 21°C

NOTE: the minimum % observed to be not less than 90% otherwise study is void.

PLOTTED RESULTS FOR MONDAYS STUDY

Fig. 8.2.4

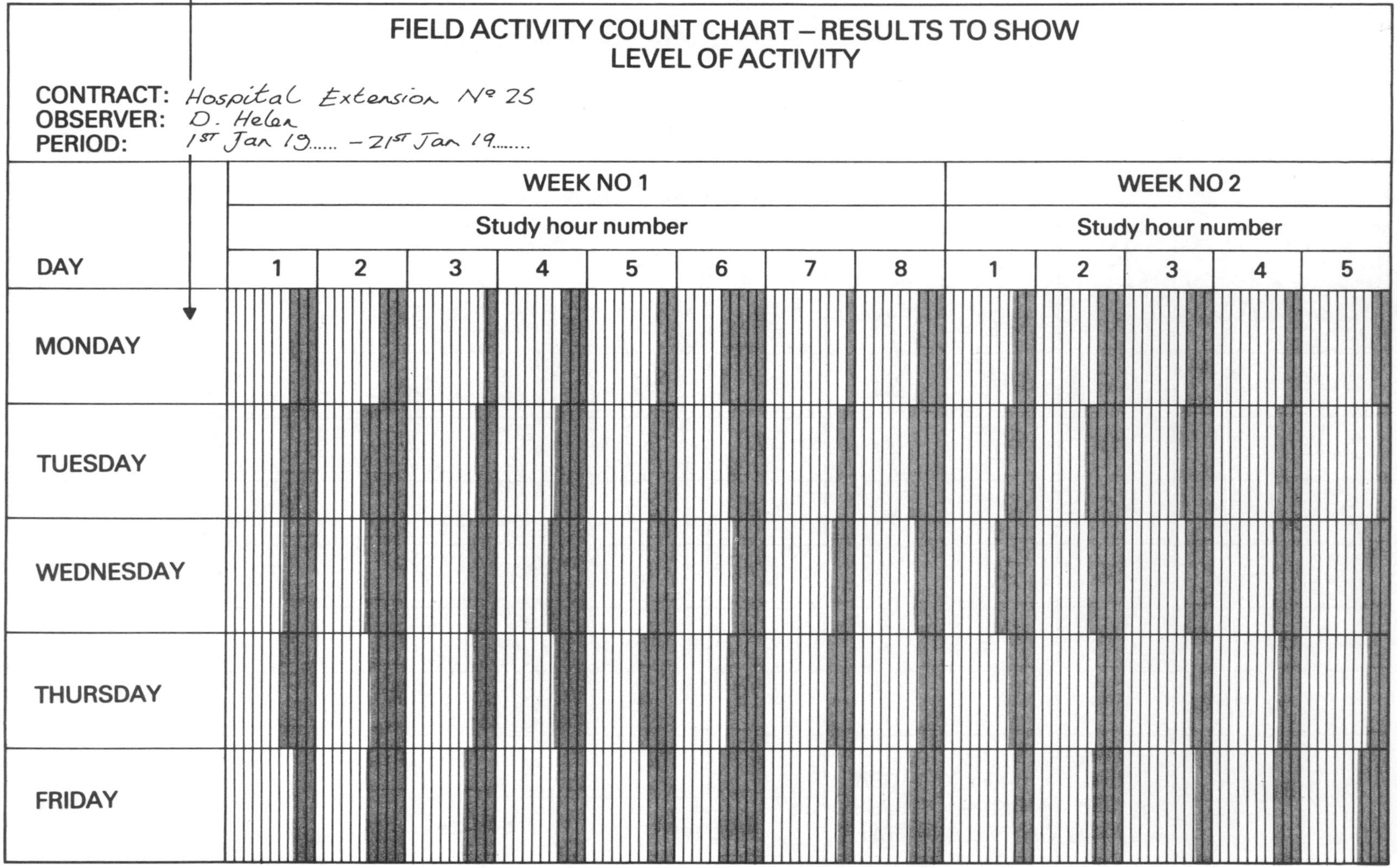

FIELD ACTIVITY COUNT CHART – RESULTS TO SHOW
LEVEL OF ACTIVITY

CONTRACT: *Hospital Extension Nº 25*
OBSERVER: *D. Helm*
PERIOD: *1st Jan 19...... – 21st Jan 19........*

	WEEK NO 1								WEEK NO 2				
	Study hour number								Study hour number				
DAY	1	2	3	4	5	6	7	8	1	2	3	4	5
MONDAY													
TUESDAY													
WEDNESDAY													
THURSDAY													
FRIDAY													

NOTE: In study 6 of the first week there is excessive inactivity because supervisor does not clear the canteen promptly each day after lunch.
The shaded portion of each hourly square (usually in red) indicates the amount of inactivity during each observation as a simple illustration to management.

Fig. 8.2.5

RANDOM OBSERVATION STUDY

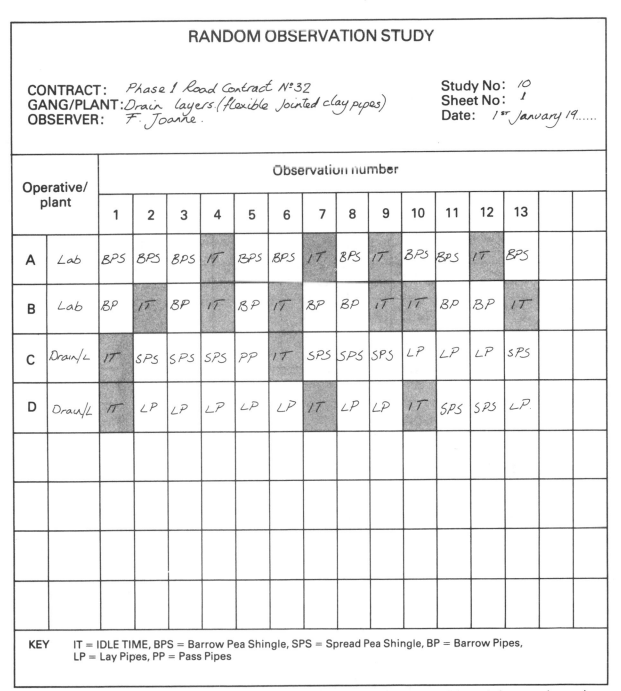

CONTRACT: *Phase 1 Road Contract N°32*
GANG/PLANT: *Drain layers (flexible jointed clay pipes)*
OBSERVER: *F. Joanne.*

Study No: *10*
Sheet No: *1*
Date: *1st January 19......*

Operative/ plant		Observation number														
		1	2	3	4	5	6	7	8	9	10	11	12	13		
A	Lab	BPS	BPS	BPS	IT	BPS	BPS	IT	BPS	IT	BPS	BPS	IT	BPS		
B	Lab	BP	IT	BP	IT	BP	IT	BP	BP	IT	IT	BP	BP	IT		
C	Drain/L	IT	SPS	SPS	SPS	PP	IT	SPS	SPS	SPS	LP	LP	LP	SPS		
D	Drain/L	IT	LP	LP	LP	LP	LP	IT	LP	LP	IT	SPS	SPS	LP.		

KEY   IT = IDLE TIME, BPS = Barrow Pea Shingle, SPS = Spread Pea Shingle, BP = Barrow Pipes, LP = Lay Pipes, PP = Pass Pipes

Note: As can be seen, this method of work measurement highlights the idle time, and due to it the gang size can be adjusted – result: a gang of 3 may be more suitable instead of 4.

Fig. 8.2.6

SYNTHESIS FROM ELEMENT DATA

PROCESS: *Hanging SW Panelled Door Size 50×1500×500 mm* STUDY NO: 72
PREPARED BY: *S. Shirley* SHEET NO: 1
DATE: *1st Jan 19......*

ELEMENT	BASIC TIME FROM SYNTHETICS DATA	ADD/SUBTRACT FOR VARIATION	TOTAL BASIC TIME	RELAXATION ALLOWANCES						TOTAL RELAX. ALLOW.	TOTAL BASIC TIME + RELAXATION ALLOWANCES.
				S	P	E	C	A	M		
1. Collect door and change work position and arrange tools	3.53	no variation	3.53								
2. Fit hanging edge of door	5.32	−1.33	3.99								
3. Fit head of door	4.22	−1.41	2.81								
4. Mark and plane door to frame width	7.49	−1.87	5.62								
5. Mark and plane bottom of door	7.03	−2.34	4.69								
6. Offer up door and mark hinge positions	1.83	no variation	1.83								
7. Cut out for one pair of hinges (75mm)	6.88	''	6.88								
8. Fit hinges to door and door to frame.	6.23	''	6.23								
9. Adjust door	2.12	''	2.12								
10. Clean up shavings	1.84	''	1.84								

GRAND TOTAL	
CONTINGENCY ALLOWANCES: (0–5%)	
PERFORMANCE/STANDARD TIME: per door	
OUTPUT PER HOUR: (if required)	

NOTE: The basic time column shows element basic times obtained from SYNTHETICS DATA (stored information of identical work elements from previous time studies) which may need adjusting because of various sizes of new work. Note that the element no. 2 basic time is 5.32 for 'fit hanging edge of door', but this door, although identical in construction and thickness, has an edge which is bigger than the door now being studied by about ¼; therefore, ¼ (1.33 mins) has been deducted to give the Total Basic Time.

Relaxation, standard and contingency allowances are next assessed, as previously explained in Time Studies.

Fig. 8.2.7

basic time of 5.32 minutes, making a basic time for the smaller door of 3.99 minutes.

Analytical estimating

This is yet another work measurement technique but which is used in calculating standard times for non-repetitive processes or jobs. It may be too expensive and time consuming to carry out time studies for a one-off job. Also, there may not be sufficient recorded previous elements or standard times in a firm's synthetic data, making the technique of synthesis inappropriate.

Analytical estimating should be undertaken by someone with considerable experience in work study, or a person (site supervisor) who may have had vast practical experience and with a knowledge of how long trade operations take.

The technique, where a one-off job's basic times are to be prepared, is to break the job down into elements, as in all time studies. Each element's basic time is assessed at a standard rate, and where it is possible synthetic data is used; this is a less precise method of obtaining standard times because it relies on judgement, but it is assessed as being satisfactory in such situations as described.

The example in Fig. 8.2.8 deals with the removal of an existing overflow and warning pipe which is too small in diameter, and which is replaced with a bigger plastic one. Some of the basic times are obtained from synthetic data, and others by analytical estimating (assessment from experience).

In using this technique a rate for undertaking the work can be fixed so that the operative, if there is a bonus scheme in operation, can assess his/her bonus earnings.

8.3 Incentive schemes

There are two areas within which financial incentive schemes fall; those of operatives and those of managers. Each area is dealt with differently by companies either as a method of reward, or inducement to higher productivity.

Schemes for managers

1. Extra payments or increases in salaries for supervisors, managers and executives for any increased output. These usually take the form of: payments by the allocation of extra shares; a percentage of any profits earned by the company; or free holidays as a bonus for well-managed, profitable schemes.
2. Fringe benefits covering such items as: a car; sickness benefits; free education for a manager's children, especially if the manager has to frequently spend time away from home and his wife, periodically, is required to accompany him.

Schemes for operatives

1. Geared incentive bonus schemes which are based on measured levels of productivity at site or shop floor level.

2. Standing bonus schemes where the cost of a bonus department set-up is either too expensive and the bonus is used as a bribe to ensure adequate numbers of employees attend work promptly and regularly during a work week.

Financial incentives for executives/managers

While some of the aforementioned are strictly financial incentive schemes used to coax the workers, staff and managers to higher productivity, others are in recognition for past efforts, or to retain the services of key personnel vital to a firm's continued success in the relevant industrial or commercial field.

It appears to be increasingly necessary for companies to add to the fringe benefits of managers/executives as a way round governments' prices and incomes policies, and which is used as a basis for appeasing vital executives, etc., who hold the keys to a company's success against competitors from home and abroad. In this regard 'alternative benefits' or 'perks' are the rewards for faithful service, and in many instances are tax free, which assists in protecting incomes against inflation.

These alternative benefits (perks) revolve around such examples as:

(*a*) free lunches;
(*b*) firm's issue of suits, with a nominal hire charge of a few pounds per year;
(*c*) free contributions to a private medical scheme such as BUPA;
(*d*) free travel abroad for wife and family while accompanying husband/father on business, or more especially to conventions;
(*e*) free advice on personal affairs;
(*f*) professional fees paid for;
(*g*) access to newspapers and periodicals;
(*h*) non-contributory pension schemes;
(*i*) non-contributory insurance schemes;
(*j*) free bridging loads, new fitted carpets, solicitor's fees when moving house;
(*k*) low interest loans;
(*l*) low mortgage rates;
(*m*) free generous supply of wine for possible entertainment of prospective customers;
(*n*) service contracts – in the event of the manager losing his/her job due to the firm eventually going into liquidation; or in a redundancy situation, an agreed sum of, say, £20 000 would be made available each year for up to 20 years (firm insures against this).

The 'perks' go as far as: free hairdressing for a wife who assists in the entertainment of prospective clients; the provision of food freezers to store, so called, special food for entertaining clients; stereo and TV sets. There are numerous entertainment 'perks' many of which are bona fide, but many more that abuse the system and which are used as a means of boosting executives' salaries. In offsetting these additional costs as overheads in the ac-

160

ANALYTICAL ESTIMATION SHEET

PROCESS: *Replacing Overflow or Warning Pipe to Cistern in Roof Space with Plastic Pipe.*

PREPARED BY: *D Helen*

STUDY NO: *10*
SHEET NO: *1*
DATE: *1ˢᵗ January 19....*

ELEMENT	BASIC TIME (IN MINUTES)	STANDARD AND RELAXATION ALLOWANCES	TOTAL: BASIC TIME PLUS ALLOWANCES	NOTES
1. Climb loft and prepare	3·25			From synthetic data
2. Cut off copper pipe which extents through fascia board and remove	1·72			From synthetic data
3. Slacken off fitting and remove pipe at junction to cistern	4·00			Estimated
4. Remove cistern fitting	2·00			Estimated
5. Enlarge hole for new fitting in fibreglass cistern	4·00			Estimated
6. Fit new fitting to cistern	2·20			From synthetic data
7. Enlarge hole through fascia board	7·00			Estimated
8. Measure new plastic pipe and cut to length	3·80			From synthetic data
9. Bend pipe to fit through fascia	6·85			From synthetic data
10. Fit pipe and attach to cistern	3·10			From synthetic data
11. Clean up and climb from loft.	8·00			Estimated
GRAND TOTAL TIME:				
CONTINGENCY ALLOWANCE:				
PERFORMANCE/STANDARD TIME: per pipe				
OUTPUT PER HOUR: (if required)				

NOTE: This example assumes that only one man will be undertaking the work. Also it may be necessary to add the elements 'travelling to and from the job.'
See the Notes Column which shows from where the Basic Times were extracted.

Fig. 8.2.8

counts a company would not then be breaching any government pay code. This means that executives effectively are worth more, and the company pays less than it would if an equivalent amount of money had to be paid because the PAYE undermines the earnings. To give executives an additional £10 000 take-home pay, normally £20 000 (to allow for tax) would have to be added to the salaries bill. The former system is cheaper.

Financial incentives for operatives

Less-enlightened managers have regarded financial incentives as the only means for motivating workers, and that by organising the work-force with close supervision and strict discipline optimum performances and productivity is ensured. Gradually, however, there has been a noticeable change, and whereas financial incentive schemes are thought to be the biggest contributor to higher productivity, other factors are now borne in mind.

Managers must show a personal interest in their work-force by allowing those with ability more responsibility and scope for creativity, and for realising their true potential.

A sense of achievement is the aim of most workers, and each one must be encouraged to strive for this. Their ideas and advice should be sought as often as possible to give them a sense of being needed and a feeling of belonging. It is a principle which fosters goodwill and enhances the company's position due to the workers' participation.

Non-financial incentive schemes also motivate the operatives and staff generally, some of which are discussed in section 5.3.

Financial incentive schemes for operatives are equated to the following:

(a) cost plus rates (standing bonus);
(b) good holiday pay;
(c) extra pay for shift work and uncomfortable, dirty and unhealthy conditions;
(d) profit sharing;
(e) long service allowances;
(f) overtime availability when operatives require it;
(g) pension fund contributions by the company;
(h) death benefits to dependants;
(i) employer's liability insurance;
(j) good sick pay.

Geared incentive bonus schemes

These types of incentive schemes allow for the making of extra payments to a work-force for increased productivity – the schemes are geared to predetermined levels of output.

Extra payments (bonus) are made to operatives provided that they produce more units of work beyond the levels (target) fixed by the Production/Bonus Department, which are normally determined from work measurement standard times obtained by work study.

There are occasions when bonus payments are made where targets are not reached, which are:

1. Where delays occur through the operatives receiving late instructions.

2. Where materials are not made available on time.
3. Other delays beyond the operatives control, e.g. sub-contractors' work which prevents continuity of main contractor's work.
4. Inclement weather delays.
5. Management's inability to cope with labour disputes which causes stoppages and which affects others on site.

Although, under the National Working Rules, operatives are entitled to a standing bonus particularly when a firm does not operate a proper scheme, this does not supplement employees' pay to the same level as a 'geared' incentive bonus scheme.

Geared incentive schemes, help to:

(a) reduce unit costs for the employer due to increased output;
(b) make supervisors more efficiency conscious and better organised;
(c) increase the levels of earnings of employees;
(d) accelerate the work and complete contracts on time;
(e) satisfy the client because the structure can be used earlier;
(f) please the architect because his supervision time is reduced.

The main criticism of incentive bonus schemes are levelled at the quality of work produced by operatives while working more quickly to beat the target; these can lead to disputes if the scheme is mismanaged due to the target being unacceptable to the operatives. The schemes have been operative for many years now, and the advantages appear to far outweigh the disadvantages.

Features of a sound financial incentive bonus scheme

1. Methods by which operatives can calculate bonus earned should be simple.
2. Bonus should be paid weekly, usually with a maximum of two weeks in arrears.
3. There should be no maximum limit to how much bonus can be earned.
4. Excessive bonus should not be reduced if the operatives have earned it.
5. Bonus earnings should not be affected by delays caused by late deliveries of materials or poor management organisation.
6. A good bonus earned one week should not be averaged with that of a bad week.
7. Man hours should be used as a target.
8. Targets should be issued to the operatives before they commence operations, and the type of scheme should be agreed before a contract gets under way – usually 50 per cent is preferred in construction work.
9. Bonus schemes should be self financing with operatives receiving a percentage of the savings in cost, and the remainder should be retained by the firm to pay for the bonus department set-up.

10. If a standard rate of work is maintained (100 work rating) operatives should be allowed to increase their normal earnings by $33\frac{1}{3}$ per cent without over-exertion.

It must be appreciated that if a bonus system is fair, and can be seen as such by the operatives, very little time would be wasted on their supervision; also they tend to be good timekeepers. Supervisors would be able to spend more time on planning and ensuring continuity of work, and providing materials when and where the operatives require them.

Under the various geared financial incentive bonus schemes the operatives are allowed the hours saved above the target hours at the following rates, depending on which scheme is agreed at commencement of a contract:

100% scheme – all the time value saved.
50% scheme – $\frac{1}{2}$ of the time value saved.
25% scheme – $\frac{1}{4}$ of the time value saved.

(remembering that the employer retains the remaining parts).

Under a properly augmented scheme the work study officer supplies the bonus surveyor sufficient standard times from the 'synthetic data' to enable the surveyor to fix appropriate targets to units of work.

Effects of each scheme

The operatives should be allowed to earn $33\frac{1}{3}$ per cent extra when working at 100 rating. In the following examples the operation to be undertaken is assumed to take 36 hours.

1. *100 per cent scheme.* Add $\frac{1}{3}$ for 100 rating, so there can be a bonus earning of 100 per cent of the time saved. Therefore 36 hours + 12 hours = 48 hours allowed for the operation, and there is 48 hours of earnings.
2. *50 per cent.* Add $\frac{2}{3}$ for 100 rating so there can be a bonus earning of 50 per cent of the time saved, therefore, 36 hours + 24 hours = 60 hours allowed for the operation, and there is 36 hours + 50 per cent of 24 hours =
   36 hours + 12 hours = 48 hours of earnings.
3. *25 per cent scheme.* Add $\frac{4}{3}$ for 100 rating so there can be a bonus earnings of 25 per cent of the time saved. Therefore, 36 hours + 48 hours = 84 hours allowed for the operation, and there is 36 hours + 25 per cent of 48 hours =
   36 hours + 12 hours = 48 hours of earnings.

There is a 75 per cent scheme where $\frac{4}{3}$ is added in the same way as the above examples.

In each of the schemes 48 hours of earnings would be allowed, but where the difference in the schemes can be seen is when an operative works either less than, or more than, 100 rating. (See Fig. 8.3.1.)

Differences: (*a*) In the 25 per cent scheme a slower worker (less than 100 rating) is better off than if working on a 50 per cent scheme.

(*b*) In the 25 per cent scheme faster workers (over 100 rating) are worse off than fast workers on 50 per cent scheme.

It is, therefore, better (for the operatives) to use a 25 per cent scheme in the early stages of the work while operatives are learning about the job, perhaps introducing a 50% scheme later when they have become accustomed to the work.

The Production Control Department/Bonus Department use various company designed forms for calculating and recording bonus earnings and which can also be utilised to show a week's cost and value, and ultimately the actual Unit Costs compared to the Standard Costs (see Fig. 8.3.2).

8.4 Quality and quantity relating to performance

Incentive schemes have sometimes been maligned as a licence for bribery to maintain a suitable level, in numbers of operatives in the workplace, and that very few benefits accrue to either the contractor or the client; also, that too many financial incentive bonus schemes are not being run on the lines for which they were intended, which are, to increase productivity or output per operative with the intention of making cost savings.

Unfortunately, it may be claimed that in fact some of what has been said is true, and that in addition it leads to a breakdown in the quality of the work produced, even though marginally the quantity is increased.

Large firms in the construction industry would contest these statements, because, the very nature of the implementation and running of bonus schemes since the early 1940s has proved in most cases to be both viable and profitable.

Supervisors are the individuals who have to ensure that while such schemes operate the quality, in particular, conforms in some degree with the specifications to maintain good working relation with the client's representative, the architect – or his subordinate, the clerk of works. Therefore, it is usual to refuse bonus payments to operatives who have completed sub-standard work although targets have been met. Perhaps, better still, the supervisor should insist that the operatives who have produced the poor work should rectify it, so that in the future the standards set would be maintained.

Supervisors have a particular duty and responsibility to ensure work is conducted in such a way so as to be maintained at a recognised standard while still trying to maintain a suitable quantity of output.

The clerk of works

This person is also referred to as the resident engineer, and to give a site supervisor some idea of how careful he/she has to be in maintaining quality control over

163

GEARED INCENTIVE BONUS SCHEMES – COMPARISONS

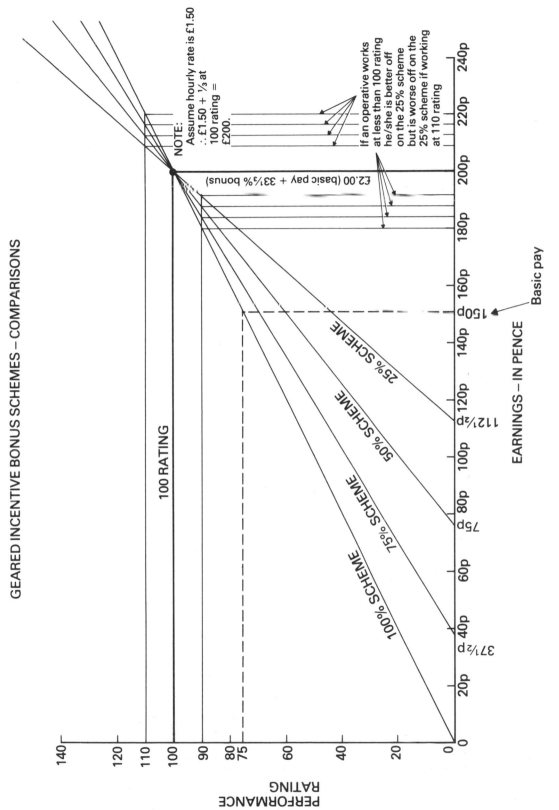

Fig. 8.3.1 Geared incentive bonus schemes — comparisons

164

COST AND BONUS SHEET

B. BUILDER LTD.

CONTRACT:　　　WEEK ENDING:　　　WEEK NO.　　　TRADE:　　　SHEET NO:

Operation hours + earnings					Allocation of hours									Bonus achievement					Week's cost + value					Inform—ation	
No.	Name	Trade	Total hours	Wages at tender rate	Bonus paid	Stage	Description of work	Mon.	Tues.	Weds.	Thurs.	Fri.	Sat.	Sun.	Total	Measure	Cost Rate	Bonus	Target Hours	Hours taken	Labour cost	Mech plant & trans. cost.	Total cost	Total value	Unit value	Unit cost	Remarks

Fig. 8.3.2

production, it is necessary to understand the reponsibilities and duties of this client/architect representative. The clerk of works therefore usually:

1. Inspects all work in progress and compares with the standards laid down in the specifications.
2. Checks site grids and other setting out undertaken by the contractor as work proceeds.
3. Inspects products in workshops off-site, before delivery, as a check on quality (sometimes the architect does this).
4. Takes samples of materials delivered to site as a check on their suitability, and either tests them him/herself, or sends them to the architect or to an independent testing laboratory.
5. Tests samples of mixes of concrete for slump and sends sample concrete cubes to independent testing laboratories (or arranges for the contractor to do so).
6. Checks that storage of materials is adequate.

It is usual for the clerk of works to have the following equipment available for simple on-site testing:
(*a*) rain gauge; (*b*) maximum and minimum thermometer; (*c*) measuring tape and rod; (*d*) moisture meter; (*e*) spirit level; (*f*) surveyor's level, tripod and staff – including a level book, pencil and calculator; (*g*) theodolite and ranging rods; (*b*) cover meter – to check concrete cover to reinforcement; (*j*) concrete testing equipment – silt testing, cube mould, concrete curing tank, sieves, scales.

It is expected that after carrying out any tests which are deemed necessary records of results are maintained, and copies would be sent to the architect for his/her inspection and approval or disapproval, along with any reports, i.e.

(i) site diary (supervisor maintains his own copy);
(ii) labour returns (to show whether suitable numbers are being maintained according to the programe – this may also include subcontractors);
(iii) details of delays, strikes, etc., (clerk of works to remain impartial in labour disputes and should not get involved);
(iv) progress reports of main contractor and sub-contractors;
(v) inspection records of materials;
(vi) reports on any tests carried out;

(vii) reports on any discrepancies in the drawings or other contract documents to the architect;
(viii) notifies the architect where outstanding information is required by the contractor;
(ix) prepares a report for the architect prior to project meetings;
(x) records measurements of work which has to be covered up (for and on behalf of the quantity surveyor);
(xi) checks and signs daywork claims;
(xii) reports to the architect any variation from the specifications by the contractor (materials or workmanship).

A clerk of works should not give instructions which will incur extra costs, without the architect's consent. Any variations from the contract particulars by the contractor would have to pass through the usual channels of communications until a variation order, duly signed by the architect, is finally received authorising the work, and which will usually entitle the contractor to extra payments in addition to the contract sum.

Clerks of works have a duty to prevent, if possible, but more usually to discover, bad workmanship at execution stage and to prevent further breaches of the specifications. He/she must inspect all stages of work each day and report any misgivings about quality to the architect. Where work is condemned and before it is pulled down or removed written confirmation must come from the architect, and any time lost due to this action should be recorded by the clerk of works, details of which would by analysed if the contractor wishes to apply for an extension-of-time later. The clerk of works is empowered to give directions, where he deems it necessary, but confirmation must follow within two days from the architect under the conditions laid down in the Joint Contract Tribunal Standard Form of Building Contract (see Fig. 8.4.1).

With the clerk of works duties in mind, the supervisor would be well advised to constantly be alert to all aspects of production, and insist that his/her site team maintains suitable standards of workmanship throughout the contract duration even though a financial incentive bonus scheme is in operation. Quality of work should not be sacrificed to appease the operatives who wish for higher bonus payments.

Architect's name
and address:

**Clerk of works'
direction**

for the works:

situate at:

Serial no:

Date:

to Contractor:

Under the terms of the Contract,

dated _____ and under Clause 12,

I issue this day the following direction on behalf of the Architect.

If this direction is not confirmed by an Architect's Instruction

received within two working days, please refer to the Architect.

This order does not authorise any extra payment.

Direction	For use in the Architect's office
Item number:	
	Covered by A1 number:

Signed _____ Clerk of works

☐ Architect   ☐ Contractor   ☐ File copy

29

Fig. 8.4.1   (*Reproduced by kind permission of RIBA Publications Ltd.*)

Chapter 9
Supervisor's effective programming and progressing of production

9.1 Programming techniques: bar charts

The bar chart, as a programme and progression technique, is the most commonly used chart on-site. It is adopted because most site supervisors find it easier to understand compared with other charts available to the construction industry, e.g. networks-critical path diagrams, precedence diagrams and line of balance charts.

With more effort on the part of site supervisors the other charts could be mastered to good effect, because, generally speaking, the latter charts/diagrams can be applied to different types of construction processes better than the bar chart, but the bar chart is used to supplement the other charts if thought desirable for the trade foremen's use. The other charts, particularly the critical path and precedence diagram, allow for a better control to be maintained on the operations on-site than can be expected from a bar chart, particularly for the most complex jobs. The line-or-balance chart is a better control document for repetitive construction jobs, such as housing estates, or a multi-storey block where each floor level is almost identical.

Site supervisors, in the main, are expected to control contracts for which they are responsible by using bar charts, so it is of some value to understand how such charts are prepared by the programmer or contracts manager. In the preparation of this and other progressing/processing charts it is necessary first to have a clear idea of how the work is to be executed, which is best achieved by draughting out a method statement, perhaps with two or more individuals discussing the problem together, i.e. programmer, contracts manager and even the site manager if it can be arranged. As an example contract for the details to follow, see Fig. 9.1.1.

Method statement

This document is prepared first which arranges one's ideas of how a contract would best be executed, with the minimum of cost, and with the optimum of resources to give a suitable level of production flow. Therefore the usual details included in a method statement relate to the following:

1. The key operations – a complete breakdown of the contract.

2. The method and sequence for carrying out each operation.
3. The plant required to carry out each operation.
4. The labour necessary to undertake each operation.
5. The operations to be undertaken by ordinary, labour-only, and nominated subcontractors (see Fig. 9.1.2).

This method statement is next used to assist in the preparation of a calculation sheet which is a prerequisite to the draughting out of a contract programme (bar chart). Later, a copy of the statement is issued as an additional control document to the site supervisor along with the other essential contract documents issued by the architect:

1. Production drawings.
2. Bills of quantities.
3. Specifications.
4. Schedules.
5. Joint Contracts Tribunal Form of Contract.

While the method statement preparation is a worthwhile exercise before attempting the programme stage, most programmers dispense with the actual document preparation and instead draw up their ideas directly on to the bar chart programme; this may show symbols beside the relevant time/duration bars for hand-over stages and other important information as instructions to the site supervisor.

Calculation sheet

When the operations, plant and gang sizes are determined and included in the method statement, the calculation sheet is used to arrive at suitable and acceptable durations for each key operation. To do this it is necessary to extract quantities for each operation from the bills of quantities, or by measuring directly from the production or other working drawings.

In the average rate per hour column of the calculation sheet (see Fig. 9.1.3) rates are included which are obtained from past personal records, synthetics data from the work study section or the estimating department. These rates are next divided into the quantities to give total hours per key operative or plant. The duration, in hours, can be

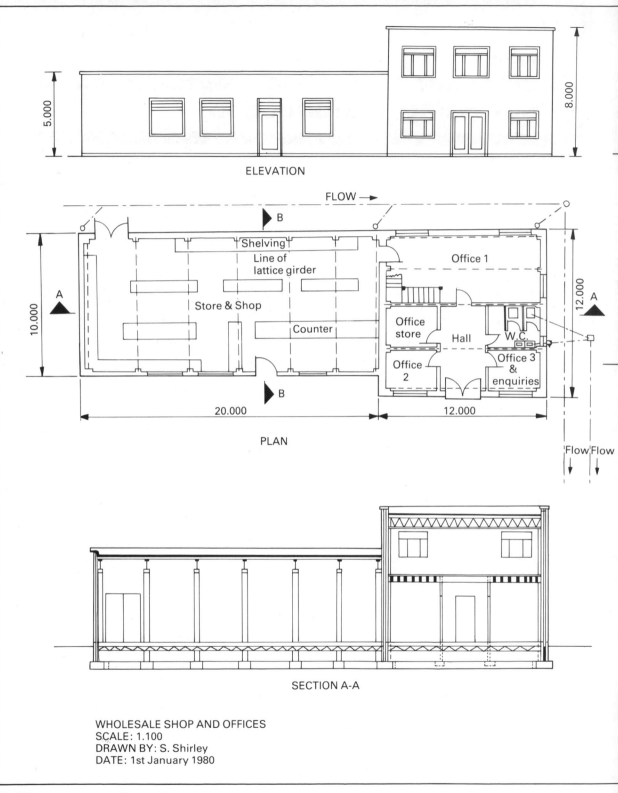

ELEVATION

FLOW →

B

Shelving

Line of
lattice girder

Office 1

A

Store & Shop

Office
store

Hall

W.C.

Counter

Office
2

Office 3
&
enquiries

A

10.000

12.000

B

20.000

12.000

PLAN

Flow Flow

SECTION A-A

WHOLESALE SHOP AND OFFICES
SCALE: 1.100
DRAWN BY: S. Shirley
DATE: 1st January 1980

Fig. 9.1.1

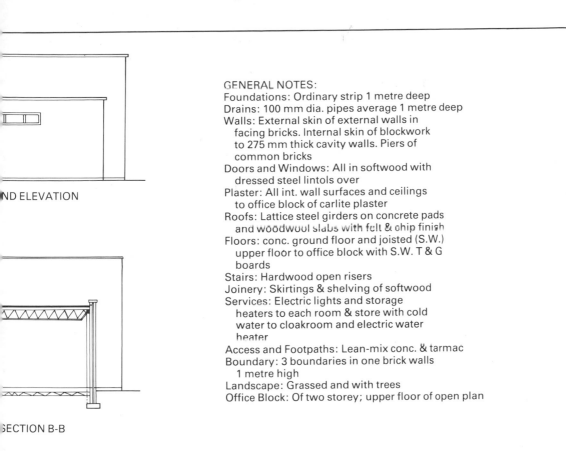

ND ELEVATION

SECTION B-B

GENERAL NOTES:
Foundations: Ordinary strip 1 metre deep
Drains: 100 mm dia. pipes average 1 metre deep
Walls: External skin of external walls in
  facing bricks. Internal skin of blockwork
  to 275 mm thick cavity walls. Piers of
  common bricks
Doors and Windows: All in softwood with
  dressed steel lintols over
Plaster: All int. wall surfaces and ceilings
  to office block of carlite plaster
Roofs: Lattice steel girders on concrete pads
  and woodwool slabs with felt & chip finish
Floors: conc. ground floor and joisted (S.W.)
  upper floor to office block with S.W. T & G
  boards
Stairs: Hardwood open risers
Joinery: Skirtings & shelving of softwood
Services: Electric lights and storage
  heaters to each room & store with cold
  water to cloakroom and electric water
  heater
Access and Footpaths: Lean-mix conc. & tarmac
Boundary: 3 boundaries in one brick walls
  1 metre high
Landscape: Grassed and with trees
Office Block: Of two storey; upper floor of open plan

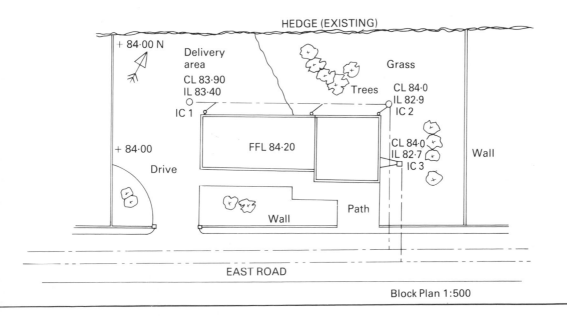

Block Plan 1:500

METHOD STATEMENT

CONTRACT: Wholesale Shop + Office
CLIENT: J.A. Aberson

SHEET NO: 1/2
PREPARED BY: D Helen
DATE 1st /.........

OP. NO.	OPERATION	METHOD AND SEQUENCE.	PLANT	LABOUR	NOTES
1.0	*Preliminaries*				
	Access + temp road	See site layout plan for access position and remove existing pavement slabs and lay hardcore ballast for vehicle access	JCB 7B	3 lab	Selection of buckets for JCB.
	Hut accommodation and Security fence	As on site layout plan. Fence to be in line with proposed wall position with one set of double access doors opening into site	Standard t/man shut canteen and store	3 lab	Mess hut to accommodate max 2 cooperative
	Temp. Services	Standpipe as per site layout and lagged/protected against frost above ground level w.c. etc. connected to main drains as soon as possible	—	Sub Con. GF/2 lab	—
	Setting out	Building first, then drainage			new supply of pegs and boards
	Surface strip	Building area first then RL. dig for drive. Spoil heap in position as shown on site layout plan	JCB 7B	2 lab	—
2.0	*Substructure*				
	Exc. fdn trench	Excess material removed from site immediately by lorry to tip at London Road	JCB 7B 2 Lorries	1 lab.	Timbering used throughout.
	Main drains	Commence immediately after trench (found) is completed.	JCB 7B	2d/lay 2 lab 4 lab	—
	Conc. founds	Use ready-mix conc laid immediately excavations completed	—	4 lab	—
	Brickwork to D.P.C.	concentrate on office block first.	100l mixer barrows	8b/lay 4 lab	—
	H/core and membrane to floor	office block first and consolidate by vibrator roller	vib. roller	4 lab	—
	Conc ground floors	Office block first - see spec for bay sizes using ready-mix conc.	Steel formers and pegs	4 lab	—
3.0	*Superstructure*				
	Bwk, piers, blockwork	Office block to 1st floor a priority. External door and window frames built with brick work	Scaffold 100l mixer	8b/lay 4 lab	Steel scaffolder outside only

Item	Description	Plant	Labour	Remarks
Padstones	Cast in-situ (office area first) and with ragg bolts cast in for steel beams.	—	3 C+J	—
4.0 Roof				
Lattice girders	Lifted into position by mobile crane: office block first	Mobile Crane	3 C+J	Firrings fixed when in position
Decking of Firrings, woodwool slabs	All nailed complete office block first	Mobile crane	3 C+J	Stack onto the girders but evenly distributed
Roof covering in felt + chips	Both roofs to be completed in one operation starting with office block	—	SubCon	—
Rainwater goods + flashings	Start immediately both roof coverings completed to prevent staining of brickwork	—	Sub. Con	Fixed to gulley at earliest opportunity

NOTE The remaining important operations would continue to be listed as dealt with above. Method Statements should be prepared before dealing with the contract programme in order to decide on correct method and sequence for doing the work on site. This document could be used on site by the site manager.

Fig. 9.1.2

established by dividing the number of key trade or other operatives into the total hours, e.g.

Substructure   618 m² cavity brickwork – Average rate or output per hour of 1 bricklayer is, say, 0.5 m².

Therefore, $\dfrac{618\,m^2}{0.5\,m^2} = 1236$ hours.

So, for 1 bricklayer there is 1236 hours of work.

Then, for 8 bricklayers there would be $\dfrac{1236}{8} = 154\frac{1}{2}$ hours of work which is $19\frac{1}{2}$ days.

If a duration for an operation appears to be excessive the programmer could reduce the duration by allowing a bigger gang size, or by introducing an extra piece of plant. The next stage is to prepare a contract programme using the information extracted from the calculation sheet.

Contract programme (bar chart)

The programme is best drawn up on tracing paper to enable extra copies to be produced later using dyeline photocopying process.

The operation numbers, operations, gang/plant size and durations are next copied from the calculation sheet before attempting the lining in of the duration bars. Any adjustments to the sequence of operations should be finally made at this stage (remembering, also, to correct the method statement if alterations are made).

When the duration bars are finally included on the programme they should, as far as possible, follow in close succession so as to provide continuity of operations; flowing from the top left-hand corner of the chart, and concluding at the bottom right-hand corner (see Fig. 9.1.4).

By studying the Fig. 9.1.4 there are a number of important details of which a site supervisor should be aware, such as:

1:   Minor, non-key operations are excluded from the operations column, but their durations are included with the associated key operations, i.e. polythene d.p.c. is part of hardcore to ground floor.
2.   Operations such as hutting follows immediately after access, etc., to provide continuity of work for the four labourers – otherwise, two gangs of four labourers would have to be introduced.
3.   Operations such as temporary services can start before the hutting and security fence is completed.
4.   An attempt should be made to balance the gangs correctly so that the fluctuation of labour is not excessive each week. Note on Fig. 9.1.4 that the general labour content each week in the early stages of the contract fluctuate between two and four in numbers. Ideally it would be best if, say, four labourers could be used throughout, and a further effort could be made to rearrange the operation bars to achieve this.

A good programme is one which has the following attributes:

(a)   easy to produce;
(b)   simple to understand;
(c)   simple to up-date due to changing circumstances;
(d)   well illustrated to show labour and plant requirements'
(e)   easily adapted to monitor progress;
(f)   has simple symbols as a warning to the site supervisor for chasing information, orders, etc;
(g)   provides as far as possible for all resources to be used continuously whilst on-site.

By studying the aforementioned on how a bar chart programme is prepared, the site supervisor can see that most of the details included are not by guesswork but as a scientific approach to the problem of controlling construction projects. It also takes into consideration the experience of the programmer and those who may periodically assist him/her; and most certainly past records of previous similar work greatly adds to the accuracy by which such charts can be prepared.

9.2   Programming techniques – networks (critical path methods)

Bar charts tended to be the only types of control charts used on construction sites until the late 1950s when the critical path network method was introduced for controlling the progress of production in the construction and other industries.

The network chart/diagram was quite a revolutionary system when first introduced. It defines, from all the operations and activities in a project, those which are critical ones, and therefore ones which require careful progressing to prevent delays which would extend a contract duration, so using up the contractor's valuable resources.

The use of this network chart is mainly limited to the more complex construction projects for which the major firms appear to be more successful in tendering. The application, however, is far reaching, and can be applied to the simplest of work, but which, unfortunately, too few individuals have attempted to apply in a real situation, or even bothered to learn – the attitude being that the bar chart is an easy technique to learn and apply, and, anyway, most individuals especially those at site level find it simpler to understand.

The most cost-conscious managers see the critical path network as a superior control document to that of bar charts, because the effects of a delay in one activity over successive ones can be seen more readily. Instead of relying entirely on the experience of a planner/programmer to arrange the job operations into a suitable sequence which he/she feels will suit the firm, instead, the operations are arranged with logical reasoning uppermost in the mind of the programmer which overcomes the possibility of making wrong assumptions many site supervisors complain about regularly.

Symbols used in networks-critical path method

1. *Activities or operations*

These are denoted by arrows, the length of which, generally, is immaterial. One activity (operation) which takes two days could have a representative arrow whose length is the same as an activity requiring 20 days to complete. Usually the activity name is printed above the arrow while the realistic expected duration is printed below (see Fig. 9.2.1).

2. *Nodes or events*

These denote the start and finish of a planned activity and are shown at the beginning and end of each arrow. In order that an activity can easily be identified each node/event is allotted a number. In Fig. 9.2.2 the activity is referred to as activity 3–4, which identifies it from a different excavation activity on a large job.

3. *The dummy arrow*

A dummy is a dotted arrow (see Figs 9.2.3 and 9.2.4) but which is not allocated a duration and which is not an activity. It is used, in the main, to ensure that no two activities have the same identifying numbers, or where one activity relies on another being completed before it is started in a parallel arrow routes situation.

4. *Lead and lag arrows*

These are almost similar to dummies but are allotted a duration although they are not strictly representing an acitivity. They are used where a short delay is required in carrying out a subsequent activity from a previous one, and where this would lead to a short lag or delay in completing the final activity in a cycle (see Fig. 9.2.5).

5. *Faulty activity arrangements*

An activity commences from a node event and the only activities which start and end without activity joining at either the tail event or head event respectively, are the very first and last activities on a critical path network (see Fig. 9.2.6).

Network (CPM) preparation

As in bar charts, it would be advisable to prepare a method statement to show, at least, the job broken down into operations which would indicate the extent of the work to be undertaken. Similarly, durations for each of the operations would be determined by the preparation and use of a calculation sheet.

When the two previously mentioned documents are finalised (which would generally take into consideration plant and other resources) the following steps are taken in the preparation of a network – critical path method (CPM).

Step 1: Using logical reasoning an arrow network, which represents all the activities relating to the construction work, is drawn bearing in mind the sequences in which the work 'could' be, and not how it should be, carried out by the contractor, unless there are some particular handing-over stages or special requests on the sequence of work by the client which affects a present adjoining structure, or production in the client's factory, etc. (see Fig. 9.2.7).

The network (CPM) differs from the bar chart because the programming officer prepares the latter by listing the operations in the order in which they are to be executed to suit the firm's resources, e.g. relying on labour and plant availability. Networks are prepared, generally, by considering which operations must be completed before another one can start. Drainage operations may be delayed until most of the structural work of a building has been completed, but some firms complete the same operation as early as possible while the excavator is still available on-site.

If a logical approach was made, which is how a network is draughted out, an arrow representing the operation (drainage) would show when it is possible to start and finish the activity, which may be any time between the setting out activity and the second fix plumbing activity 23–26 (see Fig. 9.2.12). Now compare this network (CPM) with the bar chart for the same job in Fig. 9.1.4. The bar chart determines when the drainage is to start and finish to suit the contractor's resources, leaving little flexibility for what in effect is a non-critical activity.

Step 2: When a logical, satisfactory network has been compiled, as in Fig. 9.2.7., the event numbers are inserted to identify each activity. Also, the durations in days or weeks are extracted from the calculation sheets, along with the proposed gang and plant sizes, and these are included with the appropriate activities (see Fig. 9.2.8).

Step 3: It is now essential to calculate the earliest starting time for each activity by commencing at activity 1–2 (see Fig. 9.2.9).

(a) The earliest start is the beginning of day 1, which is 0. This is inserted in the left-hand box above the first event. Therefore the earliest finish of activity 1–2 is calculated by adding its duration and earliest starting time, i.e.
earliest finish = 0 + 3 = day 3 (inscribe this in the second event).

(b) The earliest start for activity 2–3 is the earliest finish of the preceding activity, which is day 3. Therefore, earliest finish = 3 + 6 = day 9.

(c) The earliest start for activity 3–4 is the earliest finish for the preceding activity, which is day 9. Therefore, earliest finish = 9 + 4 = day 13.

(d.) The earliest start for activity 3–5 is the earliest finish of the preceding activity, which is day 9. Therefore, earliest finish = 9 + 2 = day 11.

Special note: As the activity 5–6 cannot start until after both activities 3–4 and 3–5 have been completed, the

CALCULATION SHEET

CONTRACT: Wholesale Shop + Offices
CLIENT: J G Aberson

SHEET NO: 1
PREPARED BY: D.Helen
DATE: 1st/.........

OP. NO.	OPERATION	QUANTITIES	AVERAGE RATE/HR.	TOTAL HOURS	PLANT/GANG SIZE	DURATION IN HOURS	DURATION IN DAYS
1.0	PRELIMINARIES.						
	Access and temp road		—	—	JCB 7B 4 lab	1	2
	Hut accommodation and security fence	3 huts	—	—	4 lab	1	5
	Temporary services	—	—	—	Sub.Cont.	1	2
	Setting out	—	—	—	Gen/foreman 2 lab	1	3
	Surface strip	920 m²	10m²/hr	13	JCB 7B.	13	1½
2.0	SUBSTRUCTURE						
	Excavate foundation trenches.	120 m³	10m³/hr	12	JCB 7B	12	1½
	Drainage	86 m	1m/hr	86	JCB 7B 2d/lay 4 lab	43	5½
	Concrete to foundations	32 m³	—	—	4 lab	8	1
	Brickwork to D.P.C.	130 m²	0.5m²/hr	260	4 8/c 2 lab	65	8¼
	Hardcore + membrane to ground floors	318 m²	3m²/hr	106	4 lab. vib roller	26½	3¼
	Concrete to ground Floor.	318 m²	—	—	4 lab.	—	3
3.0	SUPERSTRUCTURE						
	Brickwork and blockwork	618 m²	0.5m²/hr	1236	8 B/L 4 kb 100k mixer	154½	19½
	External doors frame and windows						
	1st Floor joists to office block		1hr.each		3 c+J	—	4
	Padstones on piers	24 no	1hr.each	24	3 c+J	8	1
4.0	ROOF						
	Lattice girders	12 No	—	—	Mobile crane 3 c+J	16	2
	Decking Slabs (woodwool)	318 m²	5m²/hr	63½	Mobile crane 8 c+J	21¼	2¾
	Felt roof covering	318 m²	—	—	Sub.Cont.	—	3½
	Rain water goods and flashings.		—	—	Sub Cont.	—	2
5.0	SERVICES						
	Utility Services	—	—	—	Sub cont	—	1st Fix 2nd Fix 4
	Plumbing	—	—	—	Sub cont	—	5 + 3

Item						
Carpentry and joinery + W.C. partitions	—	—	1	3 c+J	1	6+10
Plastering (offices)	318m²	—	1	Sub. Cont.	1	8
Grano and other screeds	—	—	1	Sub. Cont.	1	6
Linoleum floor finishes (offices)	262m²	—	1	Sub. Cont	1	4
Decorations and glazing	—	—	1	Sub. cont.	1	14
7.0 EXTERNAL WORKS						
External brick boundary wall (foundations incl)	122 m	—	1	46/lay 2lab	1	15
Lean-mix conc base to drive and paths.	1200m²	10 m²/hr	120	4L.vib.roll	30	3¾
Tarmac	—	—	1	Sub. Cont	1	4
Landscape	—	—	1	Sub. Cont	1	10
Clean up and clear site	—	—	1	4 lab.	1	6

NOTE The quantities shown have been measured off from the drawing. True quantities for key or major operations could be extracted from the Bills of Quantities. The average rates per hour would be issued by the work study section, the estimator or own personal records retained from the feed back of previous jobs.

Fig. 9.1.3

CONTRACT: Wholesale Shop and Office
JOB NO: 242
DATE: 1st January 19......

BUILDING CONTRACTOR
CONTRAC

OP. NO.	OPERATION	GANG/ PLANT SIZE	DUR IN DAYS	MONTH FEB 19... MARCH
				WEEK COMM 18th / 25th / 3rd / 10th / 17th / 24th — WEEK NO 1 2 3 4 5 6
1.0	PRELIMINARIES			
1.1	ACCESS & TEMPORARY ROAD	JCB 7B 4 LAB		
1.2	HUTTING & SECURITY FENCE	4 LAB		
1.3	TEMPORARY SERVICES	S.C.		
1.4	SETTING OUT	G/F 2 LAB		
1.5	SURFACE STRIP	JCB 7B 1 LAB		
2.0	SUBSTRUCTURE			
2.1	EXCAVATE FOUNDATION TRENCH	JCB 7B 1 LAB		2.1
2.2	DRAINAGE	JCB 7B 2 d/lay		
2.3	CONCRETE FOUNDATIONS	4/LAB		
2.4	BRICKWORK TO D.P.C.	4 B/L 2 LAB		
2.5	HARDCORE, ETC TO GROUND FLOOR	4 LAB VIB ROLL		
2.6	CONCRETE TO GROUND FLOOR	4 LAB		
3.0	SUPERSTRUCTURE			
3.1	BRICK & BLOCK WORK (WINDOW FRAMES)	8 B/L 8 LAB		3.1
3.2	1st FLOOR JOISTS TO OFFICE BLOCK	100L MIXER 3 C+J		
3.3	PADSTONES ON PIERS	3 C+J		
4.0	ROOF			
4.1	LATTICE GIRDER (OFFICES & SHOP)	CRANE 3 C+J		
4.2	DECKING (WOODWOOL SLABS)	CRANE 3 C+J		
4.3	FELT ROOF COVERING	SC		
4.4	RAINWATER GOODS & FLASHINGS	SC		
5.0	SERVICES			
5.1	UTILITY SERVICES (WATER/ELEC)	SC		
5.2	PLUMBING	SC		
5.3	ELECTRICAL WORK & HEATERS	SC		
6.4	G.P.O.	SC		
6.0	FINISHES			
6.1	CARPENTRY & JOINERY (WC + PARTITIONS)	3 C+J		
6.2	PLASTERING (OFFICES)	SC		
6.3	GRANO + OTHER SCREEDS	SC		
6.4	LINOLEUM FLOOR FINISHES (OFFICE)	SC		
6.5	DECORATIONS + GLAZING	SC		
7.0	EXTERNAL WORKS			
7.1	EXTERNAL BRICK BOUND. WALL	4 B/LAY 2 LAB		
7.2	LEAN-MIX CONC. BASE (ROAD & DRIVE)	VIB ROLL 4 LAB		
7.3	TARMAC ROADS, ETC.	SC		
7.4	LANDSCAPE	SC		
7.5	CLEAN & CLEAR SITE.	4 LAB		

CLIENT: J. Aberson Ltd.,
Summerhill
Guildford.
Architect U. Tom, F.R.I.B.A.
Q.S. A. Alma, FRICS.
SYMBOLS
Delivery ●   Notify ➡
Information required ◆

LABOUR:	1	2	3	4	5	6		
GENERAL LAB	4	4	4	2	4	2	2	4 4 4 4
DRAIN LAYER					2	2		
BRICK LAYERS					4/2	4/2	4/2	4/2 ... 8/4 8/4 8/
C + J								
PLANT: JCB 7B								
100L MIXER								
SCAFFOLDING								
VIB ROLLER								

Fig. 9.1.4

ROW, GUILDFORD 2657

AMME

PREPARED BY: D. Helen

	MAY				JUNE				JULY		
14th	21st	28th	5th	12th	19th	26th	2nd	9th	16th	23rd	30th	7th	14th
9	10	11	12	13	14	15	16	17	18	19	20	21	22

STUDENT NOTE:–

This chart could also be used suitably as a Pre-Tender Programme.

CONTRACT COMPLETION

4·1

5·1 1ST FIX / 1ST FIX / 2ND FIX / 2ND FIX

6·1 1ST FIX / 2ND FIX

GLAZE

EX CN., CONC., BRICKWORK

7·1

| | 4 | 4 | | 4 | 4 | 4 | 4 | 4 |

| 8/4 | 8/4 | 8/4 | 8/4 | | | | | | | 4/2 | 4/2 | 4/2 | 4/2 |
| 3 | 3 | ---- | 3 | 3 | 3 | 3 | 3 | 3 | ---- | 3 | 3 | 3 | 3 | 3 |

NETWORKS – CPM

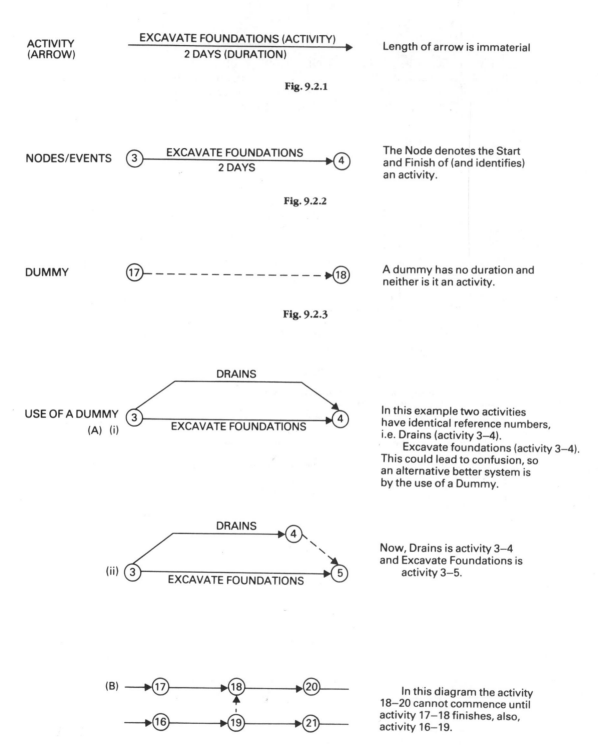

ACTIVITY
(ARROW)

EXCAVATE FOUNDATIONS (ACTIVITY)
2 DAYS (DURATION)

Length of arrow is immaterial

Fig. 9.2.1

NODES/EVENTS ③ EXCAVATE FOUNDATIONS 2 DAYS ④

The Node denotes the Start
and Finish of (and identifies)
an activity.

Fig. 9.2.2

DUMMY ⑰ ────────────► ⑱

A dummy has no duration and
neither is it an activity.

Fig. 9.2.3

USE OF A DUMMY
(A) (i)

DRAINS
③ EXCAVATE FOUNDATIONS ④

In this example two activities
have identical reference numbers,
i.e. Drains (activity 3–4).
Excavate foundations (activity 3–4).
This could lead to confusion, so
an alternative better system is
by the use of a Dummy.

(ii)

DRAINS ④
③ EXCAVATE FOUNDATIONS ⑤

Now, Drains is activity 3–4
and Excavate Foundations is
activity 3–5.

(B) ⑰ ⑱ ⑳
⑯ ⑲ ㉑

In this diagram the activity
18–20 cannot commence until
activity 17–18 finishes, also,
activity 16–19.

Fig. 9.2.4

NETWORKS – CPM

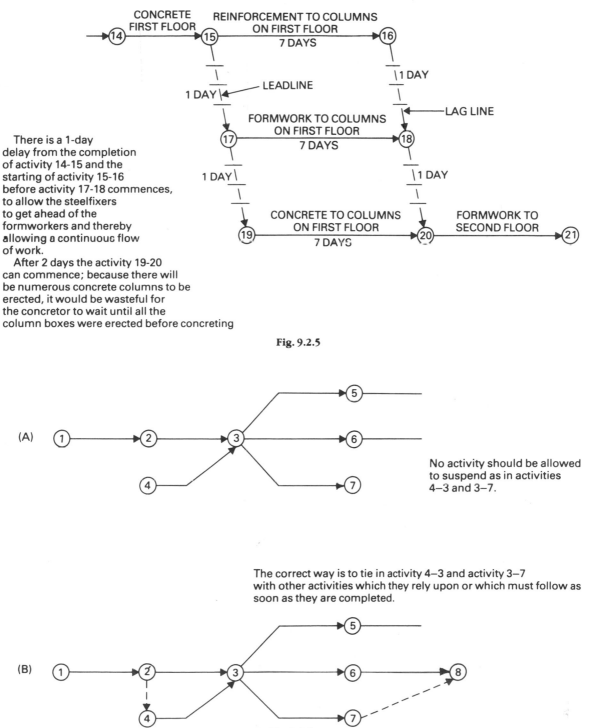

CONCRETE FIRST FLOOR

REINFORCEMENT TO COLUMNS ON FIRST FLOOR
7 DAYS

LEADLINE

1 DAY

1 DAY

LAG LINE

FORMWORK TO COLUMNS ON FIRST FLOOR
7 DAYS

1 DAY

1 DAY

CONCRETE TO COLUMNS ON FIRST FLOOR
7 DAYS

FORMWORK TO SECOND FLOOR

There is a 1-day delay from the completion of activity 14-15 and the starting of activity 15-16 before activity 17-18 commences, to allow the steelfixers to get ahead of the formworkers and thereby allowing a continuous flow of work.

After 2 days the activity 19-20 can commence; because there will be numerous concrete columns to be erected, it would be wasteful for the concretor to wait until all the column boxes were erected before concreting

Fig. 9.2.5

(A)

No activity should be allowed to suspend as in activities 4–3 and 3–7.

The correct way is to tie in activity 4–3 and activity 3–7 with other activities which they rely upon or which must follow as soon as they are completed.

(B)

Fig. 9.2.6

NETWORKS – C.P.M.

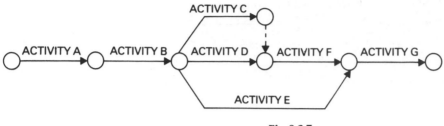

Fig. 9.2.7

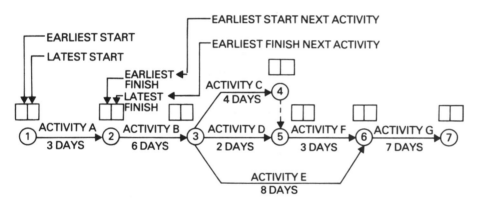

Fig. 9.2.8

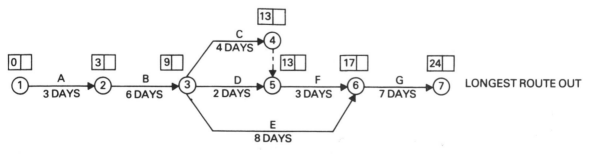

Fig. 9.2.9

earliest that activity 5–6 can commence is day 13. This is inserted in the event for the earliest finish of activity 3–5. Bearing this example in mind one always considers the longest route which has to be taken until the last activity is reached.

Step 4: Now, calculate the latest starting and latest finishing times by commencing at the last activity and working back to the first activity, remembering this time that the shortest route back should be taken, as follows:

(a) In Fig. 9.2.10 the latest finish for activity 6–7 will be the same as the earliest finish i.e. day 24. Therefore, the latest start for activity 6–7 will be the duration deducted from the latest finish, i.e.

day 24–7 = day 17 (insert this figure in the second square above event 6)

(b) The latest finish for activity 5–6 will be the latest start of activity 6–7. Therefore, the latest start of activity 5–6 is the latest finish less the duration, which is:

day 17–3 = day 14 (insert this figure in the second square above event 5).

(c) The latest finish for activity 3–5 will be the latest start of activity 5–6. Therefore, the latest start of activity 3–5 is the latest finish, less the duration, which is:

day 14–2 = day 12.

Special note: As activity 3–6 has the latest start of day 17–8 = day 9, this figure fixes the latest start above event 3, Bearing this example in mind one always considers the shortest route back until the first activity is reached.

Step 5: The critical path can be drawn on to the chart through the arrows whose starting and finishing times are the same (see Fig. 9.2.11).

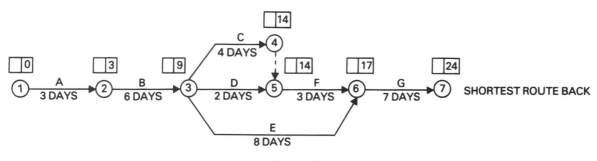

Fig. 9.2.10

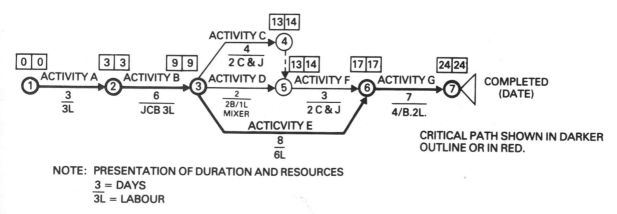

NOTE: PRESENTATION OF DURATION AND RESOURCES
3 = DAYS
3L = LABOUR

Fig. 9.2.11

The darker line on the network denotes the critical path. It is the route which shows the activities/operations on a job which are critical. Under no circumstances must there be any delays in completing these activities, otherwise the contract duration will be extended.

The critical activities would have to be monitored closely by the site supervisor, and if difficulties can be foreseen or delays take place the contract could be brought back onto target by introducing:

(*a*)   bigger gang sizes;
(*b*)   more gangs of workers;
(*c*)   more plant;
(*d*)   overtime working (starting earlier and finishing later);
(*e*)   weekend working;
(*f*)   better bonus scheme;
(*g*)   better techniques through work study (method study);
(*h*)   subcontractors.

Usually, as a check on the plotted critical path on the network, a 'network–event times and float sheet' is prepared to verify the route or routes of the critical path, and to record how much float (total and free) each activity has.

Network–event times and float sheet

Using a properly worked out network, as in Fig. 9.2.12, this network sheet (Fig. 9.2.13) is prepared using the information contained on the network.

1.   Firstly, the activities are listed in sequence in column 1 and the durations are recorded against each in column 2.
2.   The earliest starting times and latest finishing times are copied from the network and are recorded in columns 3 and 6 respectively.
3.   The earliest finishing times are obtained in column 5 by adding the duration to the earliest starting time for each activity.
4.   Now, the latest starting times are worked out by subtracting the durations from the latest finishing times, which are inserted in column 4.

5.   *Floats*
(*a*)   Total float. This is the total amount of time an activity can be delayed before it becomes a critical one. It is calculated by deducting the earliest starting time, plus the duration from the latest finishing time. See Fig. 9.2.13 and column 7. It is important to work from the actual drawn network times and not those on the event times and float sheet for total floats.
(*b*)   Free float. This is the amount of time the start or finish of an activity entering an event can be delayed without affecting the start of any activity leaving that event. It is calculated by deducting the earliest starting time of an activity plus its duration from the earliest finishing time (see column 8). It is important to work from the actual network times and not those on the event times and float sheet for free floats. (See Fig. 9.2.14 for illustration of the meaning of floats.)

The critical path is now determined by looking down the total float and free float columns 7 and 8, and if they are both equal to zero the 9th column is ticked to show the critical path. The network critical path is then finally checked by this method.

This network event times and float sheet can be retained for use by the site supervisor in conjunction with the network (CPM) chart.

Contract programme for network (bar chart)

Although a network (CPM) may be prepared as a control document on-site many programmers also illustrate the operations to be undertaken by producing a bar chart for the information determined on the network–event times and float sheet for the less enlightened supervisors and operatives on-site. A bar chart produced from network information tends to be a more accurate and logical chart than one prepared by the normal procedure.

Preparation
If one studies the bar chart in Fig. 9.2.15 the activities have been listed from the network–event times and float sheet. In addition, the durations and gang/plant sizes are recorded. Next, the time bars are drawn in from the earliest starting times and latest finishing times. The durations are then darkened in where the resources most suitably fit within the earliest starting times and the latest finishing times.

As an example, the earliest starting time and latest finishing times of the activity 5–23, drainage, is between day 10 and day 74. The duration for drainage can then be darkened in (6 days) at the most convenient time to do the operation, bearing in mind that the resources near the bottom of the chart are kept as constant and to the most optimum level to prevent high peaks and deep depressions. Drainage in this example is a non-critical activity. Most non-critical activities can be arranged in an attempt to level off the use of resources.

Where an activity is a critical one the dark bar representing the duration would be exactly the same length as between the earliest starting time and latest finishing time.

9.3   Programming techniques: line-of-balance methods

While bar chart programmes of varying types are used successfully on repetitive process contracts, such as: housing projects, or multi-storey buildings whose floors are almost identical in layout, it is argued that the line-of-balance charts are more suited in such situations.

The line-of-balance programming system was developed in this country by the National Building Agency, and assumes that each unit (house or floor level) is to be constructed repetitively, and, it is hoped, each trade's work will be done in a continuous cycle until the whole of a project is completed.

Many housing projects contain designs which are identical throughout the new estate, and where this is so the assessment or calculation of durations for the completion of each presents few problems. Where, however, estates are to be developed which include a series of differently designed units, the preparation of a line-of-balance chart requires an additional calculation process for the activities which are necessary to complete a unit. The average time for the completion of identical activities is calculated and used for each unit. To do this for each house type a list of operations/activities is prepared and compared, and a list of operations which are common to each are then recorded so that they can be incorporated into a line-of-balance chart as if each house was identical.

If the houses are of different designs and a list of operations have been prepared the durations for each operation would be calculated as follows:

Foundations (excavation and concreting):

Type A house, 20 in number at 10 hours each.
Type B house, 10 in number at 12 hours each.
Type C house, 8 in number at 8 hours each.

Therefore, total hours for foundation work for:

Type A houses = 20 × 10 = 200 hours,
Type B houses = 10 × 12 = 120 hours,
Type C houses = 8 × 8 = 64 hours.
No. of houses 38 Total 384 hours.

The average hours required for foundation work and which are used per house is, therefore, $\frac{384}{38} = 10.1$ hours approximately.

The duration for foundation work, irrespective of type of house is taken as 10.1 hours. This same technique is used for all the remaining operations in the construction of a house. In effect, it is the average duration for each operation which is used in the preparation of the line-of-balance chart or, indeed, for the ordinary bar chart in the event of one being used.

Details required for line-of-balance chart

1. List of the main operations/activities in the building of a house (or floor level when constructing a multi-storey building).
2. Durations for the completion for each operation.
3. Labour/gang sizes.
4. Plant to be used.
   All the above are extracted from the calculation sheet and method statements.
5. The time buffers between main activities to allow for minimum of delays.
6. Total time to build one unit and, therefore, the first handover stage.
7. Rate of handover for subsequent units after the first one is completed.

This all depends on how many gangs or machines (resources) are to be used, and also the date of completion if a contract exists between the builder and client. When the chart is drawn up due consideration must be given to the client's requirements where the work is undertaken through a contract, or the rate of completion of units in a speculative building situation depending on the demand for the units/houses. In either case the steeper the activity lines on the chart the quicker the work is expected to be completed. Resources, such as labour, plant, materials, etc., should be scheduled to arrive on-site at the correct times and in sequence to prevent delays, and storage space must be adequate to maintain correct levels of material to prevent shortages arising which could further contribute to delays.

Sequence of operations in the preparation of a line-of-balance chart

1. Extract the durations from a previously prepared calculation sheet for each activity, remembering that minor work such as timbering to trenches would be grouped with foundation works, so that if timbering is calculated to take 1 day and excavation, etc., are to take 3 days — foundation excavations would be 1 + 3 days — 4 days.

   One could now list, by network or connecting arrows, the operations in sequence as shown in Fig. 9.3.1 (as an example only the four main stages for constructing a house are shown).

2. Draft out similar activities for the remaining houses/units, the slopes of which depend on the handing-over stages, or the resources in the form of labour (number of gangs) and plant to be used or is required to be used (see Fig. 9.3.2).

3. Include buffers between either the major stages in the construction of a house, or between certain operations whose durations in the past have fluctuated considerably, such as, foundation work, when the work is undertaken during the winter. These buffers range in duration between one day and ten days depending on the contractor's previous experiences, particularly where recruitment of labour/subcontractors is concerned (see Fig. 9.3.3). If one looks at Fig. 9.3.3 the dotted line gives an indication of why buffers are used because it shows how the actual performance compares with the planned performance. The buffer absorbs deviations of previous activities.

Figure 9.3.4 indicates the rhythm in which the work is to be executed by either one gang completing an operation on, say, the first house and then proceeding without delay on to the second house, and so on, until the similar operation on all houses is finished (see roof operation in Fig. 9.3.4).

By studying Fig. 9.3.4 one can also observe that it would be essential to use two separate gangs on the sub-structure to provide continuity of work and to prevent delays occurring. The vertical dotted lines indicated at what times (day) the various operations on each unit can be started and where the gang proceed to

BUILDING CONTRACTORS
CONTRACT PROGRA

Contract: *Wholesale Shop and Offices*
Job no: *242*
Date: *1ST January 19.....*

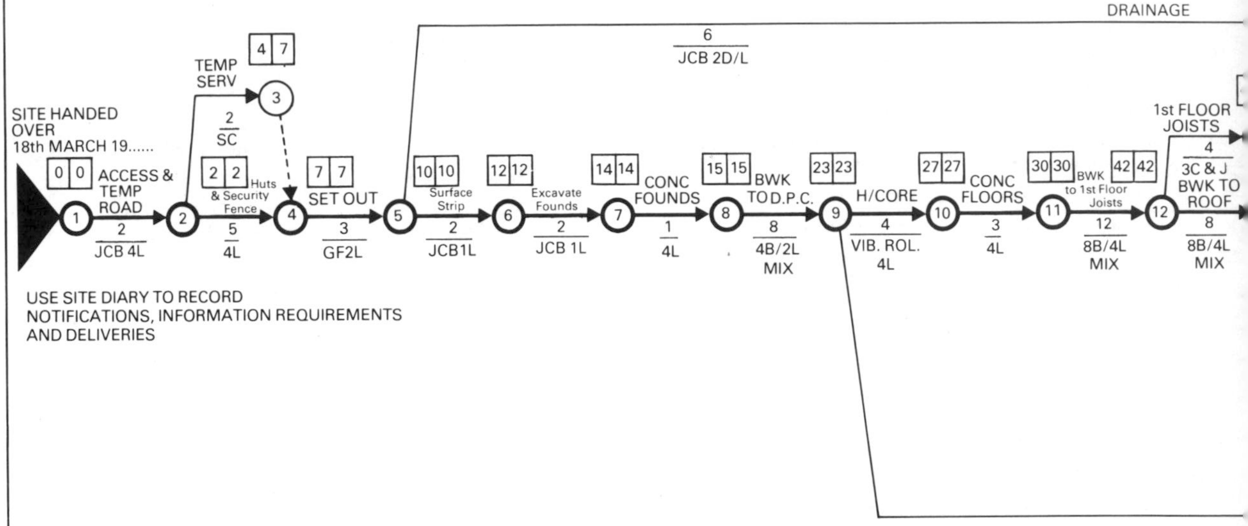

DRAINAGE

6
―――
JCB 2D/L

TEMP
SERV   4 7

2   (3)
―
SC

1st FLOOR
JOISTS

SITE HANDED
OVER
18th MARCH 19......

0 0   ACCESS &
         TEMP   2 2   Huts   7 7   10 10   12 12   14 14   15 15 BWK   23 23   27 27   30 30 BWK   42 42   4
(1)   ROAD   (2)   & Security (4) SET OUT (5) Surface (6) Excavate (7) CONC   (8) TO D.P.C. (9) H/CORE (10) CONC (11) to 1st Floor (12) 3C & J
                            Fence   Strip   Founds   FOUNDS   FLOORS   Joists   BWK TO
    2   5   3   2   2   1   8   4   3   12   8   ROOF
 ―――――   ―――   ―――――   ―――――   ―――――   ―――   ―――――   ―――――   ―――   ―――――   ―――――
 JCB 4L   4L   GF2L   JCB1L   JCB 1L   4L   4B/2L   VIB. ROL.   4L   8B/4L   8B/4L
                                                                                                 MIX   4L   MIX   MIX

USE SITE DIARY TO RECORD
NOTIFICATIONS, INFORMATION REQUIREMENTS
AND DELIVERIES

CLIENT: *J. Aberson Ltd*
         Summerhill
         Guildford

ARCHITECT: *U. Tom. F.R.I.BA.*

Q.S. *A. Alma. F.R.I.C.S.*

SYMBOLS

                    EARLIEST   14 17   LATEST
         (1)   START   START
                    TIME   TIME (IN DAYS)

                            (DURATION IN DAYS)

         (2)   2   (PLANT & LABOUR)
                   ―――――   (REQUIREMENTS)
                   JCB 4L

Fig. 9.2.12

ROW, GUILDFORD. 2657

NETWORK – C.P.M.

Prepared by: *D Helen*

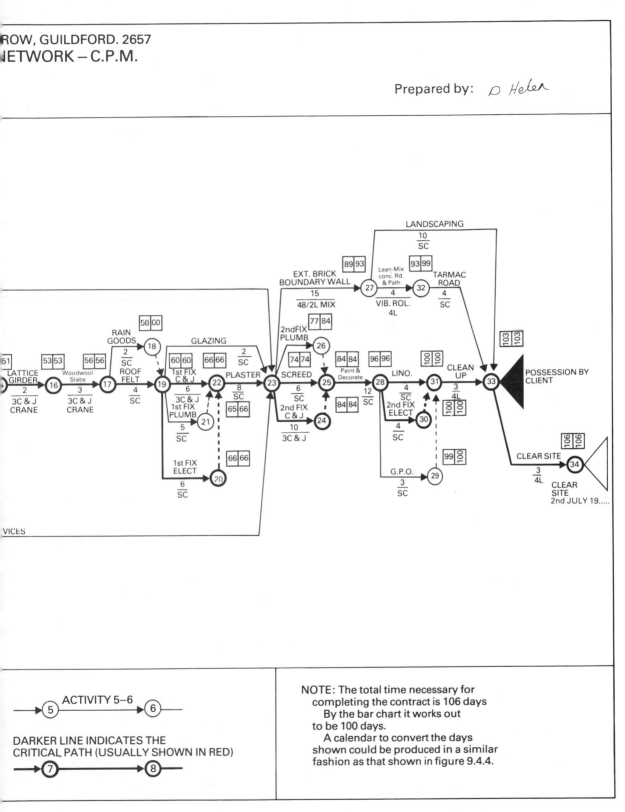

ACTIVITY 5–6

⑤ ———————→ ⑥

DARKER LINE INDICATES THE
CRITICAL PATH (USUALLY SHOWN IN RED)

⑦ ———————→ ⑧

NOTE: The total time necessary for
completing the contract is 106 days
By the bar chart it works out
to be 100 days.
A calendar to convert the days
shown could be produced in a similar
fashion as that shown in figure 9.4.4.

NETWORK-EVENT TIMES AND FLOAT SHEET

Contract: *Wholesale Shop and Offices*

Contract no: 242

Date: *1st January 19.......*

Prepared by: *D. Clare*

Activity ①	Duration (in days) ②	Start (day no.) Earliest ③	Start (day no.) Latest ④	Finish (day no.) Earliest ⑤	Finish (day no.) Latest ⑥	Float (in days) Total ⑦	Float (in days) Free ⑧	Critical activity ⑨
1-2	2	0	0	2	2	0	0	✓
2-3	2	2	5	4	7	3	0	
2-4	5	2	2	7	7	0	0	✓
3-4	—	4	7	4	7	3	3	
4-5	3	7	7	10	10	0	0	✓
5-6	2	10	10	12	12	0	0	✓
5-23	6	10	68	16	74	58	58	
6-7	2	12	12	14	14	0	0	✓
7-8	1	14	14	15	15	0	0	✓
8-9	8	15	15	23	23	0	0	✓
9-10	4	23	23	27	27	0	0	✓
9-23	4	23	70	27	74	47	47	
10-11	3	27	27	30	30	0	0	✓
11-12	12	30	30	42	42	0	0	✓
12-13	4	42	46	46	50	4	0	
12-14	8	42	42	50	50	0	0	✓
13-14	—	46	50	46	50	4	4	

17-18	2	56	58	58	60	2	0	
17-19	4	56	56	60	60	0	0	✓
18-19	1	58	60	58	60	2	2	✓
19-20	6	60	60	66	66	0	0	
19-21	5	60	61	65	66	1	0	✓
19-22	6	60	60	66	66	0	0	✓
19-23	2	60	72	62	74	12	12	
20-22	–	66	66	66	66	0	0	✓
21-22	–	65	66	65	66	1	1	✓
22-23	8	66	66	74	74	0	0	
23-24	10	74	74	84	84	0	0	
23-25	6	74	78	80	84	4	4	
23-26	3	74	81	77	84	7	0	✓
23-27	15	74	78	89	93	4	0	✓
24-25	–	84	84	84	84	0	0	
26-25	–	77	84	77	84	7	7	
25-28	12	84	84	96	96	0	0	✓✓
27-32	4	89	95	93	99	6	6	
27-33	10	89	93	99	103	4	4	
28-29	3	96	97	99	100	1	0	✓✓
28-30	4	96	96	100	100	0	0	
28-31	4	96	96	100	100	0	0	
29-31	–	99	100	99	100	1	1	
30-31	–	100	100	100	100	0	0	
31-33	3	100	100	103	103	0	0	
32-33	4	93	99	97	103	6	6	✓
33-34	3	103	103	106	106	0	0	

Fig. 9.2.13

INTERPRETATION OF FLOATS AND EVENT TIMES

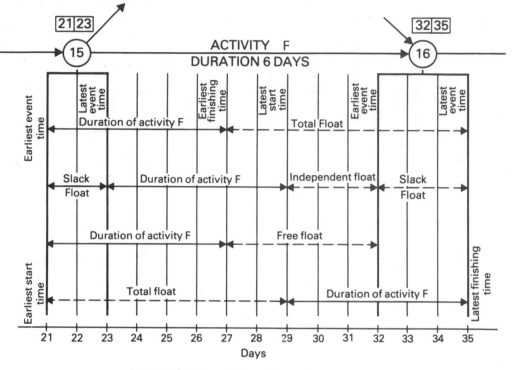

Fig. 9.2.14 Interpretation of floats and event times

after it has completed the operation. When the sub-structure for house number one is completed by the tenth day, the gang proceeds to house number three. The second gang starts sub-structure for house number two on day five, and when it is completed by day 15 would proceed to house number four and so on.

The gang that starts the superstructure on house number one would proceed to house number five when one is completed. Four gangs, however, would be required for superstructure work.

Five gangs would be necessary for services and finishes unless the slope of the line could be levelled off slightly, which means a wider buffer at house number eight, but it would reduce the gang numbers and, unfortunately in this case, would extend the contract duration (see Fig. 9.3.5) from day 107–108 until day 115. So, in reducing the gang numbers and increasing the contract duration cost comparisons should be made before a final decision to accept the modified programme could be taken.

When arranging the lines or bars on a line-of-balance chart care must be exercised to obtain the best slopes to allow correct handing-over stages of units, and that the buffers (whether stage or activity buffers) are sufficient to allow for:

1. Irregular levels of output of gangs or plant.
2. Delays caused by inclement weather.
3. Delays in delivery of materials, etc.
4. Rest breaks of the operatives.

5. Holiday periods.
6. Illnesses or absenteeism.
7. Disputes, etc.

Natural rhythm should be the aim at all times, and where possible the same gang should be given the responsibility to carry out similar operations to each unit, and not be changed from one activity to another different type. By continuing on one type of activity they can increase their bonus earnings by greater output as they learn a process thoroughly. A good bonus structure leading to increased earnings creates stability of labour on-site, thereby reducing labour wastage and time spent on recruitment.

Unbalanced activity sequences should be avoided where possible, although this is sometimes difficult to achieve in practice. An unbalanced activity is one which when completed by a gang falls at a time when there is no immediate subsequent work, or the start of the following similar activity on the next unit is passed, and if then commenced later than planned a consequence will be to eventually run into the buffer (see Fig. 9.3.3) or extension to the contract period will result.

Figure 9.3.6 shows a typical line-of-balance chart which has been prepared for a housing contract of 20 units. It may be of interest to note how the resources (labour and plant) can be illustrated, and that the lines are shaded in as the work is completed.

9.4 Programming techniques: precedence network diagrams

The precedence diagram is an extension and, perhaps, is an improvement on the network–CPM and is most suited to complex building and engineering work. A critical path can still be highlighted on such a chart and, indeed, earliest and latest start and finish times are calculated in a similar manner to those of the network–CPM. Also, total floats and free floats can be determined as a check on how long activities/operations can be delayed before they become critical to a project, because critical operations, if delayed, may extend the finishing time and thereby incur additional expenditure for the contractors. The client would also be dissatisfied.

Of all the programming charts the precedence diagram is the one which contains the most control information, and a site supervisor needs to be able to interpret the many meanings, symbols and details contained thereon.

As a first step in the learning process regarding precedence diagrams it is essential to first study and achieve an ability to understand the network–CPM. The surest way to be proficient at understanding networks is by preparing examples. Once networks–CPM are understood the basic procedure for the preparation of precedence diagrams will be easier to grasp, and, therefore, it is strongly recommended that by way of learning the network–CPM is studied first.

The basic information as required by all other programming methods is needed. This information can be extracted from the calculation sheet. The actual draughting out of the network is by the use of logic: this is to say, one considers the earliest each operation could start and the latest it could finish if the correct level of resources were available. One never considers when it is most convenient for the firm to do the work as one does in bar chart programming.

In Fig. 9.4.1 one can study the most common symbols used in precedence diagrams and learn their meanings. Activities are represented by boxes, as compared to networks–CPM where arrows are used instead. The box is divided into sections so that the appropriate data can be inserted, i.e. activity, activity member, duration and earliest and latest start and finish times.

The arrows generally have no time value but serve to indicate the sequence and relationships of operations. If, however, a programming officer wishes to show a delay between the completion of one activity to the start of another – say, after completion of the glazing the putty should be allowed to harden a little before painting commences (known as lag start) – then a duration may be shown in this situation (see Figs 9.4.1 and 9.4.2).

When draughting out the network of the precedence diagram the arrows can be applied in a variety of ways to indicate clearly when activities should start and finish. (See Fig. 9.4.2 for typical activity relationships.)

Earliest and latest start and finish time are next calculated after the network is finalised and the activity, activity number and duration has been inserted in the appropriate box representing each operation.

The earliest start and finish times are calculated first. This is done by starting at the first activity at the beginning of day 1, which is zero, and the earliest finish is day 0 plus the duration of six days, which equals the end of day six. The second activity's earliest start is the same day as the earliest finish of the previous activity – day six. Therefore, the earliest finish of the second activity is day six plus the duration of three days, which equals the end of day nine, and so on (see Fig. 9.4.3(A)).

The latest start and finish of each activity is calculated as shown in Fig. 9.4.3(B). The calculations commence from the last activity. The latest finish of activity G is the same as the earliest finish, which is the end of day 36. The latest start will be day 36 less the duration of three days, which equals day 33. The immediate preceding activities F and E have their latest start and finish times calculated by inserting activity G's latest start of 33 in the last box. Then by deducting their durations of day seven and nine respectively, the latest starts will be day 26 and day 24 respectively, and so on until activity A is reached.

The critical path is usually shown by encircling the activity rectangles in red and drawing the arrows in similarly (see Fig. 9.4.3(B)). This route follows the activities whose earliest and latest start and finish times are the same.

Figure 9.4.4 (on pages 198–9) shows a typical precedence network diagram for the wholesale shop and offices project shown in the drawings in Fig. 9.1.1. As an additional aid for the site supervisor the dates for the earliest and latest start and finish times can be seen at the lower part of the diagram. Remember that where a working day shows, say, day 25, this means the end of day 25 and the beginning of day 26.

For total floats and free floats calculations see method adopted for networks–CPM in section 9.2.

Summary

Precedence network diagrams are prepared using the following sequence.

1. Prepare a list of the main activities using the project drawings, bills of quantities and specifications.
2. A method statement could be prepared to give some indication of the sequence of construction of the projects. The labour and plant may also be determined at this stage (see Fig. 9.1.2).
3. Compile details for use on a calculation sheet, quantities being extracted from the scale drawings or bills of quantities. The durations for each activity/ operation are then determined (see Fig. 9.1.3).
4. Using logic, draught out the network.
5. Insert all durations and other particulars on to the network.
6. Calculate earliest start and finish times.
7. Calculate latest start and finish times.
8. Determine the critical paths.
9. Calculate total floats and free floats.
10. Prepare a calendar of working days.
11. Show holiday periods and other relevant information.

190

CONTRACT: Wholesale Shop + Offices
JOB NO: 242
DATE: 1st January 19........

BUILDING CONTRACTO
CONTRACT PROGRAMME FRO

EVENT NO.	Activity	GANG/ PLANT SIZE	DUR IN DAYS
1-2	ACCESS + TEMP ROAD	JCB 4L	2
2-4	HUTS + SECURITY FENCE	4L	5
2-3	TEMP. SERVICES	SC	2
4-5	SETTING OUT	GF 2L	3
5-6	SURFACE STRIP	JCB 1L	2
5-23	DRAINAGE	JCB 20L	6
6-7	EXCAV. FOUNDS.	JCB 1L	2
7-8	CONC. FOUNDS.	4L	1
8-9	B.W.K. TO D.P.C.	4B/2L	8
9-10	HARDCORE TO FLOORS	4L	4
9-23	UTILITY SERVICES	SC	4
10-11	CONC. FLOOR	4L	3
11-12	B.W.K. TO 1ST FLOOR	8B/4L	12
12-13	JOISTS TO 1ST FLOOR	3 C+J	4
12-14	B.W.K. TO ROOF	8B/4L	8
14-15	PADSTONES ON PIERS	3 C+J	1
15-16	LATTICE GIRDERS	CRANE 3 C+J	2
16-17	WOODWOOL DECKING	3 C+J	3
17-18	RAINWATER GOODS	SC	2
17-19	ROOF FELT	SC	4
19-20	ELECTRIC 1ST FIX	SC	6
19-21	PLUMBING 1ST FIX	SC	5
19-22	CARPENTRY 1ST FIX	3 C+J	6
19-23	GLAZING	SC	2
22-23	PLASTER	SC	8
23-24	CARPENTER 2ND FIX	3 C+J	10
23-25	SCREED.	SC	6
23-26	PLUMBING 2ND FIX	SC	3
23-27	BRICK BOUNDARY WALL	m.h. 4B/2L	15
25-28	PAINT + DECORATE	SC	12
27-32	CONC. ROAD + PATH	VIB ROLL 4L	4
27-33	LANDSCAPING	SC	10
28-29	G.P.O.	SC	3
28-30	ELECTRIC 2ND FIX	SC	4
28-31	LINO	SC	4
31-33	CLEAN UP	4L	3
32-33	TARMAC ROAD.	SC	4
33-34	CLEAR SITE	4L	3

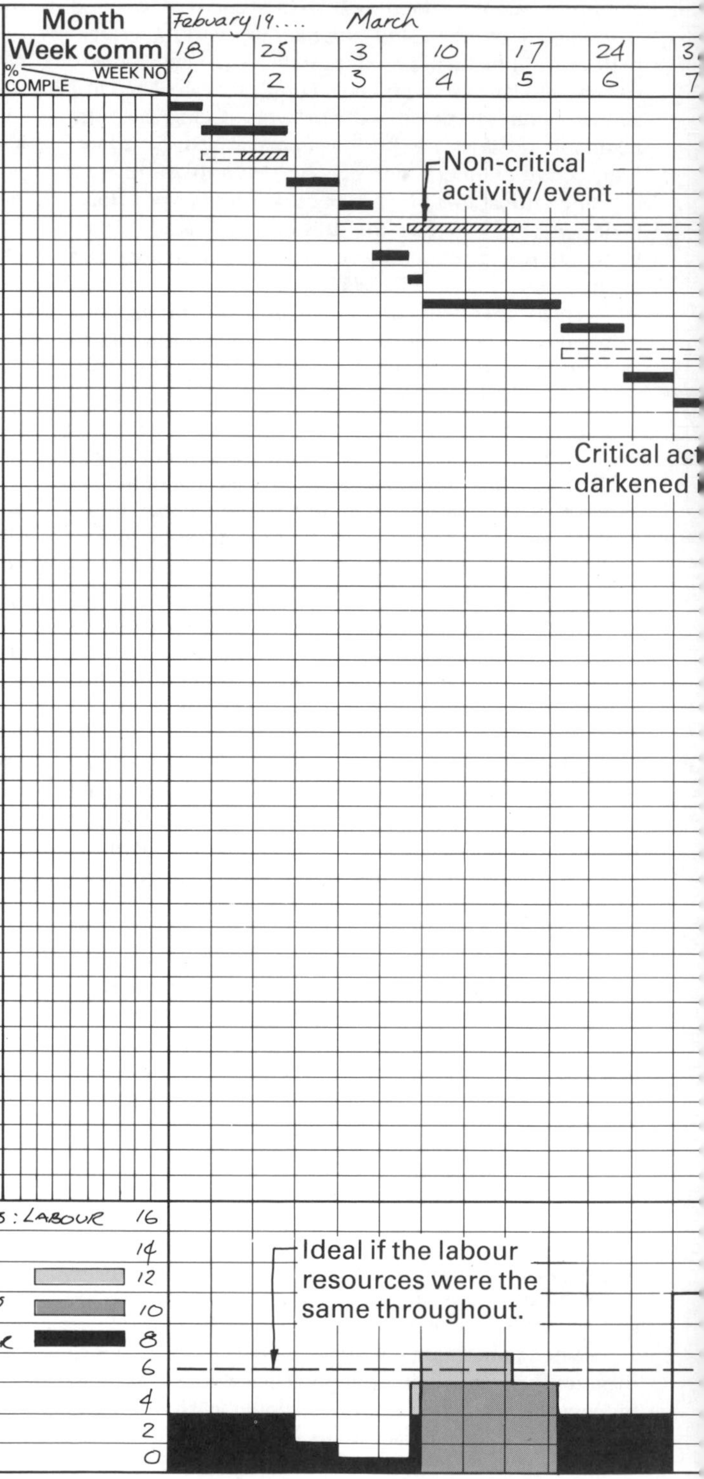

Month — Febuary 19.... March

Week comm	18	25	3	10	17	24	3.
WEEK NO	1	2	3	4	5	6	7

% COMPLE

─ Non-critical activity/event ─

Critical act darkened

CLIENT: J. Aberson Ltd.
Summerhill
Guildford.
ARCHITECT: U. Tom. FRIBA
Q.S. A. Alma, FRICS

SYMBOLS:
- • Delivery
- ◆ Information
- ➡ Notification

RESOURCES: LABOUR	16
	14
C + J	12
BRICKLAYERS + LAB	10
GEN LABOUR	8
	6
	4
	2
	0

┌ Ideal if the labour resources were the same throughout.

Fig. 9.2.15

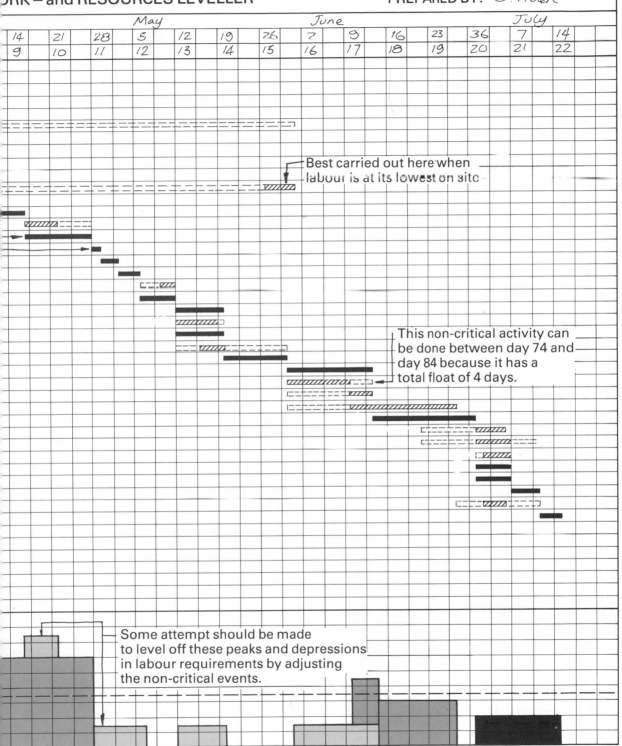

LINE-OF-BALANCE

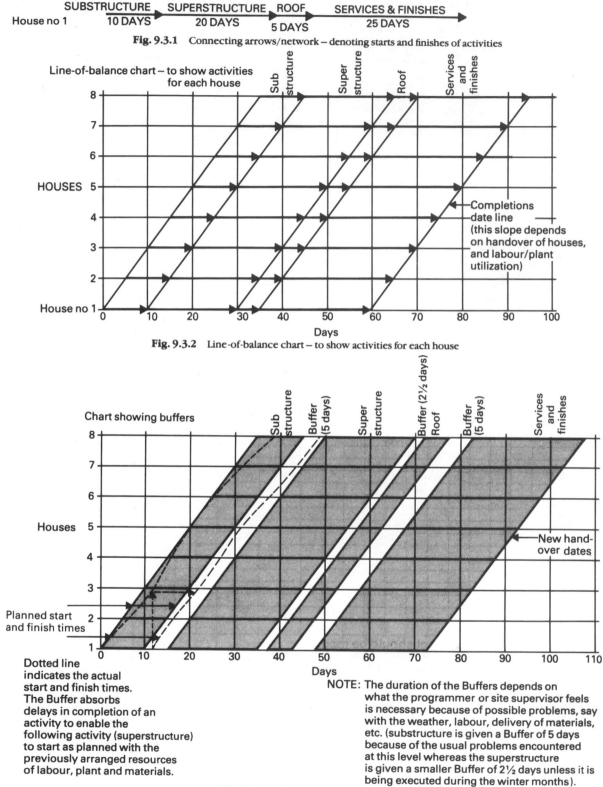

House no 1 SUBSTRUCTURE 10 DAYS → SUPERSTRUCTURE 20 DAYS → ROOF 5 DAYS → SERVICES & FINISHES 25 DAYS →

Fig. 9.3.1 Connecting arrows/network – denoting starts and finishes of activities

Line-of-balance chart – to show activities for each house

Completions date line (this slope depends on handover of houses, and labour/plant utilization)

Fig. 9.3.2 Line-of-balance chart – to show activities for each house

Chart showing buffers

New hand-over dates

Planned start and finish times

Dotted line indicates the actual start and finish times. The Buffer absorbs delays in completion of an activity to enable the following activity (superstructure) to start as planned with the previously arranged resources of labour, plant and materials.

NOTE: The duration of the Buffers depends on what the programmer or site supervisor feels is necessary because of possible problems, say with the weather, labour, delivery of materials, etc. (substructure is given a Buffer of 5 days because of the usual problems encountered at this level whereas the superstructure is given a smaller Buffer of 2½ days unless it is being executed during the winter months).

Fig. 9.3.3 Chart showing buffers

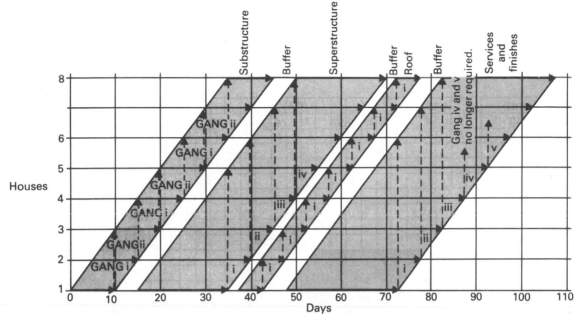

NOTE: In Stage 1 (Substructure) 2 gangs are required – assuming the gangs do all the different types of work under this heading.

In Stage 2 (Superstructure) 4 gangs are required (referred to as natural rhythm × 4)

In Stage 3 (Roof) one gang only is required to give a purely natural rhythm.

In Stage 4 (Services & Finishes) 5 gangs are required to give continuity of work and to prevent delays.

Fig. 9.3.4 Chart to show rhythm (natural and multiples of natural rhythm)

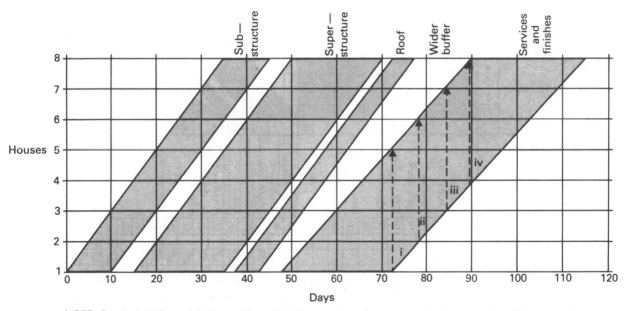

NOTE: By sloping the activity/stage line a little the number of gangs required to complete all the units is reduced from 5 to 4, but the last house will be handed over on day 115 compared to the previous attempt in figure 9.3.4 of day 107–108. As the first house will finish on day 72 and the last house finishes on day 115 a house will be ready for occupation every 5½ days approximately. (115 − 72 = 43 ∴ 43 days/8 houses = 5.38 days/house). It is usual at commencement to determine completion periods to assess cash-flow.

Fig. 9.3.5 Chart to show widening buffer (which reduces gang numbers but increases contract duration)

BUILDIN⬤

PROGRAMME OF

Contract: *Private Detached Houses, Gleeblands.*
Contract no: *× 257*

Year & month	19....	APRIL					MAY				JUNE				JULY	
Week ending		2	9	16	23	30	7	14	21	28	4	11	18	25	2	9	16
Week no.		1	2	3	4	5	6	7	8	9	10	11	12	13	14	15	16

Number of dwellings houses

Curser to indicate date of observance

Shaded to show work completed.

Commence work on-site

| WEEK NO | | 1 | 2 | 3 | 4 | 5 | 6 | 7 | 8 | 9 | 10 | 11 | 12 | 13 | 14 | 15 | 16 |

OPERATIONS: Prelims / Access huts / Fences / Site set out / Main drains / Utility services / Temp roads / Excavate Foundations / Brickwork to D.P.C. / Drainage / Hardcore + conc. floors / Brickwork to first floor and frames / First floor joist / Brickwork to roof / Roof structure / Tiling

RESOURCES

LABOUR	LAB	4	4	4	4	4	8	8	8	8	8	8	8	4	4	10	10	15	15	15	15	15	15	15	15	15	15	15	15	10	10	5	5
	GJ																									2	2	2	2	2	2	4	4
	B.L.												3/1	3/1	3/1	3/1	12/4	12/4	12/4	12/4	12/4	12/4	12/4	12/4	12/4	12/4	12/4	12/4	12/4	12/4	12/4	21/7	21/7
	SC																																
PLANT	JCB1																																
	JCB2																																
	MIX1																																
	MIX2																																
	SCAF																																

Fig. 9.3.6

ORS LTD.

NE OF BALANCE

CLIENT: R. Developer.
ARCH/DESIGNER: R. I. Barker.
DATE: 23rd Feb 19......
PREPARED BY: S. Smith.

AUG			SEPT				OCT					NOV						
13	20	27	3	10	17	24	1	8	15	22	29	5	12	19				
20	21	22	23	24	25	26	27	28	29	30	31	32	33	34	35	36	37	38

Student note:– Holiday periods are not shown but would be necessary to give a true picture of the contract duration

— Buffers

← Slope allows for 2 houses completed each week (20 handed over in 10 weeks)

Completion date.

| | 20 | 21 | 22 | 23 | 24 | 25 | 26 | 27 | 28 | 29 | 30 | 31 | 32 | 33 | 34 | 35 | 36 | 37 | 38 |

Screed | All second fixings cct/elect Plumb. etc | Painting + commence ext work | Floor finishes | Clear out snags | Road surface finishes and clear site

				5	5	5	5	5	5	5	5	5	5	5	5	5	8	8	8	8	8	8	8	8	3	3	3	3
6	6	9	9	9	9	7	7	7	7	7	5	5	5	3	3	3	3	3	3	3	3	3	3	3	3
2½/7	2½/7	2½/7	2½/7	2½/7	12/4	12/4	12/4	12/4											

(A) ACTIVITIES

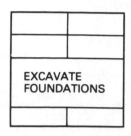

This shaped box indicates the many operations or activites to be undertaken on a project.

(C) DURATION

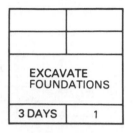

This is the period of time which has been assessed/calculated for undertaking the activity.

(B) ACTIVITY NUMBER

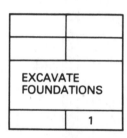

This is used to identify and differentiate one activity with another.

(D) EARLIEST AND LATEST START AND FINISH TIMES BOXES

The earliest and latest times are inserted in the appropriate boxes after calculation are made as shown in Fig. 9.4.3 (A).

(E) ARROWS OR LINES

1.

There is no time value given to an arrow in this type of network diagram except where there is to be a lag (delay) start or finish of an activity see Fig. 9.4.2 – F and G

2.

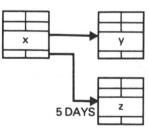

Arrows or lines also show relationships of one activity with other activities. In example activity **x** starts before both activities y and z.
Activity z cannot start until 5 days after activity x has finished. Example of this lag start in construction are:
1. To allow for the curing of concrete.
2. Mortar setting in brickwork before roof construction.
3. Screed curing and drying before laying floor tiles.
4. Putty setting on windows before painting.

Fig. 9.4.1 Precedence networks – symbols

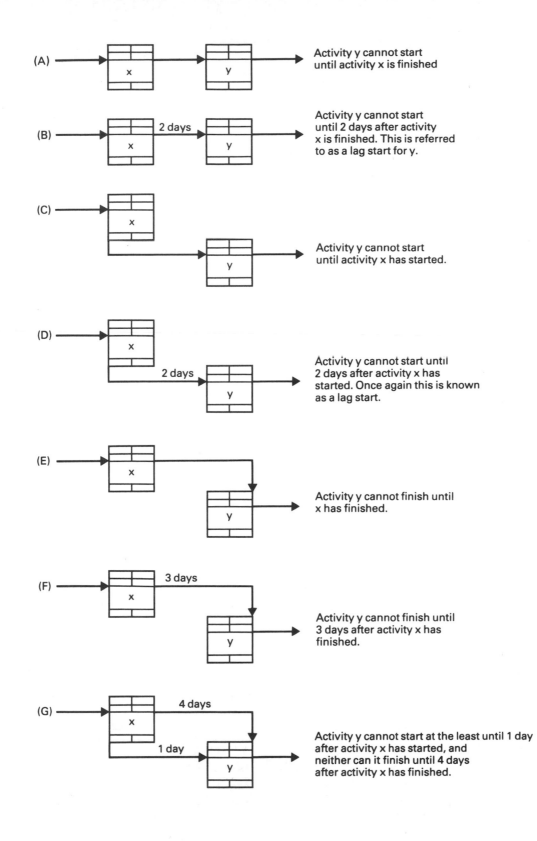

Fig. 9.4.2   Precedence network – activity relationships

198

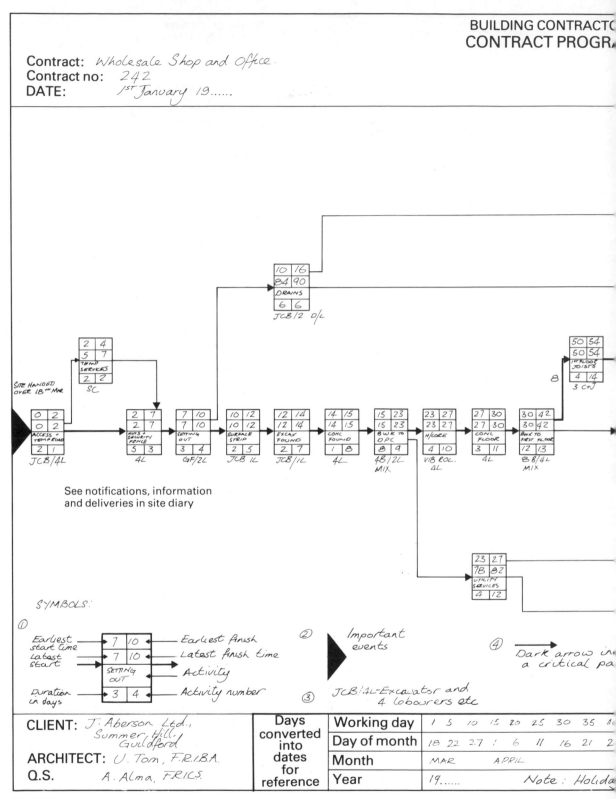

Contract: *Wholesale Shop and Office.*
Contract no: *242*
DATE: *1ST January 19......*

See notifications, information
and deliveries in site diary

SYMBOLS:

① Earliest start time → Earliest finish
Latest start → Latest finish time
→ Activity
Duration in days → Activity number

② ▶ Important events

③ ▶ JCB/4L-Excavator and 4 labourers etc

④ → Dark arrow in a critical pa

CLIENT: *J. Aberson Ltd., Summer Hill, Guildford*
ARCHITECT: *U. Tom, F.R.I.B.A.*
Q.S. *A. Alma, F.R.I.C.S.*

Days converted into dates for reference	Working day	1	5	10	15	20	25	30	35
	Day of month	18	22	27	1	6	11	16	21
	Month	MAR		APRIL					
	Year	19......			Note: Holida				

Fig. 9.4.4

RROW, GUILDFORD. 2657
ECEDENCE DIAGRAM

Prepared by: *D. Clare*

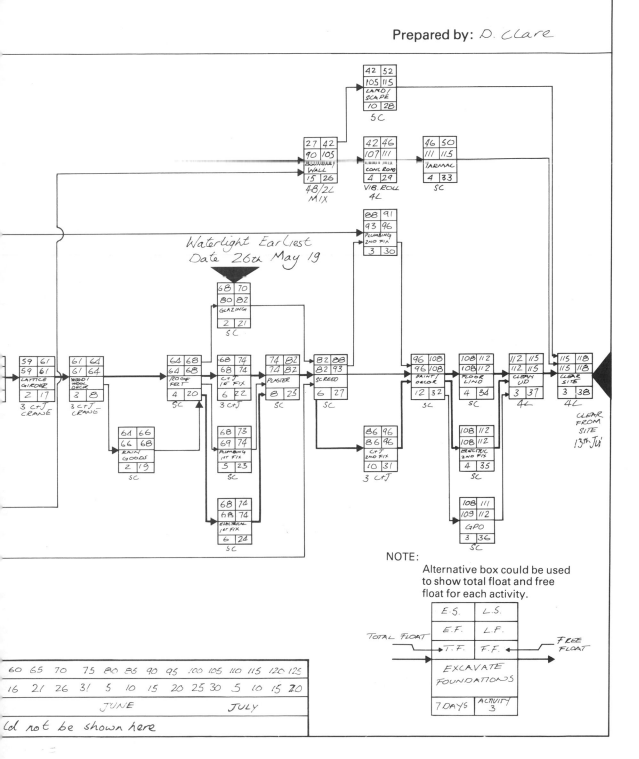

NOTE:
Alternative box could be used
to show total float and free
float for each activity.

E.S.	L.S.
E.F.	L.F.
T.F.	F.F.
EXCAVATE FOUNDATIONS	
7 DAYS	ACTIVITY 3

TOTAL FLOAT

FREE FLOAT

60	65	70	75	80	85	90	95	100	105	110	115	120	125
16	21	26	31	5	10	15	20	25	30	5	10	15	20

JUNE   JULY

ld not be shown here

(A) EARLIEST START AND FINISH TIMES

Calculations start from the
first activity remembering to take
the longest route out to the last activity
and adding the durations together

Earliest finish is the same as
earliest start of next activity.

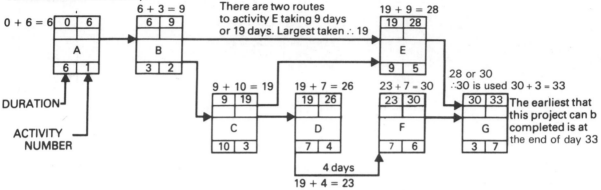

(B) LATEST START AND FINISH TIMES

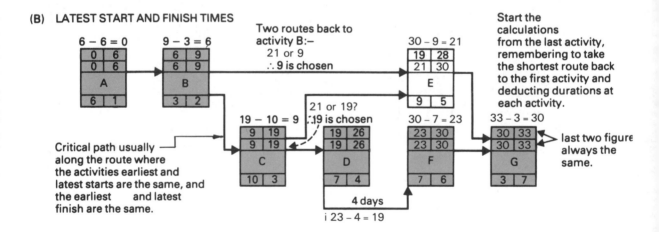

Fig. 9.4.3 Precedence diagram – earliest and latest start and finish times

Chapter 10
Supervisor's role relating to site layout and health and safety requirements

10.1 Site layout

As in the layout of workshops or mass production lines in factories, critical analysis of proposed site layouts should be undertaken before attempting to establish the positions of hutting, static machinery, plant, workshops areas, compounds and various bulk or key materials, thereby utilising the space available economically. It also assists in minimising plant, labour and materials movement during construction work, and ensures that work on building and other construction positions is not impeded by the thoughtless storage of materials, etc., on these locations.

The most economically laid out site is the one which enables the optimum of work to be undertaken without unnecessary delays occurring due to poor distribution of labour and materials. The contractor's planning officer attempts to find the best general site layout in consultation with the estimator, contracts manager and plant manager, if time allows before work actually commences on-site. Some consideration may be given to this problem even at the pre-tender stage, i.e. before a contract has been tendered for.

Neat, methodical and efficient site layouts convey to all concerned, including the general public, the sound standards by which an efficiently organised contractor operates. In addition to the concern shown with efficiency the following should be foremost in the minds of the contractor's planning officer and site supervisor.

1. Labour relations:
    (a) safety at the work places and routes to and from them;
    (b) health of the employees;
    (c) welfare of the employees;
    (d) remuneration of the employees.
2. Subcontractor's interests.
3. Public relations, to reduce or prevent a nuisance from dust, smoke, fumes and noise, and to maintain the safety of the general public.
4. Statutory conditions relating to, in particular, the highways (roads, pavements, verges, footpaths and bridleways).

A site layout plan should be furnished by the contractor's head office to the site supervisor. Where a contractor fails in this the supervisor would have to shoulder the responsibility and prepare one, bearing in mind the points previously mentioned.

The following requires some consideration when preparing site layout plans (see Fig. 10.1.1).

Existing services

The positions of each public utility services must be established as soon as possible and should be shown on the site layout plant or be marked out immediately on-site with pegs or other suitable means to prevent damages occurring during construction work (see the example in Fig. 7.3.3). The utility services are:

1. Existing sewers/drains (foul, surface or land) and any manholes/inspection chambers, rodding eyes, gulleys, cesspits, septic tanks, catchpits or soak-a-ways.
2. Water mains or distribution pipes, including any control valves or hydrant connections.
3. Electricity cables above or below ground.
4. Gas mains.
5. Telecommunications cables above or below ground.

The depths and exact locations are most important details which would be obtainable from the separate utility services authorities. For example: British Telecom request a copy of the site plan to be sent to them on application and they will show the approximate position of their underground cables by a hatched area instead of lines for cables. They also stamp the drawing with an Important Notice, before returning to the applicant, stating that the cable positions are within 3 metres and that precise positions can be located on-site by one of their engineers with special equipment provided that two clear working days notice are given.

Hoardings, etc

Under this heading one considers the following, always taking account of accessibility, cost and whether or not permission will be granted for their use by the local authority:
1. Hoardings – secures site, prevents noise and dust nuisance to adjoining properties.
2. Fences.
3. Gantries.
4. Shorings to adjacent buildings.

SITE LAYOUT PLAN

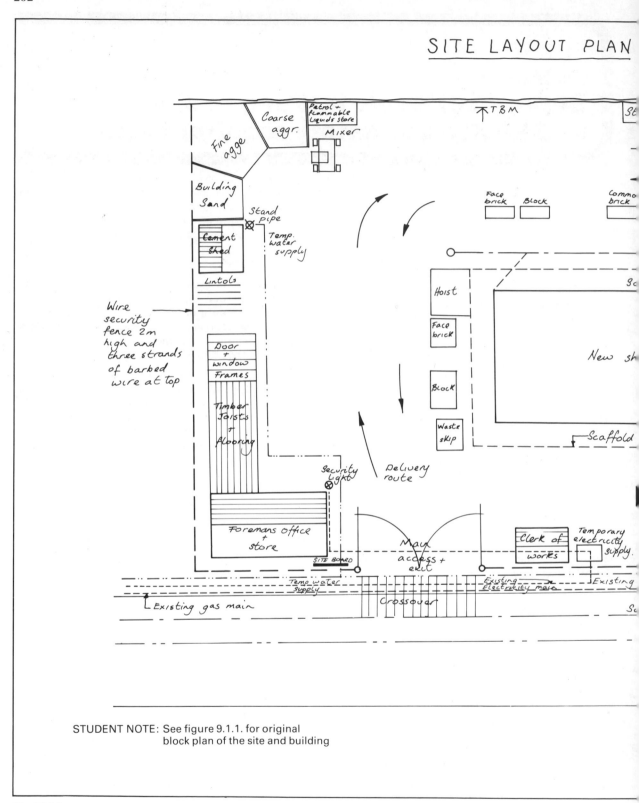

Coarse aggr.

Fine aggr.

Petrol + flammable Liquids store

Mixer

↑ TBM

Building Sand

Stand pipe

Temp. water supply

Cement Shed

Lintols

Wire security fence 2m high and three strands of barbed wire at top

Door + window Frames

Timber Joists + Flooring

Foremans office + store

SITE BOARD

Security Light

Delivery route

Face brick

Block

Common brick

Hoist

Face brick

Block

Waste skip

New sh

Scaffold

Clerk of works

Temporary electricity supply

Main access + exit

Temp water supply

Existing Electricity main

Existing

Existing gas main

Crossover

STUDENT NOTE: See figure 9.1.1. for original block plan of the site and building

Fig. 10.1.1

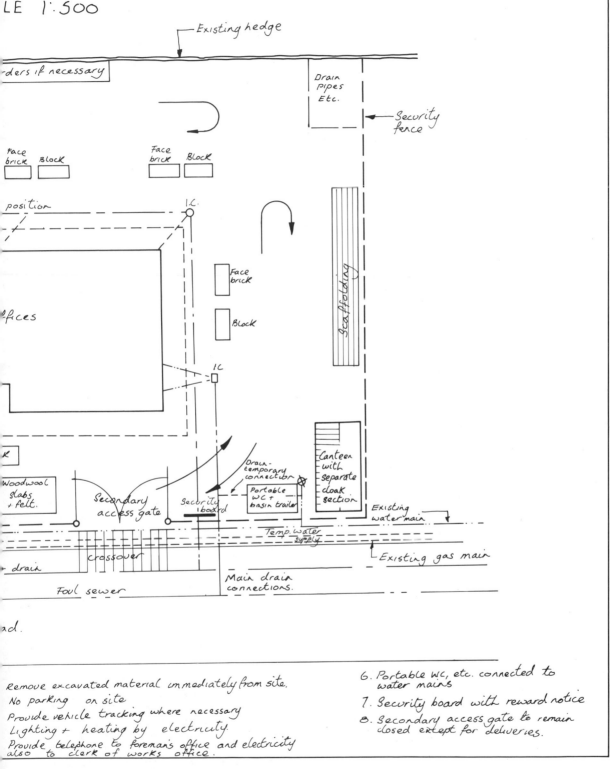

LE 1:500

Existing hedge

...ders if necessary

Drain pipes Etc.

Security fence

Face brick   Block   Face brick   Block

position   I.C.

Face brick

Block

...fices

I.C.

Scaffolding

Woodwool slabs + felt.

Secondary access gate

Security board

Drain temporary connection

Portable WC + basin trailer

Canteen with separate cloak section

Existing water main

Temp. water supply

Existing gas main

...drain

crossover

Main drain connections.

Foul sewer

...ad.

Remove excavated material immediately from site.
No parking on site
Provide vehicle tracking where necessary
Lighting + heating by electricity.
Provide telephone to foreman's office and electricity also to clerk of works office.

6. Portable WC, etc. connected to water mains
7. Security board with reward notice
8. Secondary access gate to remain closed except for deliveries.

The heights, types of materials to be used, sizes of members and methods of construction could be indicated and should be shown on the site layout plan. Careful positioning of each is essential to enable work to flow unhindered to completion before their eventual removal. See Fig. 10.1.2 for hoarding licence application form.

Access and exits

The size of access would depend on how restricted the site is or the constrainment of the locality of the site. It also depends on the probable sizes of the delivery vehicles and the type of plant to be used on-site. Under the heading of 'access on site' the following must inevitably be taken into consideration:

1. Number of access-to-site points allowed by the highways authority so as to interrupt the general public traffic minimally.
2. Access and exit road sizes for vehicles and plant movement and the thicknesses of the base depending whether it is permanant or temporary, i.e: (a) timber sleepers; (b) proprietary track; (c) hardcore, hogging, etc; (d) concrete; (e) tarmac/asphalt.
3. Crossovers and methods of protecting the existing footpaths, etc., or possible reinstatement later.
4. Ramps which are necessary for delivery of materials where access would otherwise be difficult.
5. Traffic flow directions should be shown to prevent congestion and afford satisfactory flow of vehicles.

Signs and notices

There should be some attempt by the contractor's planning officer to show the positions of the more important signs and notices which should be displayed on-site. Other signs can be erected at the discretion of the site supervisor for the following reasons:

1. To assist in directing plant and materials deliveries.
2. For security and to act as warnings to the public and employees.
3. For ease of location of administration, safety, health and welfare facilities.
4. To show the names and organisations who are party to the contract.
5. To help maintain good public relations.

Strategic positions are chosen for the siting of the notices so that maximum display is achieved. Example of signs and notices are as follows:

1. Contractor's site and name board stating the client, architect and others involved with the work. Permission should normally be obtained as to the size and shape from the client or architect.
2. Direction finger boards, including access/entry and exit boards.
3. Transportation routes to prevent heavy lorries using adjacent owners' roads thereby causing a nuisance/disturbance (public relations board).

4. 'Trespassers will be prosecuted' – warning notices (see Fig. 7.4.1).
5. Reward notices (see Fig. 7.4.1).
6. Administration area and offices signs for visitors to report to.
7. Parking area signs.
8. Bulk materials stacking/storing areas.
9. Reporting notices – warning all delivery drivers to report to main or checker's office before proceeding on to the site.
10. Danger notices.
11. Warning notices attached to incomplete scaffolding or uninspected trench timbering.
12. Fragile roof notices.
13. Overhead cable warning notices.
14. Position and depth of existing underground services pipes/cables.
15. Safety switches.
16. Fire alarm.
17. Hard-hat area notices.
18. Unsafe plant or machinery notices.

One can see that the site supervisor has a wide responsibility in this area and should attempt not only to use notices to maintain a safe site, but also to help in creating an efficient site, and to warn off possible unauthorised visitors.

Administration and other accommodation

Consideration should always be given to the possibility of ensuring that the administration block/huts are kept at a suitable distance from the construction work and be in such a position to enable the administrative staff to conduct their duties without having a background of excessive noise and distractions. The contractor's planning officer should show the careful positioning of the facilities on a copy of the project block plan (site layout plan) drawn to a suitable scale, then once the accommodation is erected it may stay until the end of the contract. Sometimes where there is a phased handover of work, especially on a very large housing contract, the huts are dismantled and moved to a new area. Naturally, portable/caravan type accommodation is best suited where a disturbance is inevitable. Maximum and minimum labour and site staff numbers must be assessed so that the optimum facilities can be provided to satisfy the safety, health and welfare Acts and Regulations, and to conform to the agreements incorporated in the appropriate Working Rule Agreements.

Accommodation should be provided for the following:

1. *Site, staff, etc.:* (a) Site manager; (b) clerk of works; (c) site engineer; (d) quantity surveyor; (e) timekeeper and wages clerk; (f) typists and telephonists; (g) production control and bonus surveyors; (h) meetings and conference room; (i) toilets and cloakroom.
2. *Operatives and subcontractors* (a) canteen and rest room; (b) drying room; '(c) first-aid room; (d) toilets and washing facilities.

THE BOROUGH OF KINGSFIELD ON SEA

HIGHWAYS ACT 1980

APPLICATION FOR HOARDING LICENCE

TO: The Borough Surveyor,
    TOWN CROSS,
    KINGSFIELD ON SEA   Date . . . .

1. I/We wish to (erect) (take down) (alter) (repair) (the outside of) a building known or to be known as

   within the Borough and hereby make application for a licence to erect a close boarded hoarding or fence to the satisfaction of the local authority so as to separate the building from the street.

2. I/We understand. I/We must indemnify and keep indemnified the Council against all actions demands claims and costs arising by the erection use and subsequent removal of the said hoarding or fence and in order to secure the due performance of such indemnity to enter into such policy or policies of insurance as the Council may require and approve and to produce to the Council for inspection the said policy or policies of insurance and the receipt or receipts for the last premium or premiums relating thereto.

3. I/We enclose deposit of £   having arranged for the Town Clerk at the Guildhall Kingsfield on Sea to inspect my/our insurance documents as required in 2 above.

Signed . . . .

NAME & ADDRESS (Block Capitals)

. . . .

. . . .

. . . .

Fig. 10.1.2

The provision of each depends on the size of the contract and the numbers employed. Accommodation can, of course, be shared with the subcontractors if agreed at commencement of the contract and the appropriate Health and Welfare Certificate, F2202, has been prepared.

Stores, storage facilities, and compound

Areas should be designated on the site layout plan for each of the following, and the sizes of each may be shown:

1. Hardstands for bulk, heavy materials.
2. Compounds (size, height and type of fence, and gate width).
3. Stores shed (with steel lock-up cupboards for valuable items).
4. Curing shed for concrete test cubes.
5. Cement and other essential sheds, e.g. for petroleum, oil and explosives.

Workshops

Positions should be indicated to enable the site supervisor to arrange for the erection of the following types of workshops:

1. Fitter's shops and work area.
2. Joinery shop and machinery area.
3. Reinforcement and bar bending area.
4. Concrete mixing area.

Temporary services

1. *Water supply:* stand pipe positions are normally indicated on the site layout plan, but in addition water is required temporarily for WCs, urinal stalls, wash basins and shower units.

2. *Electricity supply:* an indication should be given for transformers positions for plant and power tool operations. Lighting for the works and hutting is usually essential.

3. *Gas:* if portable Calor gas or other types of gas are to be used for heating processes, North Sea or other piped gas may not be necessary and should be clearly stated.

4. *Telephones:* a decision must be given on which offices are to be provided with a telephone. The clerk of works/resident engineer may have to be provided with this method of communication. Additionally the site operatives may request a pay-phone close to the canteen. Details should be given on the site layout plan.

5. *Drainage:* the provision of chemical toilets may be dictated in the initial stage, but where it is possible connections should be made to the main drainage at the earliest convenience for health and welfare reasons. The number of toilets required are laid down in the construction regulations.

If the site is a substantial one then the provision of separate WCs, etc. are essential for the site staff and operatives.

Plant and equipment positions

Depending on the complexity, size of site and the rate of output expected one would consider the positions of plant and equipment under two headings – static and movable.

Static
(a) mixers and weigh-batching points; (b) cranes – tower; (c) woodworking mechanical saws and other machines; (d) scaffolding; (e) hoists (passenger and materials); (f) weigh bridges; (g) cleaning and wash points for movable plant/vehicles.

Moveable
(a) vehicles – lorries, fork-lift trucks and dumper; (b) compressors; (c) skips; (d) cranes – tracked or rubber tyred; (e) pumps.

Other considerations

1. Hardstands for parking of staff and operatives' vehicles.
2. Material stockpile areas for optimum storage to prevent congestion and shortages on site occurring.
3. Security lighting positions.
4. Pedestrian passageways and footpaths.
5. Stop-go lights (temporary traffic lights).
6. Viewing platforms.
7. Temporary bench mark position (TBMs) and other control points.
8. Spoil heaps or dumps.
9. Roads to be sealed off or temporarily closed (no entry).
10. Skip positions.
11. Protective measures for existing trees, pavements, gardens and other features which are not to be disturbed.

As much information as possible should be displayed on a site layout plan to inform the site supervisor of the decisions of head office and architect (see Fig. 10.1.1).

10.2 Healthy and safe working places

Each site supervisor must by necessity be aware of, and be able to read with understanding, the contents of the many Acts of Parliament and subsequent Regulations which governs the health and safety of individuals on or near to workplaces. It is evident in the principal piece of legislation, the Health and Safety at Work, etc., Act 1974, that management is charged with ensuring that health and safety aspects of a company or small firm permeate down to all employees. Since the inception of the Act a Policy Statement and a Statement of Intent should by now have been prepared and displayed, stating the aims of the company and responsibilities of individuals regarding safety, health and wealth. Most of the experienced operatives are aware of the many problems and dangers associated with the construction industry, but generally it is with regards to the inexperienced worker (new entrants, including apprentices) that every conceivable type of hazard should be highlighted to prevent these people from injuring themselves or causing others to be injured by their carelessness. Induction courses would benefit new employees to the industry to cover such points.

Hazards exist and accidents take place even on the most health and safety conscious site because of:

1. Carelessness by individuals.
2. The variable nature of construction work.
3. The high labour turnover experienced by many contractors.
4. The variable and changing weather conditions.

As a result 16 million working days were lost to the construction industry in 1978.

The provision of healthy and safe working places is a necessity on-site, but equally workshops, factories and offices must be places where work can be done without risk to employees.

The workers in each trade are not altogether familiar with the practices of other trades and do not readily appreciate the dangers of certain processes. Therefore, under the Health and Safety at Work, etc., Act 1974, each employee is charged with ensuring that his/her actions do not present undue hazards to other workers or third parties. It is then obvious that a site supervisor should see that untrained and inquisitive individuals are not allowed to do work which they were not trained to do, especially where their work/action have a direct bearing on the health or safety of others e.g.:

1. Erection of scaffolding: This should be undertaken by persons experienced in this kind of work. Trained scaffolders are the obvious answer – those who have attended a special course, or courses, in scaffolding approved by the Construction Industry Training Board. There are three grades of scaffolder (receiving remuneration accordingly) – Trainee, Basic and Advanced.

2. Trench timbering or sheeting: Competent and experienced operatives are expected to do this type of work, as laid down in the Construction Regulations. If there are no experienced operatives then the work must be supervised by a competent and experienced supervisor.

3. Maintenance of electrical power hand-tools: Inspection and maintenance should be undertaken on a regular basis (once a week) by a competent electrician or electrical engineer.

4. Mounting of abrasive wheels: The wheel must be fitted by a trained person who is appointed in writing on Form 2 346 and whose name is recorded in the Register of Appointment of Persons to Mount Abrasive Wheels (see Fig. 10.2.1).

5. Cartridge guns: Operators must be trained and certificated in the use of such guns, unless they are working under direct supervision of someone else who is certificated.

6. Woodworking machines: Operators must be sufficiently trained in the use of the machines they are using, and must have been instructed in the inherent dangers in using such machines – unless they are under close supervision.

7. Explosives and even machinery, etc: Should only be used by trained persons.

Dangerous or unhealthy processes

Warnings should be given to operatives working with or adjacent to dangerous or unhealthy processes. Additionally, warning notices and signs can be used effectively by the different trades or supervisors where particular processes are hazardous. Take for instance the oxyacetylene equipment of a welder; if used correctly it is safe, but the resultant fumes are toxic and cause irritation to the eyes and lungs especially in confined spaces. Other workers can be protected by giving them alternative work in different locations until the welding process ceases. Adequate natural or artificial ventilation helps to protect the user of such equipment; or, alternatively respiratory equipment should be provided. Other unhealthy processes are: sand blasting, heat processes, processes emitting fumes, processes which are noisy, etc. (see section 6.5).

In attempting to achieve a healthy and safe working environment the site supervisor must take action as follows:

Before work commences
1. Keep workplaces free from debris (waste, unwanted materials and equipment) at floor level.
2. Ensure workplaces, where practicably possible, are flat, firm, and non-slippery and of reasonable size to enable operatives to undertake the process or operation safely and to prevent fatigue.
3. Construct work platforms (scaffolds, etc.) with sound materials using experienced and trained scaffolders, and inspect at least once every week or during and after adverse weather conditions.
4. Foresee hazards and take steps to reduce them by removing dangerous projections to scaffolding, protruding nails or objects, and unnecessary overhanging obstructions.
5. Obtain a clearance certificate and approval from the safety supervisor/officer before allowing electric arc welding, etc., to be undertaken to boilers, cylinders and other storage vessels or in confined spaces because residue of previously stored or used materials could cause an explosion. Thorough cleaning or ventilating of the vessels, etc. may be necessary.
6. Carefully clear away other dangerous substances before allowing the men to work, e.g. contaminated

PART 2

Appointment of persons to mount abrasive wheels (regulation 9)

	APPOINTMENT			REVOCATION	
Name of person appointed	Class or description of abrasive wheels for which appointment is made (*See Note 7*)	Date of appointment	Signature of occupier or his agent	Date of revocation of appointment (*See Note 5*)	Signature of occupier or his agent
(1)	(2)	(3)	(4)	(5)	(6)

Fig. 10.2.1

earth, asbestos substances (notify the Factory Inspectorate before removing blue asbestos), acids, red lead- coated objects and fumes, etc.

7. Fence off large holes, unattended dangerous points around the work area and machines, and if possible adequately cover to prevent operatives falling or being injured.

8. Provide raking, flying or dead shoring where necessary to prevent the collapse of walls or building in close proximity to workplaces (especially if the work undertaken may lead to the possible weakening of a structure). Additionally, provide tell-tales (glass or cement dots) as a check on structural movement.

9. Provide, where necessary, good access and egress points (tied ladders, ramp, etc.); labour/material movements communication routes should be kept clear of obstructions.

10. Stack materials safely and in correct quantities to reduce obstructions. Materials for immediate use only should be stacked on scaffolding and should be distributed around the scaffold to prevent overloading.

11. Fix warning notices/posters to highlight hazardous points or areas on-site for the benefit of employees and visitors.

12. Provide adequate lighting and ventilations when required.

13. Warn workers of the inherent hazards associated with any new work and materials they are to use. Also, instruct them on any necessary precautions.

Hazardous materials are those which are highly inflammable, toxic, corrosive, etc., and should have warning labels or posters shown on the container printed in black on a yellow or orange background, as follows:

1. Exploding bomb for explosive substances.
2. A flame over a circle for oxidising substances.
3. A skull and crossbones for toxic substances.
4. A flame for highly inflammable substances.
5. A hand and a piece of metal being dissolved by liquid dropping from two test tubes for corrosive substances.
6. A St Andrew's Cross for harmful or irritant substances (see Fig. 10.2.2).

In addition to the symbols on the labels, warnings on the way the substances should be handled must be shown (this is the responsibility of the supplier of the substances).

As work proceeds

1. Clean the work area periodically to minimise the following; tripping, ankle twisting, slipping, nails or other sharp objects from entering the feet.
2. Use sound materials and observe the manufacturers' instructions when using them.
3. Ensure tools are sound and are well maintained.
4. Ensure power hand tools are in good working order and are used by competent persons. There should be a supply of electricity with reduced voltage of 110 volts. This is a safety measure on construction

sites because of the damp conditions which generally prevail.

5. Use appropriate plant to reduce the strain on operatives. This makes work less fatiguing and, therefore, leads to fewer accidents. The plant should be operated by a trained, competent person.

6. Provide suitable protective clothing, footwear and safety gear which should be worn depending on the type of hazard to be encountered.

7. Seek advice from the structural engineering department when using formwork/falsework if there is any doubt about its adequacy.

8. Observe correct sequences of work particularly where dismantling operations are under way.

9. Leave major demolition to the experts, but for others, observe the sequence of working including the appropriate safety measures shown in either the BEC Construction Safety Handbook or BS Codes of Practice 94.

10. Abandon certain operations during severe weather conditions, e.g. in high winds do not operate tower cranes beyond the manufacturers' recommendations. Workers should remove themselves from high platform working; sheeting operations; and deep excavations where there is a possibility of flooding.

A site supervisor should have sufficient foresight, experience and training to enable him/her to take steps to prevent the thousand and one accidents which could occur on-site. As an added incentive to see that supervisors are forever vigilant to health and safety matters many firms maintain accident records so that an analysis can be made to arrive at the most recurring type of accident. The records also show which sites are the most successful in minimising the accident risk.

The form in Fig. 10.2.3 is a typical site accident report which the site supervisor is expected to complete monthly.

Finally, the site supervisor must be aware that if unsafe practices are discovered by the Factory Inspector he/she may be served with an Improvement Notice or Prohibition Notice (see section 10.3).

10.3 Safety, health and welfare arrangements

The passing of the Health and Safety at work, etc., Act 1974 saw the introduction by the government of the Health and Safety Commission which is answerable to the Secretary of State for Employment for all safety, health and certain welfare matters affecting employees work-places and establishments throughout Great Britain.

Health and Safety Commission

The Commission is mostly comprised of members which are either from the employers' or employees' organisations, with the remaining members being from public or other bodies. It is responsible for ensuring that the Act

SYMBOLS FOR LABELS USED ON DANGEROUS SUBSTANCE CONTAINERS

Fig. 10.2.2 Symbols for labels used on dangerous substance containers

B. BUILDER LTD.

SITE ACCIDENT REPORT – MONTHLY

PREPARED FOR: _ _ _ _ _ _ _ _ _ _ _ _ _

PREPARED BY: _ _ _ _ _ _ _ _ _ _ _ _ _

SITE/CONTRACT: _ _ _ _ _ _ _ _ _ _ _ _   DATE: _ _ _ _ _ _ _ _

		THIS PERIOD		OVERALL	
1.	NUMBER OF PERSONNEL (average on site)				
2.	NUMBER OF ACCIDENTS REPORTED (requiring an employee to be absent from work for more than 3 days).				
		FATAL	NON-FATAL	FATAL	NON-FATAL
	Accident types:				
	(a) Persons falling – from scaffolds, ladders, roofs, etc				
	(b) Fall of materials onto persons.				
	(c) Excavations or tunnel collapse.				
	(d) Cranes, hoists, transport or other plant.				
	(e) Powered hand tools.				
	(f) Ordinary hand tools.				
	(g) Strains, twists.				
	(h) Striking or abrasings on objects.				
	(i) Burns, scalds or electric shocks				
	(j) Others.				
3.	NUMBER OF MINOR ACCIDENTS (where employees did not have to stay off work more than 3 days).				
4.	NUMBER OF SITE-MAN-HOURS WORKED				
5.	NUMBER OF HOURS LOST BECAUSE OF ACCIDENTS				

NOTE: Accident numbers extracted from the
ACCIDENT BOOK. FORM B1 510 and from
ACCIDENT REPORTS sent to
H.M. INSPECTOR OF FACTORIES.

SIGNED: _ _ _ _ _ _ _ _ _ _ _ _ _ _ _
SITE SUPERVISOR: _ _ _ _ _ _ _ _ _ _

Student Note: The accident types can be further broken down to
give a greater analyses of the types of accidents.
The frequency rate of accidents may be calculated
to compare with the national average.

Fig. 10.2.3

is implemented fully. Therefore, the Commission is responsible for:

1. Preparing safety, etc., policies relevant to industry and commerce.
2. Arranging safety, etc., research and publishing results.
3. Providing an information and advisory service for industry and commerce.
4. Making proposals for the introduction of Regulations to the Secretary of State where and when required.
5. Consulting with and taking advice on safety, etc., from special advisory bodies set up by each industry.
6. Establishing, maintaining and directing the Health and Safety Executive.

Health and Safety Executive

The Executive implements the Commission's directives through its seven Health and Safety Inspectorates, i.e. Factory; Mines and Quarries; Explosive; Nuclear; Alkali Works; Pipe Lines; and the Employment Medical Inspectorate. Each of these Health and Safety sections of the Executive have responsibilities for health and safety matters within the construction industry at appropriate times, e.g. when pipelines are being constructed; where explosions occur – and so on, but it is mainly the Factory Inspectorate representative (Factory Inspector) who visits the various sites to inspect the works. but under the Health and Safety (Enforcing Authority) Regulations 1977 local authorities have the responsibility to see that standards of safety, health and welfare are maintained within offices and catering establishments.

Factory inspector

These inspectors are the front-line enforcers of the various safety, health and welfare legislation. They ensure that firms and individuals operate by adhering closely to the various safety, health and welfare Statutes, and the Codes and Regulations subsequently prepared by the Secretary of State for Employment. The inspectors have far-reaching powers and may enter any factory or place of work to carry out their duties. Employers and individuals must take cognisance of any of their lawful requests.

Where conditions or activities are in breach of the safety, etc., law Notices would be served on the employer or his/her representative indicating that the problem should be rectified immediately.

There are two main types of Notices under the Health and Safety at Work, etc., Act 1974 which are:

1. Improvement Notices.
2. Prohibition Notices.

Improvement Notices

These Notices would be served by the factory inspector on the individual who is responsible for the offending conditions met with on-site (the site manager/supervisor). This notice would point out the offence, and would stipulate the period by which it should be rectified.

Prohibition Notices

If an activity or process is being conducted which the inspector feels could lead to individuals being injured a Prohibition Notice would be served which normally takes effect immediately. This means that the activity or process should cease until the hazard has been eliminated.

The site supervisor, in each case, should rectify the hazard, whether it is:

(*a*) a totally inadequate or unsafe technique;
(*b*) an unsafe or dangerous material, piece of plant or equipment;
(*c*) an unsafe work area.

Failure to comply with the details laid down on either of the notices could lead eventually, on summary conviction, to a substantial fine and/or imprisonment.

Other requirements

In addition to the government and local authority bodies and individuals concerning safety, etc., there is the requirement that employers must nominate or employ trained personnel or qualified organisations who implement the safety policies of firms, and help to keep them within the law regarding safety, health and welfare matters, e.g.

1. Safety supervisors; 2. Safety officer;
3. Safety consultants; 4. Safety groups;
5. Site safety supervisors; 6. Safety representatives;
7. Safety committees.

Safety supervisor
It is the responsibility of each employer who employs more than 20 employees to appoint, in writing, a safety supervisor under the Construction (General Provisions) Regulations 1961. The safety supervisor, among other duties, would ensure that the firm complies with the various legislation. As stipulated in the Construction Regulations, this individual must be experienced and knowledgeable on safety, health and welfare matters of the construction industry, and if possible should be allowed to devote all of his/her time in this capacity. The Safety Supervisor may be given the responsibility to employ and supervise/direct safety officers if the size of a firm permits.

Safety officer
Normally firms refer to all individuals employed full time on safety, health and welfare matters as safety officers. These persons should be qualified in safety, etc., through the City and Guilds of London examinations, or preferably by having also attended courses held by the CITB, BECBAS, FCEC/TSD or take the examinations of the Institute of Industrial Safety Officers to attain Corporate Membership.

Site supervisors must observe and implement all instructions from the safety officers regarding safety, etc. Failure to act on these officers' recommendations causes conflict with the management, but more seriously, with the Factory Inspectorate. Safety officers therefore have the following duties, to:

1. Implement the firm's safety, etc., policies.
2. Advise management of legislation changes.
3. Advise on what general improvements can be made regarding safety matters throughout the firm.
4. Prepare reports on the causes of any accidents or dangerous occurrences.
5. Ensure site managers/supervisors compile reports weekly/monthly, and issue data to HM Factory Inspectorate when required.
6. Conduct site inspections and prepare reports on problems and conditions encountered, copies being issued to the site managers, for reference or corrective action to be taken, and to the contracts managers and general manager.
7. Analyse accident reports and prepare trends/statistical charts for everyone's instruction (see Fig. 10.2.3).
8. Arrange for necessary training of individuals on safety, etc., matters by films, visits, courses, discussions, etc.
9. Arrange for essential training to satisfy the various legislations' special requirements, e.g. certification for abrasive wheels mounters, etc.
10. Attend safety committee meetings and observe and act on recommendations from Site/Shop safety committees and safety representatives, and enter into discussion with trade union representatives where necessary regarding safety, health and welfare arrangements.

These points are the full responsibility of the safety supervisor as well as any safety officers where the distinction is made within a firm.

Where it is unecomonical for a small firm to employ full-time safety officers the alternative would be to either employ the services of a safety consultant, or join a safety group.

Safety consultants

These persons are qualified individuals offering a safety, etc., service to firms on a private consultancy basis – a fee being charged either per visit or on an hourly rate or by a retainer, and act in a similar capacity as safety officers.

A safety consultancy, however, is normally a firm undertaking inspections and preparing reports, etc., for businesses to ensure that they remain within the law.

Safety groups

This type of organisation is set up by a group of small businesses to look after the safety, etc., affairs of each. It would employ a general manager and a few qualified safety officers who would then service each of the groups's businesses regarding safety, etc. The directors of the group would be representatives from the firms who are members of the group. So, with the employment of a few safety officers the safety, etc., affairs of many businesses

would be preserved, which is substantially cheaper than if each business were to employ its own safety officer.

Site safety supervisor

The site manager normally acts in this capacity and is trained preferably under the BEC/CIOB Scheme, or similar scheme, which is approved by the CITB. The site manager (site safety supervisors) therefore must:

1. Understand the main principles and requirements of the various safety legislation.
2. Ensure that work is organised safely, with the minimum of risk to employees, and maintain safe working areas.
3. Ensure that all plant and equipment, including hand power tools, are inspected and maintained regularly.
4. Only allow trained operatives to operate the various plant.
5. Carry out regular inspections regarding safety health and welfare facilities, and keep appropriate records and reports.
6. Accompany and cooperate with the visiting safety officers or factory inspectors, particularly where recommendations are made to make improvements after hazards are highlighted.
7. Where necessary, see that protective clothing and equipment are made available and used.
8. Check whether subcontractors have their own safety, etc., provisions, and, if not, to arrange facilities for them.

Safety representatives

Under the Safety Representatives and Safety Committee Regulations 1977 a recognised trade union (UCATT; TGWU; GMBU; FTAT – depending on whether the building or civil engineering industries are involved) within a firm may nominate in writing to the employer the employee who will represent the union members and section of workers on such matters as safety, etc. In the construction industry it is expected that the nominated person/persons for site representative on-site will be in the employment of the main contractor with at least 12 months experience, and two years for representatives in a factory, workshop or main contractor's yard.

The safety representative, therefore:

1. Listens to complaints from operatives regarding safety, health and welfare.
2. Checks on safety of the workplaces, and is allowed to carry out a full inspection not less than every three months.
3. Represents the employees when discussing safety, etc., with employers and factory/other inspectors.
4. Raises important issues at safety committee meetings, and with management.
5. Keeps up to date with new legislation and attends approved courses on safety, etc., when required, provided that the appropriate notice is given to the employer before doing so.

Safety committees

When a request is made in writing to an employer to set up a safety committee it must be made by at least two safety representatives under the Safety Representatives and Safety Committee Regulations 1977.

The committee monitors and suggests improvements on matters relating to safety, etc., and may consist of representatives from the management as well as the unions. It is a further attempt to improve conditions for employees at the workplace.

10.4 Safety inspections notifications, notices, etc.

Safety inspection, testings and thorough examinations

In the Construction Regulations 1961 to 1966 references are made to inspections, testings and thorough examinations to working places, lifting appliances and gear, and certain structures, e.g. caissons and coffer dams. Site supervisors should appreciate that each requires a different treatment under the Regulations, examples being as follows:

Inspections

The level of inspections relate mainly to commonsense measures which anyone with any degree of concern for their own and others' safety would take before using certain equipment or undertaking special operations. Written evidence is seldom required of inspections but is essential when undertaking thorough examinations. Reports, however, of inspection to scaffolding are made weekly on form F91 Section A (see Fig. 10.4.1). Also, after inspections are carried out on special lifting appliances, anchorages and safe load indicators, a report is made on form F91 Section C. The following are typical examples where inspections are made without the need to make reports:

1. Trench excavations and timbering should be inspected for stability by a competent person at the beginning of each day or shift if operatives are to work within the trench excavations.
2. Hoists (materials and passenger) should be inspected weekly to ensure that parts are functioning safely and that there are no clearly visible or other defects.
3. Cranes, etc., should be inspected weekly.

Testing

This is another activity which should be undertaken periodically. Take for instance the previous examples:

1. There are no recommended tests on trenches.
2. Hoists should be tested before use and after substantial alterations or repairs and results should be entered on Form F91 (part II) H.
3. Cranes should have been tested sometime within the preceding four years before they are about to be used (record on Form F96).

Thorough examinations

Thorough examinations should be conducted by competent persons (this usually means trained individuals in this type of work, e.g. mechanic, fitter, etc.) after certain events have occurred or time has elapsed. After each thorough examination a report is made out to comply with the legislation requirements. The reports are recorded on statutory forms which are retained, where possible, on-site as proof to the factory inspector that the safety statutes are being observed – typical examples being as follows:

1. Trenches and trench timbering should be thoroughly examined every seven days; after the use of explosives nearby; after unexpected falls of rocks; etc.; after damage has occurred to trench timbering (see Fig. 10.4.2 for Form F91 (part I) B).
2. Hoists should be thoroughly examined at least every six months (see Fig. 10.4.3 for Form F91 (part II) H).
3. Cranes (exceeding 1 tonne safe working load) should be thoroughly examined at least every 14 months (see Fig. 10.4.4 for Form F96).

Notifications to the Health and Safety Inspectorate

These are submitted generally on standard forms, and although there are many occurrences or situations requiring notifications to be made to the inspectorate the following are the most common and important.

1. The commencement of building/civil engineering operations which are to last for more than six weeks (Form F10).
2. Accidental death of employees at work, or accidents which cause them to be absent from work for more than three days (Form F2508).
3. Where diseases occur due to lead, etc. (Form F2508A).
4. Outbreaks of fire which causes serious injuries or stoppages of work (Form F2508).
5. Where cranes or hoists are found to be unsafe or have collapsed (Form F2508).
6. Where an office is to be used by personnel for more than 21 hours per week and more than six months' duration (Form OSR1).
7. Where a power grinding or similar wheel disintegrates irrespective of whether damages have occurred or not (Form F2508).

A notification is also required to the careers officer on taking into employement or transference of a young person (Form F2404).

Prescibed notices, certificates and placards

Under the Factories Act 1961, Office Shops and Railway Premises Act 1963, and other Acts and Regulations certain notices are prescribed for display on-site or in the work area. While there are many, the most commonly displayed notices are as follows:

1. Insurance Certificates, under the Employers Liability Compulsory Insurance Act 1969, should be displayed on-site.

KEY TO SAFETY SIGNS

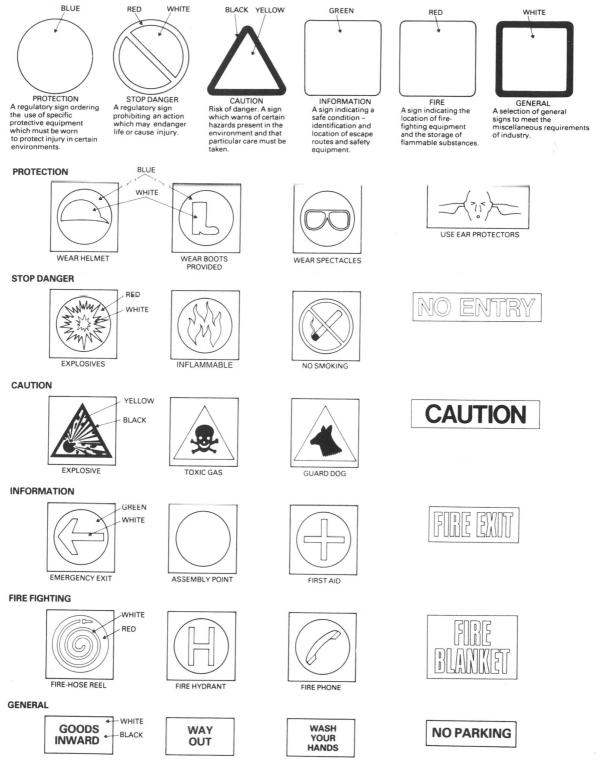

Fig. 10.2.4   British and European standard safety signs – some examples

Factories Act 1961

FORM F91

SECTION A

Construction (Working Places) Regulations 1966

SCAFFOLD INSPECTIONS

Name or title of employer or contractor

....

Address of site ....

....

Work commenced—Date ....

Reports of results of inspections under Regulations 22 of scaffolds, including boatswain's chairs, cages, skips and similar plant or equipment (and plant or equipment used for the purposes thereof)

Location and description of scaffold, etc. and other plant or equipment inspected (1)	Date of inspection (2)	Results of inspection State whether in good order (3)	Signature (or, in case where signature is not legally required name) of person who made the inspection (4)

NOTES TO SECTION A

(1) *Short check list—at each inspection check that your scaffolding does not have these faults:*

		Week		
	1	2	3	4

FOOTINGS — Soft and uneven / No base plates / No sole boards / Undermined

STANDARDS — Not plumb / Joined at same height / Wrong spacing / Damaged

LEDGERS — Not level / Joint in same bays / Loose / Damaged

BRACING 'Facade and ledger' — Some missing / Loose / Wrong fittings

PUTLOGS and TRANSOMS — Wrongly spaced / Loose / Wrongly supported

COUPLINGS — Wrong fitting / Loose / Damaged / No check couplers

BRIDLES — Wrong spacing / Wrong couplings / Weak support

TIES — Some missing / Loose

BOARDING — Bad boards / Trap boards / Incomplete / Insufficient supports

GUARD RAILS & TOE BOARDS — Wrong height / Loose / Some missing

LADDERS — Damaged / Insufficient length / Not tied

(2) *This check list is not part of the report required by Regulation 22: see also para 5 of Notes and Regulation 22 on page (ii) of cover and Notes on page 13.*

Fig. 10.4.1

Factories Act 1961

Construction (General Provisions) Regulations 1961

SECTION B

EXCAVATIONS, SHAFTS, EARTHWORKS, TUNNELS, COFFERDAMS AND CAISSONS

Name or title of employer or contractor

....

Address of site ....

....

Work commenced—Date ....

Reports of results of every thorough examination made in pursuance of Regulation 9(2) of an excavation, shaft, earthwork or tunnel or in pursuance of Regulation 18(1) of a cofferdam or caisson

Description or location (1)	Date of examination (2)	Result of thorough examination State whether in good order (3)	Signature (or, in case where signature is not legally required, name) of person who made the inspection (4)

See Notes and Regulations 9 and 18 on page (ii) of cover

Fig. 10.4.2

218

Name or title of Employer or Contractor

Address of Registered or Head Office or Address of Site

Hoists

Factories Act 1961   Section H

Reports of results of six-monthly thorough examinations

Particulars prescribed by the Secretary of State in pursuance of regulation 46 of the Construction (Lifting Operations) Regulations 1961

Description of hoist e.g., type, identification mark, capacity	Date of last previous thorough examination	Result of examination Enter details of repairs required or defects. If none enter "In good order"	Signature of person making or responsible for examination	Date of exami-nation
1	2	3	4	5

Fig. 10.4.3

HEALTH AND SAFETY EXECUTIVE

F 96

Certificate No. ....

Factories Act 1961

CONSTRUCTION (LIFTING OPERATIONS) REGULATIONS 1961

CERTIFICATE OF TEST AND THOROUGH EXAMINATION OF CRANE

(Prescribed by the Secretary of State in pursuance of paragraph (1) or (2) of regulation 28)

See Note 4, overleaf, for continuation of items 10 and 11

1. Name and address of owner of crane	
2. Name and address of maker of crane	
3. Type of crane and nature of power (e.g. Scotch derrick manual; Tower derrick electric; Rail mounted tower-electric)	
4. Date of manufacture of crane	
5. Identification number — (a) Maker's serial number	
5. Identification number — (b) Owner's distinguishing mark or number (if any)	
6. Make and type of automatic safe load indicator, if required (*See* regulation 30)	
7. Make and type of derricking interlock, if required (*See* regulation 22)	
8. Date of last previous test of crane	
9. Date of last previous thorough examination of crane	

10. Safe working load or loads In the case of a crane with a variable operating radius (including a crane with a derricking jib or with interchangeable jibs of different lengths) the safe working load at various radii of the jib, jibs, trolley or crab must be given. Test loads at various radii should be given in column (iii) and in the case of a safe working load which has been calculated without the application of a test load "NIL" should be entered in that column	(i) Length of Jib (metres)	(ii) Radius (metres)	(iii) Test load (tonnes)	(iv) Safe working load (tonnes)

11. In the case of a crane with a derricking jib or jibs the maximum radius at which the jib or jibs may be worked (in metres)	
12. Defects noted and alterations or repairs required before crane is put into service. (If none enter "None")	

I hereby certify that the crane described in this Certificate was tested and thoroughly examined on
.... and that the above particulars are correct.

Signature   Qualification

Name and address of person, company or association by whom the person conducting the test and examination is employed.

Date of Certificate ....

Fig. 10.4.4

2. Abstract of the Factories Act for Building Operations and Works of Civil Engineering Construction – Form F3.
3. Abstract of the Offices, Shops and Railway Premises Act 1963 – Form OSR 9.
4. The Woodworking Machines Regulations 1974 Prescribed, Notice F2470.
5. Electricity (Factories Act) Special Regulations 1908 and 1944, Prescribed Notice F954.
6. Asbestos Regulations 1969, Notice F2358.
7. The Abrasive Wheels Regulations 1970, Notice of Permissible Speeds Placard.
8. Dermatitis Cautionary Placard F367.
9. Electric Shock (First Aid) Placard F731.

Each different situation or activity where there is a risk to employees, visitors or the general public etc, requires a different notice, etc.: and on a single site, because there may be a multitude of activities, most if not all of the aforementioned notices may have to be displayed.

Legislation

By now site managers must realise the seriousness by which the Government and its representative, the Health and Safety Commission, treat matters relating to the safety, health and welfare of employees and third parties. As a final indication the following are most of the statutes dealing with such matters:

Acts passed by Parliament
Health and Safety at Work, etc., Act 1974.
Factories Act 1961.
Offices Shops and Railway Premises Act 1963 and 1982.
Fire Precautions Act 1971.
Control of Pollution Act 1974.
Explosives Act 1875 and 1923.
Boiler Explosions Act 1882 and 1890.
Mines and Quarries Act 1954.

Regulations introduced through the various previous Acts
Construction Regulations – General Provisions 1961 – Lifting Operations 1961 – Health and Welfare 1966 – Working Places 1966. (metricated 1984)
Abrasive Wheels Regulations 1970.
Woodworking Machines Regulations 1974.
Protection of Eyes Regulations 1974.
Asbestos Regulations 1969.
Asbestos (Licencing) Regulations 1983
Asbestos (Prohibition) Regulations 1985.
Control of Asbestos at Work Regulations 1987.
Offices at Building Operations, etc. (First Aid) Regulations 1964.
Health and Safety (First Aid) Regulations 1981.
Fire Certificates (special premises) Regulations 1976.
Electricity (Factories Act) Special Regulations 1908 and 1944.
The Work in Compressed Air Special Regulations 1958.

Diving Operations at Work Regulations 1981.
Electricity at Work Regulations 1989.
Noise at Work Regulations 1989.
The Construction (Head Protection) Regulations 1989.
Control of Lead at Work Regulations 1980.
Highly Flammable Liquids and Liquified Petroleum Gases Regulations 1972.
Food and Drugs Act 1955.
Food Hygiene (General) Regulations 1970.
The Reporting of Injuries, Diseases and Dangerous Occurrences Regulations 1985.
The Control of Substances Hazardous to Health Regulations 1989 (COSHH).
The Construction (Head Protection) Regulations 1989.
Control of Substances Hazardous to Health Regulations 1989.

10.5 Financial protection in case of fatal and other accidents or damage to health

It is of importance for site managers/supervisors and operatives alike to appreciate that there are arrangements within the industry for employees to be insured against accidents, etc., with better facilities being available through firms belonging to the BEC, FCEC, or Federation of Master Builders (FMB).

It is by agreement with the construction unions who are parties to the NJCBI and/or the CECCB that the employers have arranged for a death benefit cover of £3500 for each operative with the Building and Civil Engineering Holiday Scheme Management Company (non-profit making and set up by the industry), which also provides incapacitated workers with £8.00 per day sickness benefit (figures for 1986). A surcharge is made of 90p on each holiday stamp payment made by 18 years to pay for the Death Benefit and Retirement Scheme.

Most operatives are covered by the scheme previously mentioned except where agreements exist between employers and other workers' unions who are not party to the NJCBI or CECCB, such as plumbers and electricians. Other arrangements therefore exist in such cases.

It is expected that the new body called the Building and Allied Trades Joint Industrial Council instigated by the FMB and the TGWU will make similar arrangements to those which exist through the NJCBI.

Acts of Parliament giving special injury benefits

In addition to the Death Benefit Scheme, employers under the Employers Liability (Compulsory Insurance) Act 1969 must insure with a recognised insurance company against liability for personal injuries and diseases to their employees, and as proof of insurance cover being made under the Act a Certificate of Insurance must be displayed in a prominent place at the workplace/on-site. If the site is too small to justify the display of a certificate a sufficient number of Certificates of Insurance should be displayed at head office to enable each employee to check that they are covered under the Act. Certain employees need not be insured, however, e.g. relatives.

Under the Fatal Accidents Act 1976 the dependents of a deceased person who is killed through injuries sustained at work may claim damages from the employer. As an added safeguard for employees the National Insurance (Industrial Injuries) Act 1965 gives them the right to claim danages from an employer where industrial injuries are sustained.

Accident Book BI 510 (see Fig. 10.5.1)

This book should be retained on-site to record any form of injury sustained by employees under the Social Security Act 1975. It is therefore essential that injured individuals have the types of injuries received recorded in this book. If an injury appears minor but then later develops complications leading to the injured party becoming incapaci-tated, the Accident Book record protects not only the injured party under the Act but also the employer and is proof that the injury occurred on-site. Special benefits will then be forthcoming from the Government through its National Insurance Fund either in the form of a Disablement or Sickness Benefit.

It must be appreciated that National Insurance contributions are divided between contributions made by the employee and a larger contribution being made by the employer, the cost ultimately being passed on to the clients. National Insurance contributions are used towards:

1. National Health Benefits.
2. Industrial Injury Benefits.
3. The Redundancy Fund (claims can be made in cases of redundancy).

National Insurance Contributions are shown on each employee's PAYE Tax Deduction Card (P9 or P11) kept, usually, at the wages department of a firm's head office.

ACCIDENT BOOK

1 About the person who had the accident	2 About you, the person filling in this book	3 About the accident
▼ Give full name ▼ Give the home address ▼ Give the occupation	▼ Please sign the book and date it ▼ If you did not have the accident write your address and occupation	▼ When it happened ▼ Where it happened

Name
 	
Address
 	
Postcode
Occupation
	Your signature Date
	                    / /
	Address
	
	Postcode
	Occupation
		Date Time
		    / /
		In what room or place did the accident happen?

4 About the accident – what happened	Reporting of Injuries, Diseases and Dangerous Occurrences, RIDDOR 1985
	For the Employer only
▼ Say how the accident happened. Give the cause if you can. ▼ If any personal injury say what it is.	Please initial the box provided if the accident is reportable under RIDDOR
How did the accident happen? _____ 	Employer's initials

This book satisfies the regulations about keeping records of Accidents to people at work –
1. Under the Social Security Act 1975
2. Under the Reporting of Injuries, Diseases and Dangerous Occurrences Regulations 1985 (RIDDOR)

Fig. 10.5.1

Chapter 11
Supervisor's accurate setting out and controlling of works

11.1 Accurate dimensional instructions and other information

The person most likely to be responsible for setting out important structures on the site is the site engineer/surveyor. Site supervisors are expected to have a reasonable understanding of setting-out procedures although, generally, it is more convenient for firms to employ their own site engineers directly in a full-time capacity and who are more able to service one or more sites. Site engineers' responsibilities lie with the establishment of the critical parts and control points of a structure, which would then enable the site manager/supervisor to conduct his/her own simpler setting-out techniques, perhaps using less sensitive or precise surveying equipment.

A good site manager/supervisor should be proficient at using surveying and setting-out equipment and be conversant with setting-out procedures and other site engineering techniques. Too many of the smaller firms, however, expect the supervisor to be not only a manager of people, but versatile to a point at which there is no need to employ site engineer and other service staff. This is generally at the expense of accuracy.

Procedure before commencing setting out

One can only consider setting out when sufficient contract documents have been finalised and issued by the architect/designer. The contract starting date is noted, and every essential detail is correlated by the site supervisor which is essential to the establishment of the site and successful setting-out of the structure.

Accurate details and information is required, and to this end the person responsible for setting-out should undertake the following:

1. Check that all drawings and other documents are received to enable the key parts of the structure to be established on the site (separate foundation drawings are essential on major construction jobs). See that the drawings, etc., which are received are of the latest issue, and notify the architect/designer, through the contracts manager, where there are any deficiences.
2. Check the drawings' main dimensions and reconcile them by adding intermediate dimensions together. Make sure sufficient information is shown to allow control points to be fixed, e.g. details about bench marks (BMs); road centre lines reduced levels (RLs) and tangent points; sewer positions and their sizes, gradients, and RLs; and distances of the new structures from roads or other structures.
3. If a site grid has not already been incorporated on the designer's drawings to enable important features to be located on the site later, a grid should be superimposed on the drawings (20 m to 100 m grid, depending on the ground slope) if the size of the site and works dictates it. The site grid could also be used to establish contours or gradient of the site if so required.

   It is usually a sensible precaution to check and confirm the architect's grid of levels.
4. Where there are a number of structures observe the distances between each, from the drawings, as a check later while setting-out.
5. Record the standard of accuracy expected by the architect for setting out and controlling the works ($\pm$ m). As a guide, one should attempt to achieve the following levels of accuracy as stated in the BS5606 1978, Code of Practice for Accuracy in Buildings:
   (a) Linear measurements:
      i. General measurements:
         $\pm$ 5 mm in 5 m.
         $\pm$ 10 mm between 5 m and 25 m.
      ii. Precise measurements:
         $\pm$ 15 mm over 25 m
         $\pm$ 3 mm up to 10 m
         $\pm$ 6 mm between 10 m and 30 m
   (b) Angular (theodolite):
         $\pm$ 40 seconds, and $\pm$ 10 mm in 50 m.
   (c) Levels:
      i. spirit:
         $\pm$ 5 mm in 5 m.
      ii. optical level (builders type using single sighting):
         $\pm$ 5 mm up to 60 m.
   (d) Verticality:
      i. Spirit: $\pm$ 10 mm in 3 m.
      ii. Plumb bob: $\pm$ 5 mm in 5 m
      iii. Theodolite: $\pm$ 5 mm in 30 m.

For very high structures adjustments may have to be made for the earth's curvature.

6. When on-site check that there are the minimum of obstructions before commencing setting-out. If a good site layout plan, which attempted to minimise obstructions, was drawn up and adhered to for hutting and storage accurate setting-out would be easier.

7. Confirm the levels of the TBMs with the approved ordinance bench marks (OBMs); the expected accuracy is within ± 10 mm. TBMs should be established where possible on permanent features, e.g. existing walls, steps, etc. The positions of each TBM and other reference points can be recorded on the site plan along with their reduced levels (RL), etc.

   Also, check other levels – particularly contour and spot heights shown on the working drawings, with the actual heights on the site.

8. Check the accuracy of any tapes, levels, theodolites and other instruments before proceeding to set out. (See section 11.2.)

9. When checking and conducting the setting-out of control and other points the booked or other records should be retained in the office for future reference, which is particularly important if the setting-out accuracy comes under question later. Also, intructions regarding level points etc., given to foremen or subcontractors should be retained. The records are recorded in or on:

   ● the level books.
   ● the theodolite record sheets.
   ● sewer, road and building detail sheets.
   ● the field books (chain or tape measurements of areas).
   ● instruction sheets to foremen/supervisors and subcontractors.

   The site manager/supervisor or site engineer (depending on who sets out) is accountable for any errors in setting-out, so it is of paramount importance that constant checks are made to ensure accuracy, and that maintenance is assured of profiles and pegs which are moved or damaged from time to time by excavators, vehicles, operatives or vandals.

10. It many be necessary to check on-site boundary positions before setting-out commences, but this information should normally be issued by the architect/designer.

11. The site layout plan (produced by the firm's planning officer) should have recorded on it all the control points – coordinate positions, TBMs, etc. In selecting suitable positions the person responsible for setting-out should avoid storage and other obstructions.

12. Check on handover phases, or parts of a structure which may have to be completed first. This would determine the setting out sequence.

13. A schedule of requirements for setting-out equipment and materials may have to be submitted to the head office to ensure that a neat professional standard of work is secured. Makeshift materials for setting out not only reduces the efficiency of the site engineer/supervisor but looks untidy and is more prone to accidental damage because the pegs and profiles look unimportant to plant operators and operatives alike.

14. Before setting-out commences in earnest the site may have to be cleared of shrubs, trees, rubbish, fences and old buildings may have to be demolished. The architct's/designer's drawings would have to be checked for any tree preservation details. Protective fences would then have to be erected to not only protect the tree trunks, but also as protection for the roots and branches.

In making the ultimate decision as to who is best suited for setting out the following is a guide:

(*a*) Where the structure has to be constructed to very fine tolerances use a site engineer who is well qualified and trained. Special precise instruments should be made available for use – the whole process being, therefore, more costly. Industrialised building in concrete or steel falls into this category, especially if they are multi-complex or storey.

(*b*) Where two-/three-storey traditional buildings or simple single-storey ones are contemplated average accuracy would be expected, and a good site manager/supervisor is expected to be competent enough to set-out such works.

11.2 Checking instruments for accuracy

Site supervisors must always be aware of the limitations of the equipment and instruments at their disposal for setting-out construction work. It is of very little use relying on others for the checking of equipment, etc., because the blame for inaccurate site surveying falls squarely on the site engineer or site supervisor, depending on who actually sets the work out.

When instruments are sent to the site from the contractor's main stores or plant department they are seldom checked unless complaints have been made by the previous user. It is, therefore, essential at the commencement of setting out operations on-site, and periodically thereafter, that the instruments and equipment are checked, and if minor faults are discovered to:

● return to the yard store where they can be sent for adjustment or repair and in the mean time get replacements;
● make allowances for faults/inaccuracies during their use;
● make adjustments to the instruments, etc., on-site. (With a little training and skill adjustments can be made reasonably quickly.)

In making allowances for faults one requires a basic understanding of the instruments used, which will become apparent later in the chapter.

The instruments and equipment to which one is required to pay special attention are as follows:

1. *Tapes* – 20 metres or 30 metres long. Steel tapes must be used for accuracy and not plastic or fibreglass.

2. *Surveyors' levels* – Dumpy; – Quickset; – Automatic (self levelling).
3. Theodolite – Microptic (adjustments are similar to a Vernier instrument although the parts are encased); Vernier (the older type of design).
4. *Automatic plumbing instrument* – Optical (Hilger and Watts Autoplumb, etc); – Laser (also used for levelling).
5. *Other instruments and equipment* – Cowley level; Sitesquare; Boning rods; travellers.

Tapes

A tape can be visibly inspected for damage, but more important its actual length can be equated against what should be a correct length. This is done by checking one part of its length by another part, i.e. the 0 metres to 5 metres can be checked against the 10 metres to 15 metres. Similarly the 2 metres to 7 metres can be checked against the 13 metres to 18 metres, and so on. This procedure checks whether parts of a tape have stretched or not.

Finally, the only accurate check on a tape length is to record a true 20 metres (or 30 metres) on the side of a permanent structure, or retain a new tape for checking other tapes.

Surveyor's level

A visible inspection of the instrument can reveal many faults. The kind of faults normally encountered are associated with loose hinge screws on the tripod, and small grub screws being either missing or ineffective on such places as the eye lens and focusing screw, apart from the more obvious signs of misuse and damage caused by the instrument falling.

Adjustments to levels are of two types:

1. Temporary adjustments.
2. Permanent adjustments.

Temporary adjustments

Temporary adjustments are the normal adjustments made by the user of an instrument when first setting it up for use, i.e.

1. Standing and securing the tripod firmly on to the ground.
2. Screwing the instrument firmly to the tripod.
3. Levelling the instrument's principal bubble.
4. Eliminating parallax (cross hairs focussed so that they are clearly defined by adjusting the eye lens while sighting through the telescope on to a white piece of paper held close to the object lens to reflect light into the telescope).
5. Focusing the telescope onto a calibrated surveyor's staff.

Permanent adjustment

Surveyor's levels can be checked for accuracy but the temporary adjustments should be undertaken first before attempting to check if permanent adjustments are necessary.

There are two checks which can be made to a surveyor's level by reference to:

1. The principal bubble.
2. The diaphragm (cross hairs).

The principal bubble is checked by simply setting up the instrument and levelling the telescope while it faces in one direction; the bubble would then be expected to remain in the centre of its tube when the telescope is swung horizontally to face in the opposite direction (this check need not be conducted on automatic levels).

The diaphragm is tested while the instrument is set up, but after the bubble has been checked and adjusted, using the 'two peg test' (see Fig. 11.2.1).

The theodolite

Before the theodolite can be checked for accuracy the temporary adjustments should be carried out in a similar manner to a surveyor's level. There are, however, four permanent adjustments which could be made to certain sensitive parts if the instrument is in error, the adjustments being to:

1. The principal bubble – the method of checking the bubble is exactly the same as for a surveyor's level; see the section on the level.
2. The diaphragm alignment (horizontal movement) – see Fig. 11.2.2 (A).
3. The trunnion axis – see Fig. 11.2.2 (B).
4. The diaphragm (vertical movement) – the same checked method as used on the level. See the 'two peg test' in Fig. 11.2.1.

Note: The altitude bubble which is situated on the vertical scale should also be checked in a similar manner to that of the principal bubble, but only after the principal bubble has been checked and if necessary, corrected. This would only be necessary if the instrument is to be used for recording vertical angles.

The theodolite, therefore, should be checked for accuracy and adjustments made in the order, as outlined, starting with the principal bubble.

If the errors in an instrument are very small because the preciseness of adjustment is difficult to achieve, the instrument may be used provided that the errors are corrected by using the technique of 'equalising errors'. This is to say that change-face readings should be taken which give mean readings for each setting out point. If one checks Fig. 11.2.2 (A) where there is a diaphragm alignment problem, then peg D position cancels out errors of this kind when setting out long straight lines.

Additionally, when plumbing tall buildings and the trunnion axis is not at right angles to the vertical axis, as in Fig. 11.2.2 (B), provided that a change-face is made as a check, the mid position of the face-right and face-left readings would be taken as the true point.

Finally, to overcome small inaccuracies in the calibrations of the theodolite scales face-left and face-right readings should be taken by the 'repetition' technique (a few readings taken of the same angle with the average reading being recorded) – preferably by using a different zero, say 10°. Assume one reads an angle of 30°, deduct the zero of 10°, and the angle to be recorded would be 20°. Repeat this procedure, using a zero of 45°, take the same angle again which now reads 65° 20′ 0″ – and deduct the zero of 45°, which equals 20° 20′ 00″. The average of the two answers is recorded (20° 20 00″ + 20° 00′ 00″) ÷ 2 = 20° 10′ 00″.

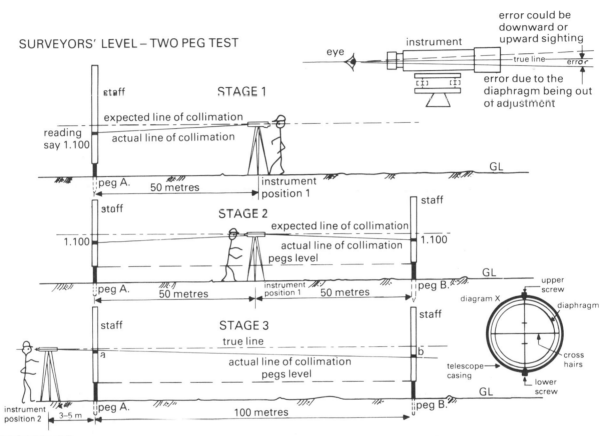

PROCEDURE:

1. Set up instrument on approximately level piece of ground (stage 1). Fix peg 'A' at 50 metres distance and take a reading on to a staff held on peg 'A'. Say the reading is 1.100 metres (the bubble check should be carried out first and if necessary should be adjusted).
2. Fix peg 'B' in opposite direction 50 metres from the instrument and knock it into the ground until the identical reading to that obtained on peg 'A' is achieved. So, although there may be an error in the instrument provided that the staff readings and the distances of the pegs from the instrument are the same pegs 'A' and 'B' will be level with each other (stage 2).
3. Now move the instrument from position 1 to position 2 (stage 3) and take readings onto a staff held on peg 'A' and peg 'B'. The readings should be the same if the instrument is true. If the readings are different, as in diagrams X, at a. and b. then an adjustment is necessary to the diaphragm by means of the screws – see diaphragm diagram.
4. Once the diaphragm is adjusted the checking procedure (stage 3) should be repeated until readings a. and b. are identical.

NOTE

a. If the instrument is seriously out of true the normal procedure is to send it to the manufacturer for repair/adjustment.
b. If there is a minor inaccuracy the instrument may be used provided sightings are kept to short distances, and the foresight and backsight readings are always the same distance from the instrument.

Fig. 11.2.1   Checking instruments for accuracy

THE THEODOLITE
(A) DIAPHRAGM ALIGNMENT

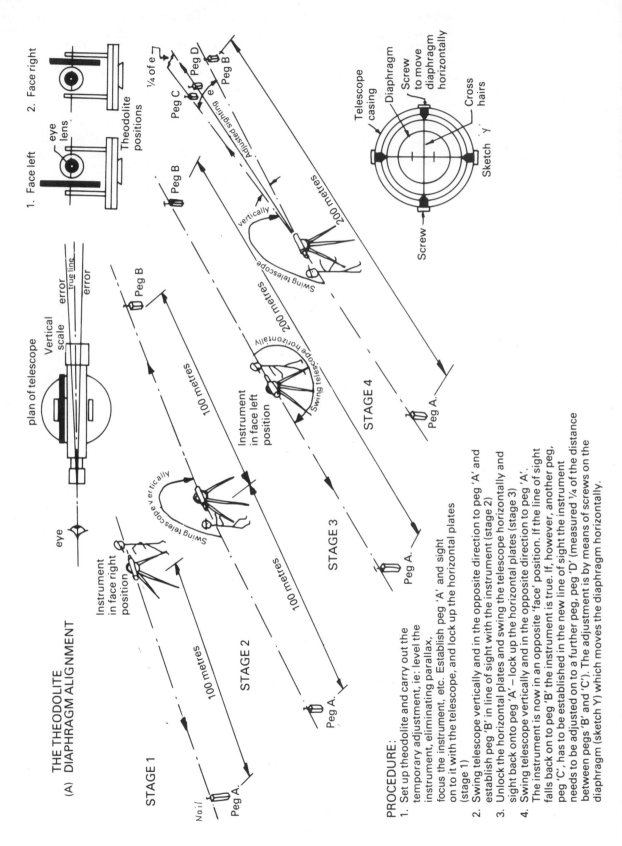

PROCEDURE:

1. Set up theodolite and carry out the temporary adjustment, ie: level the instrument, eliminating parallax, focus the instrument, etc. Establish peg 'A' and sight on to it with the telescope, and lock up the horizontal plates (stage 1)

2. Swing telescope vertically and in the opposite direction to peg 'A' and establish peg 'B' in line of sight with the instrument (stage 2)

3. Unlock the horizontal plates and swing the telescope horizontally and sight back onto peg 'A' – lock up the horizontal plates (stage 3)

4. Swing telescope vertically and in the opposite direction to peg 'A'. The instrument is now in an opposite 'face' position. If the line of sight falls back on to peg 'B' the instrument is true. If, however, another peg, peg 'C', has to be established in the new line of sight the instrument needs to be adjusted on to a further peg, peg 'D' (measured ¼ of the distance between pegs 'B' and 'C'). The adjustment is by means of screws on the diaphragm (sketch Y) which moves the diaphragm horizontally.

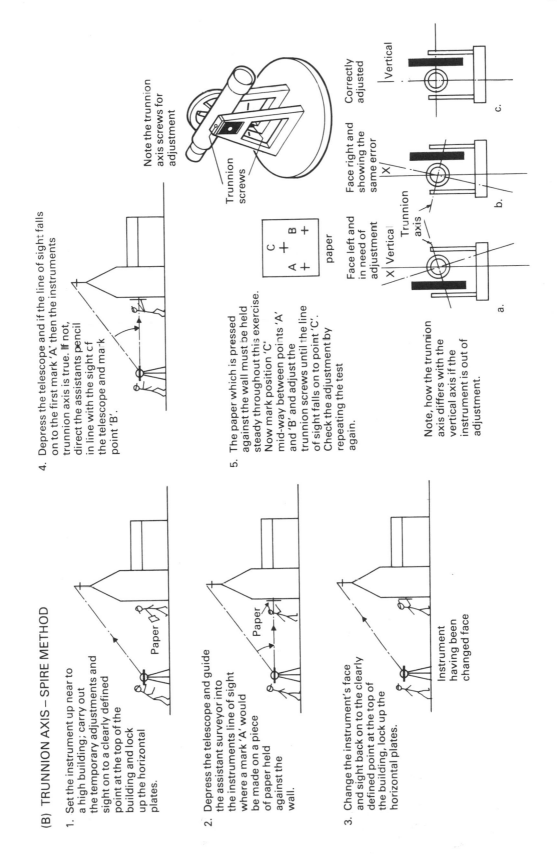

(B) TRUNNION AXIS – SPIRE METHOD

1. Set the instrument up near to a high building; carry out the temporary adjustments and sight on to a clearly defined point at the top of the building and lock up the horizontal plates.

2. Depress the telescope and guide the assistant surveyor into the instruments line of sight where a mark 'A' would be made on a piece of paper held against the wall.

3. Change the instrument's face and sight back on to the clearly defined point at the top of the building, lock up the horizontal plates.

Instrument having been changed face

4. Depress the telescope and if the line of sight falls on to the first mark 'A' then the instruments trunnion axis is true. If not, direct the assistants pencil in line with the sight of the telescope and mark point 'B'.

Note the trunnion axis screws for adjustment

Trunnion screws

5. The paper which is pressed against the wall must be held steady throughout this exercise. Now mark position 'C' mid-way between points 'A' and 'B' and adjust the trunnion screws until the line of sight falls on to point 'C'. Check the adjustment by repeating the test again.

C
+
A   B
+   +

paper

Note, how the trunnion axis differs with the vertical axis if the instrument is out of adjustment.

Face left and in need of adjustment

X Vertical

Trunnion axis

a.

Face right and showing the same error

X

Trunnion axis

b.

Correctly adjusted

Vertical

c.

Fig. 11.2.2   Checking instruments for accuracy

Automatic plumbing instruments

Small errors in these types of instruments may be allowed for while in use by rotating the instrument about its horizontal axis and determining the mean (average) of the reading. See Fig. 11.5.2, for the method adopted while plumbing high structures.

Other instruments and equipment

The Cowley level

This level would be unusable if one fails to get the instrument's two mirrors to coincide. Also one of the mirrors must float (move) and act as a compensator which cancels out errors due to inaccurate setting up by the user. This instrument was designed to be used by untrained site staff/operatives. To set up before using one should stand the tripod firmly on to the ground and then place the level on the tripod. The instrument should be swivelled around horizontally 360° while the user looks into the eye pieces. The instrument is correctly set up if the two mirrors coincide (see Fig. 11.2.3 (Ai), (Aii), (Aiii).

Sitesquare

This optical instrument was designed for setting out right angles, which is most useful for setting out the corners of square buildings and areas on flat or sloping ground (see Fig. 11.2.3 (Bi)). The instrument should be checked to find if it is adequate for setting out right angles, and if not, to enable a correction for any inaccuracies to be made by the user (see Fig. 11.2.3 (Bii)).

Boning rods

This type of equipment (three in number) is used by operatives once the main levels or control points have been established. The check is to ensure that each set of three boning rods are exactly the same length. (See Fig. 11.5.3 (B) for samples of boning rods.)

Travellers

Check that the length of any traveller agrees with the difference between the RL of the appropriately fixed sight rails and the formation level of the area or the trench which is to be excavated. (See Figs 11.6.1 (f) and 11.4.2(Ei) for samples of a traveller.)

11.3 Setting-out of buildings

Before any serious setting-out can be undertaken on-site it may be necessary to have the site cleared of shrubs, trees or derelict buildings. Care should be exercised to ensure that the trees to be removed are those indicated on the drawings by the designer. It may even be necessary to check that there are no infringements of Tree Preservation Orders relating to the site.

The setting-out of proposed structures should be done at the earliest opportunity: this normally depends on how long the site establishment takes (access, hutting, temporary services). To prevent disruption of the works careful positioning of site accommodation is essential from site layout plans which are generally produced by the contractor's planning department.

The proposed structure's position can be determined by:

1. Measuring from adjoining properties, features or centrelines of existing or proposed new roads; the sizes should be extracted from the designer's drawings (see Fig. 11.3.2 (A)).
2. Using reference points on the boundary of the site to establish the bearing of the main or front face of the proposed structure. The distances along this bearing line can be scaled from the drawing, then transferred on to site (see Fig. 11.3.2 (B)).
3. Using triangulated lines and angles which are calculated from reference pointed stations or triangulation stations situated on the drawings but which can then be transferred on to site (see Fig. 11.3.2 (C)).
4. Using a site grid which is prepared on the drawings and is later transferred around the site boundary (from 10 metres to 100 metres grid depending on the slope of the site). Any structural point, road position or inspection chamber may be simply located using coordinates found by measuring accurately from the drawings and which are then transferred on to site (see Fig. 11.3.2 (D)).

Site grid

This is the more versatile system for establishing a structure on-site and in Fig. 11.3.1 a site grid has been superimposed over the designer's drawings of the site plan. This may be done by the designer, or the site supervisor would be expected to draw out the grid if it had been omitted from the drawings. In the example (Fig. 11.3.1) a site grid spacing of 100 metres has been chosen, and has been drawn out using the same scale as the site plan, i.e. 1 : 2500.

The following list indicates the remainder of the work associated with setting out the building in the correct position on-site, and which also includes profiles.

1. Determine the coordinates of the main face of the building from the drawings; i.e. 203.00 N; 140.60 E and 224.50; 188.50 E respectively. Other positions can be determined – the inspection chambers and tangent points of curved roads. Also, the anticipated positions of the first site grid line can be located by measuring from the boundary of the site, i.e. 8 metres; 3.5 metres to site grid station A; and 3.4 metres for ranging in the site grid stations B, C, D, E and F (see Fig. 11.3.1 and 11.3.3 A).
2. Now, establish the site grid station points around the site using a theodolite and tape (see Fig. 11.3.3 (A), (B) and (C)).
3. Determine from the pegs 1, 2, 3 and 4 the proposed corners of the building A and B (see Fig. 11.3.3 (C)).
4. Set out building corners C and D using theodolite and tape (see Fig. 11.3.4 (B)).

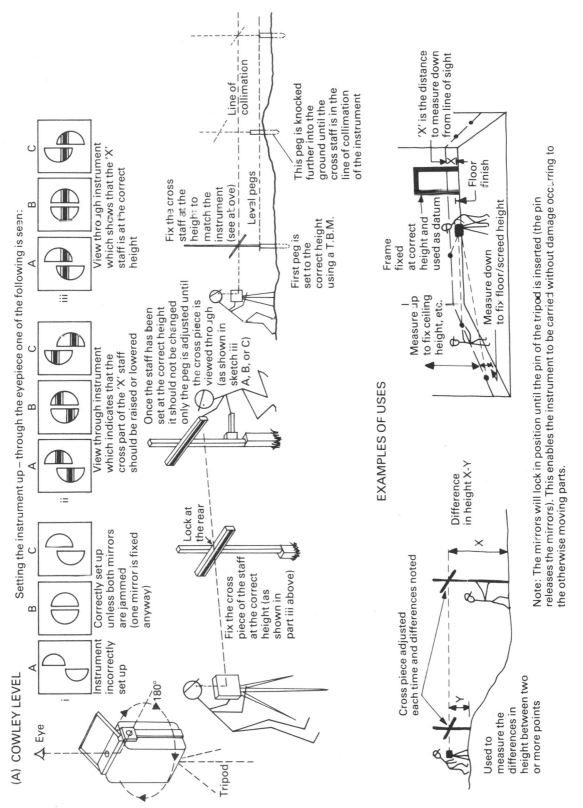

(A) COWLEY LEVEL

Setting the instrument up – through the eyepiece one of the following is seen:

Eye

180°

Tripod

Instrument incorrectly set up

Correctly set up unless both mirrors are jammed (one mirror is fixed anyway)

View through instrument which indicates that the cross part of the 'X' staff should be raised or lowered

View through instrument which shows that the 'X' staff is at the correct height

Fix the cross piece of the staff at the correct height (as shown in part iii above)

Lock at the rear

Once the staff has been set at the correct height it should not be changed only the peg is adjusted until the cross piece is viewed through the cross piece (as shown in sketch iii A, B, or C)

Line of collimation

Fix the cross staff at the height to match the instrument (see above)

Level pegs

First peg is set to the correct height using a T.B.M.

This peg is knocked further into the ground until the cross staff is in the line of collimation of the instrument

EXAMPLES OF USES

Frame fixed at correct height and used as datum

Measure up to fix ceiling height, etc.

Measure down to fix floor/screed height

'X' is the distance to measure down from line of sight

Floor finish

Cross piece adjusted each time and differences noted

Difference in height X-Y

Used to measure the differences in height between two or more points

Note: The mirrors will lock in position until the pin of the tripod is inserted (the pin releases the mirrors). This enables the instrument to be carried without damage occurring to the otherwise moving parts.

Fig. 11.2.3 Checking instruments for accuracy

(B) SITESQUARE

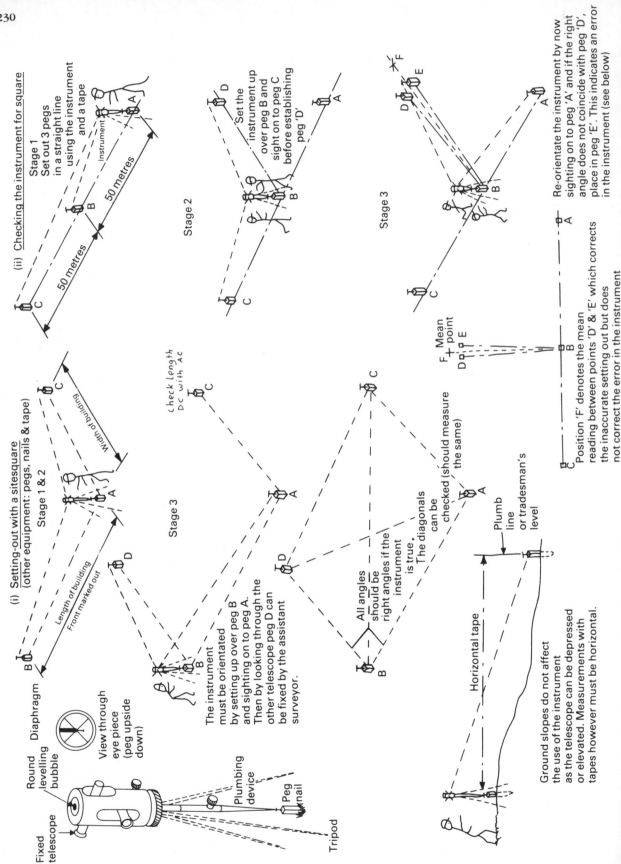

(i) Setting-out with a sitesquare
(other equipment: pegs, nails & tape)

Stage 1 & 2

Stage 3

The instrument must be orientated by setting up over peg B and sighting on to peg A. Then by looking through the other telescope peg D can be fixed by the assistant surveyor.

Width of building

Length of building
Front marked out

check length
DC with AC

All angles should be right angles if the instrument is true. The diagonals can be checked (should measure the same)

Plumb line or tradesman's level

Horizontal tape

Ground slopes do not affect the use of the instrument as the telescope can be depressed or elevated. Measurements with tapes however must be horizontal.

Fixed telescope

Round levelling bubble

Diaphragm

View through eye piece (peg upside down)

Plumbing device

Peg or nail

Tripod

(ii) Checking the instrument for square

Stage 1
Set out 3 pegs in a straight line using the instrument and a tape

Instrument

50 metres

50 metres

Stage 2

Set the instrument up over peg B and sight on to peg C before establishing peg 'D'

Stage 3

Re-orientate the instrument by now sighting on to peg 'A' and if the right angle does not coincide with peg 'D', place in peg 'E'. This indicates an error in the instrument (see below)

F Mean point E

Position 'F' denotes the mean reading between points 'D' & 'E' which corrects the inaccurate setting out but does not correct the error in the instrument

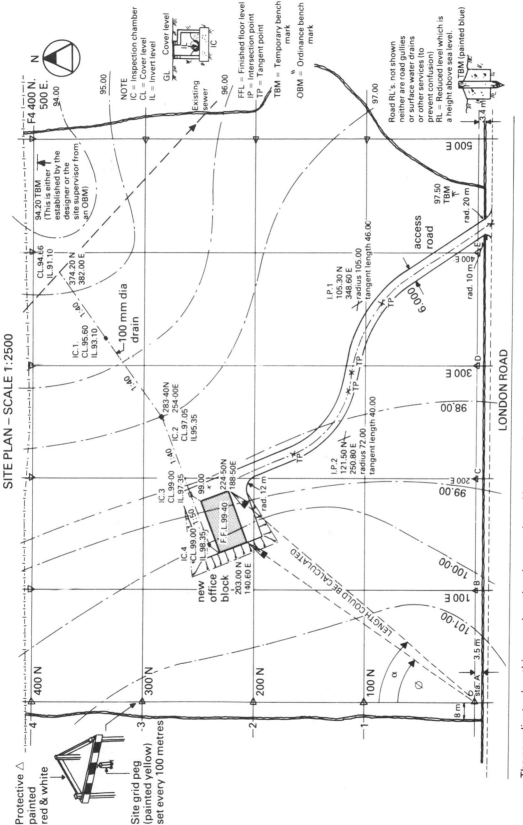

SITE PLAN – SCALE 1:2500

Fig. 11.3.1

The coordinates (northings and eastings) are evaluated by comparing the setting out points with the site grid. The site grid must first be superimposed on the site plan drawing which is later transferred on to the site. The major setting out points and centrelines may be fixed by the use of the grid or by calculating the bearings and lengths from one or more control stations, i.e: station 'A' to the first corner of the building. See evaluations in figure 11.3.3A to C.

NOTE
IC = Inspection chamber
CL = Cover level
IL = Invert level

FFL = Finished floor level
IP = Intersection point
TP = Tangent point

TBM = Temporary bench mark
OBM = Ordinance bench mark

GL Cover level

Existing sewer

Road RL's. not shown neither are road gullies or surface water drains or other services (to prevent confusion)
RL = Reduced level which is a height above sea level.

LONDON ROAD

(A) Measuring from adjoining property, etc.

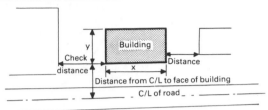

(B) Establishing face of building by projecting to the boundary and providing reference pegs

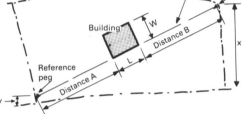

Line drawn through to boundaries from the face of the building and the distances are scaled off the drawings (A & B). Also distances x & y are scaled off so that the reference pegs can be established on-site which shows the face of the building orientation. Distances A, L & B are measured along the orientation line to fix the building's position on the site.

(C) Triangulation lines and angles

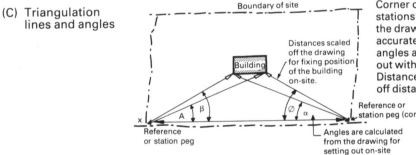

Corner of site pegs or triangulation stations' positions are determined from the drawings. They are next located accurately on-site from which the angles and triangle lines are set out with theodolite and tape. Distance xy must agree with scaled off distance from the drawings.

(D) Site grid and coordinates

(i) The front of the building is determined from the drawings first by scaling the coordinates of corners A & B from 00.
∴ corner A coordinates are 212 N and 165 E
and, corner B coordinates are 225 N and 285 E
(coordinates in metres)

(ii) The outside grid intersection points + would be set out on-site by theodolite and tape. The building's corners A & B would be set out from these grid points.

(iii) The position of the grid on the drawing would be determined on-site by measuring the first grid lines from the boundary.

Fig. 11.3.2 Setting-out buildings

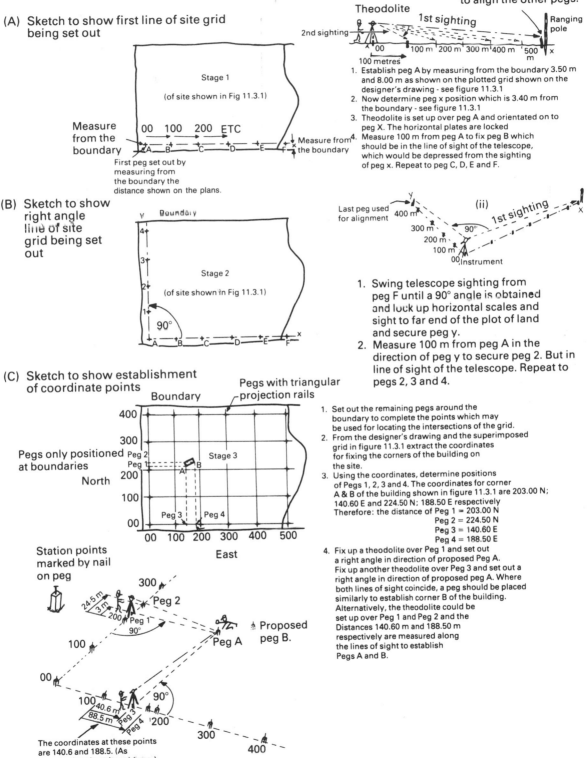

(A) Sketch to show first line of site grid being set out

Stage 1

(of site shown in Fig 11.3.1)

Measure from the boundary

00 100 200 ETC

A   B   C   D   E F

Measure from the boundary

First peg set out by measuring from the boundary the distance shown on the plans.

(i) Theodolite

This peg is only used to align the other pegs.

1st sighting

2nd sighting

Ranging pole

00 100 m 200 m 300 m 400 m 500 m x

100 metres

1. Establish peg A by measuring from the boundary 3.50 m and 8.00 m as shown on the plotted grid shown on the designer's drawing - see figure 11.3.1
2. Now determine peg x position which is 3.40 m from the boundary - see figure 11.3.1
3. Theodolite is set up over peg A and orientated on to peg X. The horizontal plates are locked
4. Measure 100 m from peg A to fix peg B which should be in the line of sight of the telescope, which would be depressed from the sighting of peg x. Repeat to peg C, D, E and F.

(B) Sketch to show right angle line of site grid being set out

Boundary

y

4
3
2
1
90°

Stage 2

(of site shown in Fig 11.3.1)

A B C D E F x

(ii)

Last peg used for alignment 400 m

y

300 m 90°
200 m
100 m
00 Instrument

1st sighting

x

1. Swing telescope sighting from peg F until a 90° angle is obtained and lock up horizontal scales and sight to far end of the plot of land and secure peg y.
2. Measure 100 m from peg A in the direction of peg y to secure peg 2. But in line of sight of the telescope. Repeat to pegs 2, 3 and 4.

(C) Sketch to show establishment of coordinate points

Pegs with triangular projection rails

Boundary

400
300
200
100
00

Pegs only positioned at boundaries

North

Peg 2
Peg 1
A B

Stage 3

Peg 3 Peg 4

00 100 200 300 400 500

East

Station points marked by nail on peg

24.5 m
3 m
200 Peg 1
90°

300 Peg 2

Proposed peg B.

100

Peg A

00

100
40.6 m
88.5 m Peg 3
Peg 4 200

90°

300

400

The coordinates at these points are 140.6 and 188.5. (As there is already a site grid peg.)
At 100 m measure 40.6 and 88.5 metres along

1. Set out the remaining pegs around the boundary to complete the points which may be used for locating the intersections of the grid.
2. From the designer's drawing and the superimposed grid in figure 11.3.1 extract the coordinates for fixing the corners of the building on the site.
3. Using the coordinates, determine positions of Pegs 1, 2, 3 and 4. The coordinates for corner A & B of the building shown in figure 11.3.1 are 203.00 N; 140.60 E and 224.50 N; 188.50 E respectively
Therefore: the distance of Peg 1 = 203.00 N
Peg 2 = 224.50 N
Peg 3 = 140.60 E
Peg 4 = 188.50 E
4. Fix up a theodolite over Peg 1 and set out a right angle in direction of proposed Peg A. Fix up another theodolite over Peg 3 and set out a right angle in direction of proposed peg A. Where both lines of sight coincide, a peg should be placed similarly to establish corner B of the building. Alternatively, the theodolite could be set up over Peg 1 and Peg 2 and the Distances 140.60 m and 188.50 m respectively are measured along the lines of sight to establish Pegs A and B.

Fig. 11.3.3 Setting-out buildings

(A) Structural grid

The structural grid follows the centrelines of the columns and walls. The points must be marked out first on to profile boards.

(B) When corners 'A' and 'B' have been established from coordinates a theodolite is set up over pegs 'A' and 'B', and pegs C and D are established at right angles.

Coordinates of point A

(C) Profiles set at suitable level and should be continuous where possible.

Profiles to be fixed parallel to corner pegs of proposed building and made of wood or tubing.

(D) A theodolite is set up over peg 'A' only. After sighting on to peg 'B' the telescope is elevated on to the profile and a point b1 is marked. The telescope is swung vertically to sight in to the opposite direction. A mark a1 is made on to the profile.

(E) Now sight on to peg 'C' and then elevate the telescope on to the profile and mark position c1: Next swing the telescope vertically and sight in the opposite direction and on to the profile and mark a2. Each profile now has the first mark from which the structural grid is measured.

(F) Centreline marks of structural grid

This plan shows how the structural grid is set out on each profile by measuring from each point a1, a2, b1 and c1 along the profiles only.

(G) Pegs A, B, C & D removed once profiles are marked out

If line (nylon) were strung between the profiles the structural grid would easily be found on the plot.

a2 Structural grid points

(H) Pad foundation base marked out using line of sight of theodolite, or by using builders' nylon line.

Profile

Structural grid

(I) Sight rails set out for reduced level dig because the site is sloping and needs to be levelled off.

Profile

Stepped continuous profile on sloping ground

Traveller used by excavator driver or banksman

Sight rails set out level with other sight rails positioned around the site

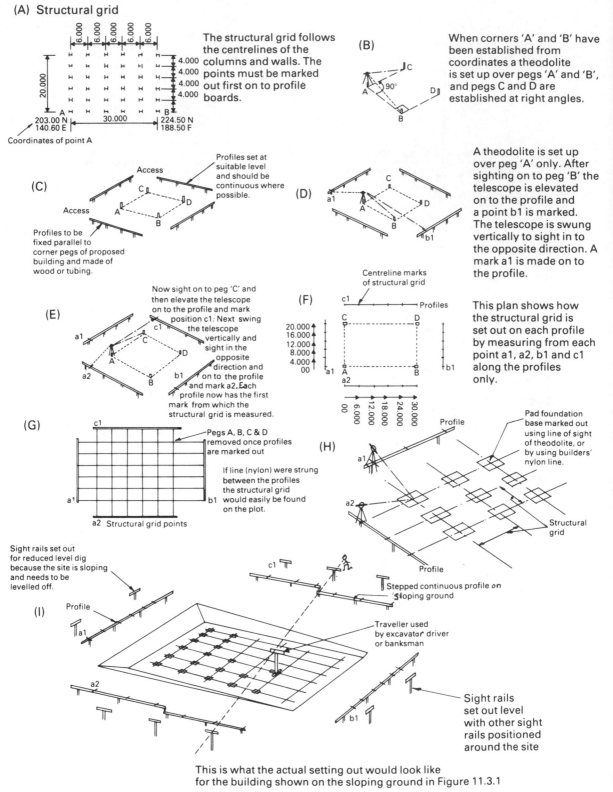

This is what the actual setting out would look like for the building shown on the sloping ground in Figure 11.3.1

Fig. 11.3.4 Setting-out buildings – steel or reinforced concrete frame

5. Determine the structural grid by checking from the drawings the centreline distances of columns and structural walls and mark this grid on the profiles. (see structural grid as follows).

Structural grid

The centrelines of the structural elements (columns and walls) forms the basis of the structural grid (see Fig. 11.3.4). The distances between the already established main corners of the building A and B will not normally be absolutely accurately positioned, The accuracy is determined when the profiles are marked out, as follows:

1. Fix pegs or posts in sufficient numbers, and which are parallel to the sides of the proposed building, to enable profile boards to be assembled level and at a suitable height from the ground. They should also be at a reasonable distance from the works to prevent excavators and vehicles from damaging them, and enable excavation work to proceed unhindered and allow access to the works.

   It is generally advised that the profile boards are continuous and are fixed at predetermined heights thereby enabling them to act as sight rails (although in the example separate sight rails are assumed).

2. Set up the theodolite over point A (which is one corner of building) and sight on to peg B (the other corner). The telescope is elevated on to the profile beyond peg B and a mark point b1, should be made.

3. Swing the telescope 180° vertically to sight into the opposite direction onto the profile beyond peg A and mark point a1.

4. Sight the telescope of the theodolite on to peg C and repeat the process to mark out points c1 and a2 (see Fig. 11.3.4 (D) and (E)).

   The profiles at this stage have one mark on each from which to mark out the remaining centrelines of the structural grid.

5. Mark out the centrelines of the structural elements (column and walls) by laying the tape along the top of the continuous profiles, which prevents the tape from sagging and allows measurements to be accurate (see Fig. 11.3.4 (F)).

6. Either builders' line or the line of sight of a theodolite is next used to determine the centreline positions of each proposed column base or wall foundation on the ground (see Fig. 11.3.4 (G) and (H)).

Note: Figure 11.3.4 (i) illustrates the profile and sight rail positions for the building on the sloping ground as shown in Fig. 11.3.1.

11.4 Setting out drains and private sewers

The site supervisor must appreciate that drains should be constructed in such a way as to conform with the Building Regulations, the main requirements being:

1. Drains should be laid straight between inspection chambers and the joints should be watertight.

2. Drains should be laid to suitable gradients to allow self cleansing.

3. Inspection chambers should be spaced not more than every 90 metres; at changes of drain direction; or at changes of gradient, but where there is a junction the distance of an inspection chamber should not exceed 12.5 metres, as shown in Fig. 11.4.2 (A).

4. Inspection chambers or rodding eyes should also be situated at the top end of drain runs.

With the above points in mind the supervisor or site engineer should next check the designer's drawings and extract the following information before the setting out process commences.

1. Sizes of the drains and types of pipes; bed construction; and details of any surround materials.

2. Sizes and numbers of the inspection chambers.

3. Cover levels of the inspection chambers.

4. Flow directions.

5. Invert levels of the drains (at the inspection chamber positions) and check if there are any back-drop inspection chambers.

6. Invert level at the discharge point, e.g. the sewer position.

7. Positions of existing services (electricity, gas, etc.) to prevent damage during excavation work.

8. Coordinate positions of the inspection chambers. If there are none given they should be determined by the superimposing of a site grid.

Finally, the Method Statement of the Contract Programme should be referred to which usually states when the various sections of drainage should be carried out. The sewers and certain drains may have to be completed at the earliest opportunity to enable roadworks to be undertaken.

The determination of sizes of drains or sewers is not the responsibility of the supervisor or site engineer, but if the wrong sizes appear to have been quoted on the drawings then query them with the designer and await instructions before proceeding with the setting out.

Procedure for setting out

In outlining the procedures to be followed when setting out drains reference is made to Fig. 11.3.1 for drain layout on a sloping site, and Fig. 11.4.1 for more precise details of the same proposed drain which shows an enlarged plan layout and longitudinal section.

It should be noted however, that longitudinal sections are generally only prepared for public sewers or large drains.

The procedure, therefore, is as follows:

1. Locate the original sewer at the point where the drain run is to connect and check the access point invert level (this is done by digging down to the sewer). The coordinates to locate the exact connection position can be determined from the superimposed site

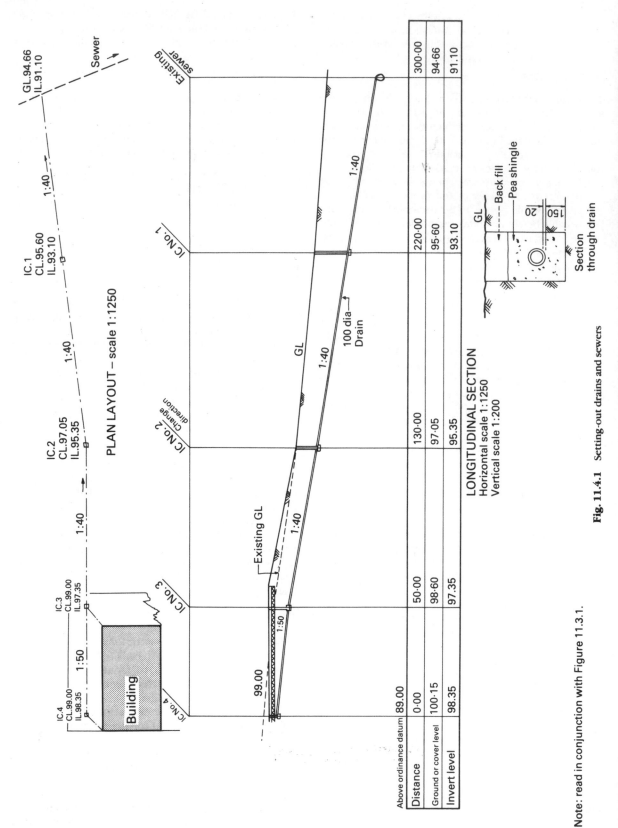

PLAN LAYOUT – scale 1:1250

LONGITUDINAL SECTION
Horizontal scale 1:1250
Vertical scale 1:200

Section through drain

Above ordinance datum 89.00						
Distance	0·00	50·00	130·00	220·00	300·00	
Ground or cover level	100·15	98·60	97·05	95·60	94·66	
Invert level	98.35	97.35	95.35	93.10	91.10	

Fig. 11.4.1 Setting-out drains and sewers

Note: read in conjunction with Figure 11.3.1.

grid, i.e. 374.20 m N and 382.00 m E (see Fig. 11.3.1) using the technique previously described in section 11.3.

2. Determine the coordinates of inspection chambers (IC2) where the change of direction of the drain takes place, i.e. 283.40 m N and 254.00 m E (see Fig. 11.3.1). A white peg would indicate the centre of the inspection chamber.

3. Determine the positions of IC3 and IC4 by measuring from the building (if it has already been set out), or by once again using coordinates. The distance between each IC can be checked from the longitudinal section (see Fig. 11.4.1).

4. The position of IC1 may simply be determined by ranging in between the peg marking the centre of IC2 and the point marking the proposed connection to the existing sewer.

   Remember that the centreline points are shown by white pegs, which also represents the centreline of the proposed drain.

5. Offset pegs are next driven into the ground at a suitable distance from the centreline pegs to enable sight rails to be accurately positioned, but additionally, to assist in the relocation of the centreline points which may be accidently damaged or moved before the drain excavations are started. Also, the distance of these offset pegs (coloured yellow) from the centreline pegs should be sufficient to allow a mechanical excavator to pass by as it carries out the trench dig; and also allows the easy use of a traveller (say from 3 to 4 metres long.) (See Fig. 11.4.2 (C).)

6. Sight rails are next erected to a suitable height above ground level and which allows for an adequate traveller to be used. The sight rails' positions are established directly over the offset pegs at each inspection chamber. Intermediate sight rails can also be used to assist in allowing excavation work and laying of drains to be executed more accurately.

   Assuming one wishes to use the same traveller length throughout (this is not always possible, therefore traveller heights should be marked on the sight rails – as shown in Fig. 11.4.2 (D)); as an example, for the calculations that are necessary see the longitudinal section in Fig. 11.4.1 and the inspection chamber details, and evaluate the differences between the existing ground levels (GLs) and the invert levels (ILs):

Calculations

	IC No. 4	IC No. 3	IC No. 2
GL	100.15 m	98.60 m	97.05 m
IL	98.35 m	97.35 m	95.35 m
Difference	1.80 m	3.25 m	1.70 m

	IC No. 1	Sewer connection
GL	95.60 m	94.66 m (point of entry into the sewer)
IL	93.10 m	91.10m
Difference	2.50 m	3.56 m

The differences now show the depths of the excavations which are necessary at the inspection chamber positions.

One always attempts to secure sight rails not less than 0.5 metres above GL. So, by choosing the deepest excavation point – which occurs at the sewer connection – one determines the length of the traveller. Therefore, the traveller is made up as shown in Fig. 11.4.2 (E), the length being determined as follows but with modifications for the pipe thickness and thickness of bed (pea shingle):

length of traveller = 3.56 + 0.5 = 4.06 metres.
sight rails heights (or RLs) can now be calculated using the following:
RL of sight rail = Invert Level (or IL) + length of traveller.
Therefore RLs of sight rails are:

at sewer connection:	91.10 + 4.06 =	95.16 m (RL)
at IC No. 1	93.10 + 4.06 =	97.16 m (RL)
at IC No. 2	95.35 + 4.06 =	99.41 m (RL)
at IC No. 3	97.35 + 4.06 =	101.41 m (RL)
at IC No. 4	98.35 + 4.06 =	102.41 m (RL)

Note: The RL means the height of a point or position above the mean sea level.

With the sight rail heights now calculated one must check whether they will be at suitable heights to enable the banksman or excavator driver to sight through easily, or too high to be of any use.

Therefore, check as follows:
Height of sight rail above GL = sight rail RL − RL of the GL.
See calculations as follows:

Position	Site rail RL	GL RL	Sight rail above GL	Comment
at sewer :	95.16 −	94.66 =	0.50 m	satisfactory
IC 1 :	97.16 −	95.60 =	1.56 m	satisfactory
IC 2 :	99.41 −	97.05 =	2.36 m	too high
IC 3 :	101.41 −	98.60 =	2.81 m	too high
IC 4 :	102.41 −	100.15 =	2.26 m	too high

If a sight rail is higher than 1.50 metres above GL operatives will find difficulty in sighting (boning) through from one to another. Therefore where sight rails are too high then one must consider lowering each of them in turn by an equal amount. An equal adjustment would also have to be made to the traveller.

Where the ground undulates considerably an alternative arrangement would be to consider each drain length separately (between inspection chambers), with different traveller heights and stepped or double sight rails being used (see Fig. 11.4.2 (F) and (G)).

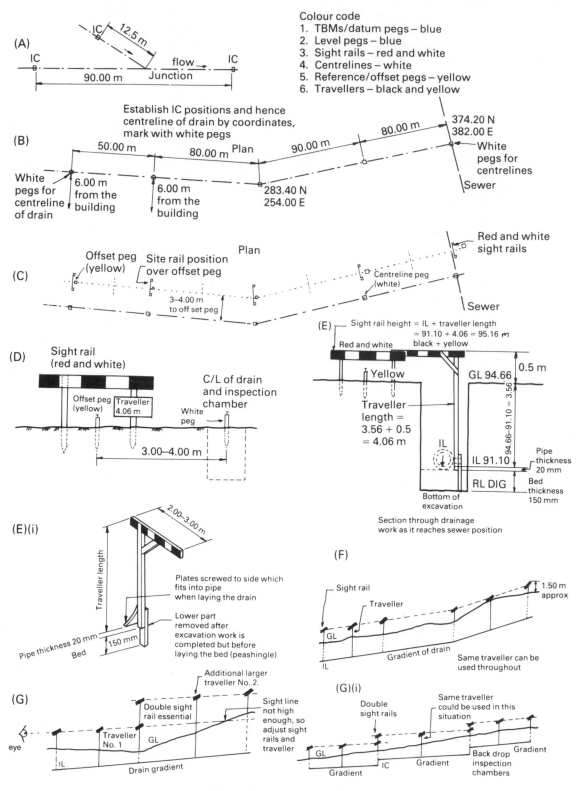

(A)

Colour code
1. TBMs/datum pegs – blue
2. Level pegs – blue
3. Sight rails – red and white
4. Centrelines – white
5. Reference/offset pegs – yellow
6. Travellers – black and yellow

Fig. 11.4.2   Setting-out drains

Now consider Fig. 11.4.3 (Ai) and (Aii) which include the sight rails set at the originally calculated reduced levels, thereby satisfying a traveller of length 4.06 metres.

The second diagram, Fig. 11.4.3 (Bi) and (Bii), shows the sight rails adjusted to a more suitable lower level to conform with the last set of calculations and which will allow an individual to bone in through the sight rails comfortably.

Procedure for fixing sight rails

Having determined the reduced levels of the sight rails at each inspection chamber (sometimes referred to as manholes) the supervisor or site engineer must now fix into the ground pegs or posts over the offset pegs. Ideally, the pegs or posts should be long enough to enable sight rails to be fixed across them at the appropriate reduced levels. If this is not possible then extension pieces to the initial pegs/posts must be fixed sufficiently high enough to accommodate the sight rails (see Fig. 11.4.4 (D)).

To fix sight rails at the correct heights:

1. Commence by setting up the surveyor's level at a convenient position overlooking, if possible, each inspection chamber position and carry out the temporary adjustments.
2. Direct the assistant with a surveying staff on to the TBM of 94.20 m, or at a position whose height is known, and observe a back-sight reading of, say, 4.30 m. This will then enable the height of instrument (HI), better known as the highest point of collimation (HPC), to be found, i.e. 94.20 m + 4.30 m = 98.50 m (see Fig. 11.4.4 (A)).
3. At each proposed sight rail position take a level reading on to a staff placed on one of the pegs/posts which will support the site rail and calculate the reduced level. In Fig. 11.4.4 two positions are being dealt with – the sewer position and the IC 1 position. Therefore, assuming the readings are 2.40 m and 1.80 m respectively, the reduced levels are calculated by subtracting the readings from the HI (98.4 m), i.e.:
    (a) sight rail peg RL at sewer = 98.5 m – 2.4 m = 96.1 m;
    (b) sight rail reg RL at IC1 = 98.5 m – 1.8 m = 96.7 m.

Because the sight rails at the sewer and IC1 need to be fixed at RLs of 95.16 m and 97.16 m respectively, at the sewer position one measures down the peg 96.1 m – 95.16 = 0.94 m, because the RL of the peg is higher than the previously calculated RL of the sight rail. Conversely, the sight rail at position IC1 needs to be fixed at a higher RL to that of the peg; therefore, the peg has to be extended by 97.16 m – 96.70 m = 0.46 m to enable the sight rail to be fixed at the previously calculated RL of 97.16 m (see Fig. 11.4.4 (B), (C) and (D)).

11.5 Controlling work on buildings

Once the main setting-out is completed and the construction work commences the site supervisor or site engineer would need to maintain control of the actual construction operations to ensure that the horizontal, alignment and verticality accuracies are according to recognised standards (BS 5606 : 1978).

Horizontal control

The depths of excavations may be controlled by one of a number of acceptable ways, the most common ones being as shown in Fig. 11.5.1 (Ai) (Aii) and (Aiii), which are:

1. By determining the peg height at ground level which can then be used to measure down from as excavation work proceeds.
2. By determining the optical height of the instrument (HI), and as the excavation work gets under way periodic checks can then be made by reading on to a surveyor's staff held resting on the bottom of the excavated hole (foundation pit) until a predetermined reading is achieved.
3. By establishing site rails at suitable reduced levels (RL) or heights above the ground level. A traveller is next made up, the length of which is determined as follows:

    Traveller length = Reduced level of the sight rail – Reduced level of excavation (bottom of excavation).

Instead of using a surveyor's level or Cowley level for determining depths of excavations, etc., a laser, whose beam is mounted horizontally, may be used which beams on to a target held inside the excavation similar to a surveyor's staff.

These systems are used primarily at ground level control, but the methods of horizontal control at first and subsequent floor levels can be achieved by the use of:

1. *Storey rods* – for each floor's level. A long rod of wood of one storey height or more (marked out in brick courses with windows, door and floor heights) may be set up on the outside or inside edge of a proposed structure, and as work proceeds upwards each trade would ensure that its courses of brickwork and floor levels, etc., coincide with the marks on the storey rod (see Fig. 11.5.2 (ii)).
2. *Tapes* – these may be used to tape up the walls or columns, stairways, lift shafts, etc., to determine the subsequent floor levels as work proceeds (see Fig. 11.5.2 (ii)).
3. *Levels* – the tape level marks are transferred at each floor level by means of a surveyor's level and staff (see Fig. 11.5.2 (ii)).

Alignment control

The accurate alignment of columns always depends on well-constructed continuous profiles which achieve the best results if erected as close to the ground level as possible. These profiles can then be used successfully to align the columns at ground level or ground floor level by means of builders lines, theodolite sightings (as shown in Fig. 11.5.1 (Bi) and (Bii) respectively), or even laser beams.

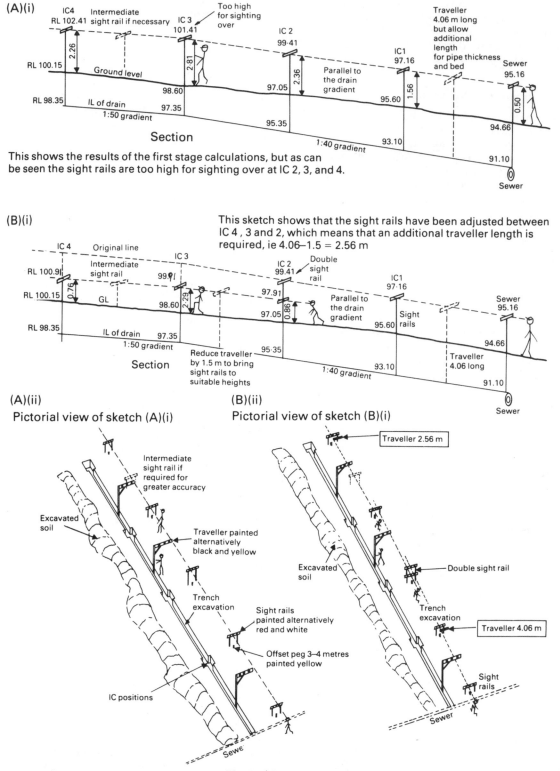

This shows the results of the first stage calculations, but as can be seen the sight rails are too high for sighting over at IC 2, 3, and 4.

This sketch shows that the sight rails have been adjusted between IC 4, 3 and 2, which means that an additional traveller length is required, ie 4.06 − 1.5 = 2.56 m

Fig. 11.4.3 Setting-out drains

The height of the instrument (HI) (or highest point of collimation (HPC)) can be determined from an ordinance bench mark (OBM) or temporary bench mark (TBM), on-site by taking a back-sight reading as shown.

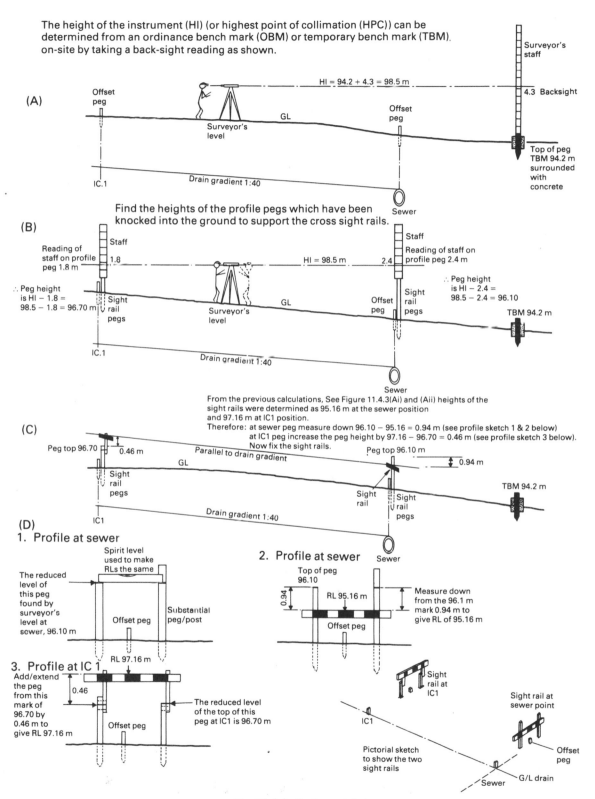

Fig. 11.4.4 Setting-out drains

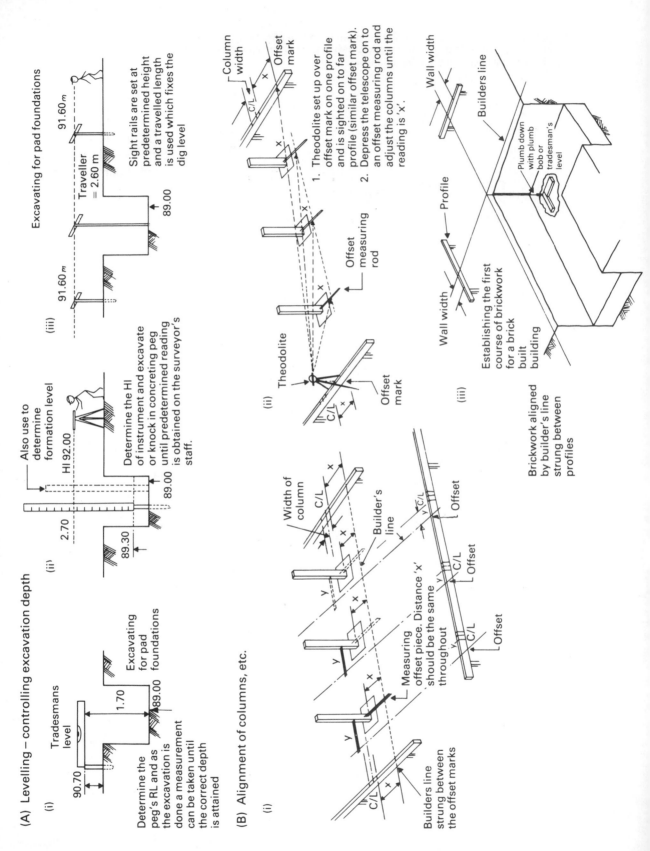

(A) Levelling – controlling excavation depth

(i)

Tradesmans level

Excavating for pad foundations

90.70
1.70
89.00

Determine the peg's RL and as the excavation is done a measurement can be taken until the correct depth is attained

(ii)

Also use to determine formation level

HI 92.00
2.70
89.00
89.30

Determine the HI of instrument and excavate or knock in concreting peg until predetermined reading is obtained on the surveyor's staff.

Excavating for pad foundations

(iii)

91.60 m
91.60 m
89.00

Traveller = 2.60 m

Sight rails are set at predetermined height and a travelled length is used which fixes the dig level

(B) Alignment of columns, etc.

(i)

Width of column
C/L
x
x
x
y
y
y
Builder's line
Measuring offset piece. Distance 'x' should be the same throughout
C/L
x
x
Offset
C/L
Offset
C/L
Offset

Builders line strung between the offset marks

(ii)

Column width
C/L
x
Offset mark
x
x
x

Offset measuring rod

Theodolite

C/L
x
Offset mark

1. Theodolite set up over offset mark on one profile and is sighted on to far profile (similar offset mark).
2. Depress the telescope on to an offset measuring rod and adjust the columns until the reading is 'x'.

(iii)

Wall width
Builders line
Profile
Wall width

Plumb down with plumb bob or tradesman's level

Establishing the first course of brickwork for a brick built building

Brickwork aligned by builder's line strung between profiles

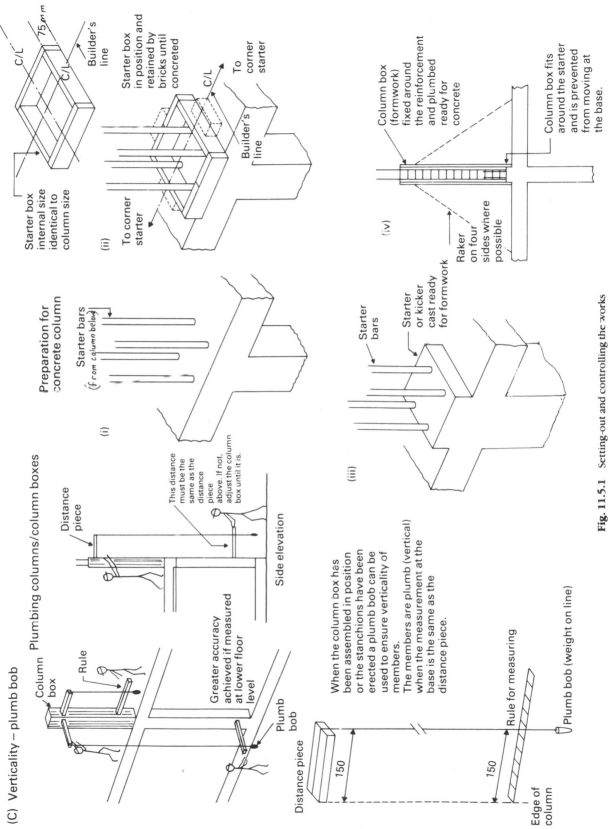

(C) Verticality – plumb bob

Plumbing columns/column boxes

Preparation for concrete column

Distance piece

Column box

Rule

Greater accuracy achieved if measured at lower floor level

Plumb bob

This distance must be the same as the distance piece above. If not, adjust the column box until it is.

Side elevation

When the column box has been assembled in position or the stanchions have been erected a plumb bob can be used to ensure verticality of members. The members are plumb (vertical) when the measurement at the base is the same as the distance piece.

Distance piece

150

Rule for measuring

150

Edge of column

Plumb bob (weight on line)

Starter bars (from column below)

(i)

Starter bars

Starter or kicker cast ready for formwork

(iii)

C/L

75 mm

C/L

Builder's line

Starter box internal size identical to column size

Starter box in position and retained by bricks until concreted

To corner starter

C/L

To corner starter

Builder's line

(ii)

Column box (formwork) fixed around the reinforcement and plumbed ready for concrete

Column box fits around the starter and is prevented from moving at the base.

Raker on four sides where possible

(iv)

Fig. 11.5.1 Setting-out and controlling the works

Horizontal control by level & tape

Verticality – theodolite

The theodolite is used to fix major points accurately from ground floor to upper floors. Each column could be plumbed with a plumb bob at each floor level, which is simpler, with accuracy checks being made by the theodolite periodically.

Coincidence point of theodolite telescope sightings

This column box or precast column needs to be pushed over in line of sight of the theodolites

Starter bars

Starter/kicker

(i)

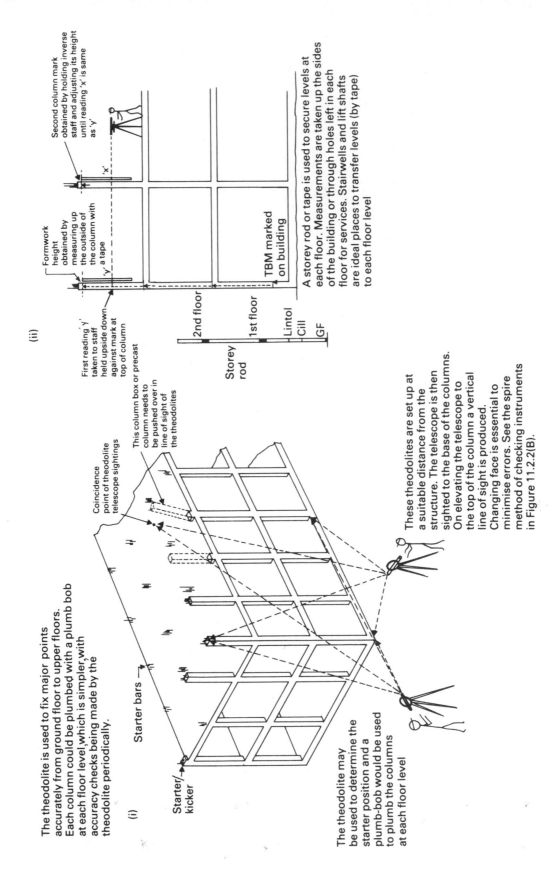

The theodolite may be used to determine the starter position and a plumb-bob would be used to plumb the columns at each floor level

These theodolites are set up at a suitable distance from the structure. The telescope is then sighted to the base of the columns. On elevating the telescope to the top of the column a vertical line of sight is produced. Changing face is essential to minimise errors. See the spire method of checking instruments in Figure 11.2.2(B).

(ii)

Formwork height obtained by measuring up the outside of the column with a tape

Second column mark obtained by holding inverse staff and adjusting its height until reading 'x' is same as 'y'

'x'

'y'

First reading 'y' taken to staff held upside down against mark at top of column

2nd floor

1st floor

Lintol
Cill
GF

Storey rod

TBM marked on building

A storey rod or tape is used to secure levels at each floor. Measurements are taken up the sides of the building or through holes left in each floor for services. Stairwells and lift shafts are ideal places to transfer levels (by tape) to each floor level

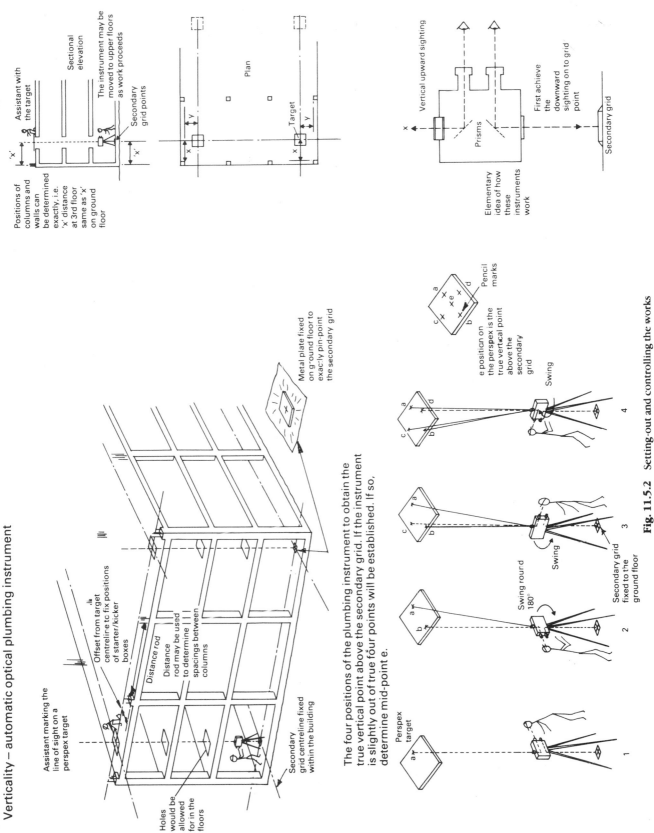

Verticality – automatic optical plumbing instrument

Assistant with the target

Sectional elevation

The instrument may be moved to upper floors as work proceeds

Secondary grid points

Positions of columns and walls can be determined exactly, i.e. 'x' distance at 3rd floor same as 'x' on ground floor

Plan

Target

Vertical upward sighting

Prisms

First achieve the downward sighting on to grid point

Secondary grid

Elementary idea of how these instruments work

Assistant marking the line of sight on a perspex target

Offset from target centreline to fix positions of starter/kicker boxes

Distance rod

Distance rod may be used to determine spacings between columns

Secondary grid centreline fixed within the building

Holes would be allowed for in the floors

Metal plate fixed on ground floor to exactly pin-point the secondary grid

The four positions of the plumbing instrument to obtain the true vertical point above the secondary grid. If the instrument is slightly out of true four points will be established. If so, determine mid-point e.

Perspex target

Pencil marks

e position on the perspex is the true vertical point above the secondary grid

Swing

Swing round 180°

Secondary grid fixed to the ground floor

Fig. 11.5.2 Setting-out and controlling the works

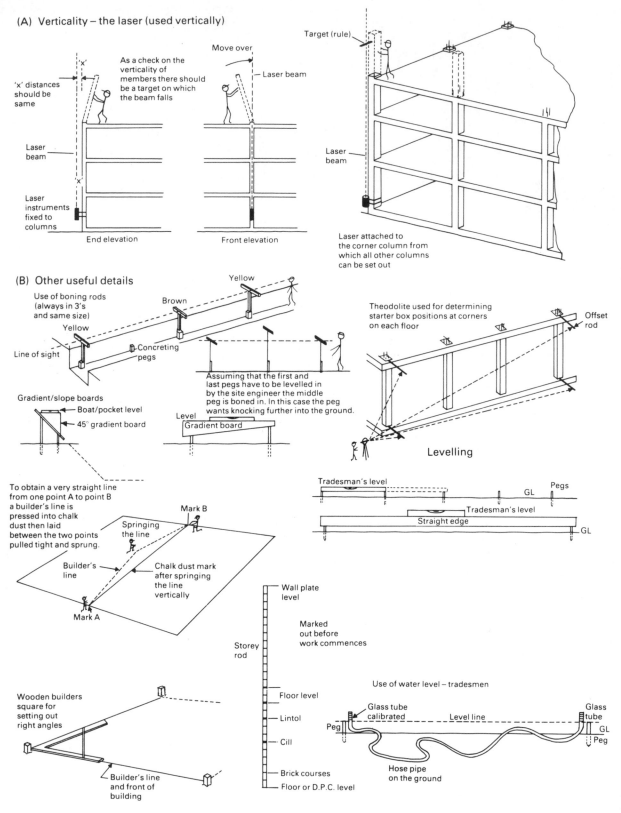

(A) Verticality – the laser (used vertically)

'x' distances should be same

As a check on the verticality of members there should be a target on which the beam falls

Move over

Laser beam

Laser beam

Laser instruments fixed to columns

End elevation

Front elevation

Target (rule)

Laser beam

Laser attached to the corner column from which all other columns can be set out

(B) Other useful details

Use of boning rods (always in 3's and same size)

Yellow

Brown

Yellow

Line of sight

Concreting pegs

Assuming that the first and last pegs have to be levelled in by the site engineer the middle peg is boned in. In this case the peg wants knocking further into the ground.

Gradient/slope boards

Boat/pocket level

45° gradient board

Level

Gradient board

Theodolite used for determining starter box positions at corners on each floor

Offset rod

Levelling

To obtain a very straight line from one point A to point B a builder's line is pressed into chalk dust then laid between the two points pulled tight and sprung.

Mark B

Springing the line

Builder's line

Chalk dust mark after springing the line vertically

Mark A

Tradesman's level

GL

Pegs

Tradesman's level

Straight edge

GL

Wooden builders square for setting out right angles

Builder's line and front of building

Storey rod

Wall plate level

Marked out before work commences

Floor level

Lintol

Cill

Brick courses

Floor or D.P.C. level

Use of water level – tradesmen

Glass tube calibrated

Level line

Glass tube

Peg

GL

Peg

Hose pipe on the ground

Fig. 11.5.3

For each of the techniques used for alignment requires the following marks to be made on the profiles for each column or load bearing wall:

1. The centreline of the columns and walls (structural grid).
2. The widths of the columns and walls.
3. An offset mark for each column and wall at a suitable distance say, 300–600 mm from the marks representing the edges of the columns or wall (see Fig. 11.5.1 (B)).

It is obvious that a builder's line cannot be strung between the centreline marks as it would impede the accurate erection of the columns. The only true and accurate way to align each column is to use offset marks with a builder's line (or theodolite sighting) strung between them because the columns would then be erected without touching the line, and the accurate alignment is achieved by using offset rods/measures, as shown in Fig. 11.5.1(Bi) and (Bii). It should be noted that the offset measure from the columns to the builder's line or theodolite sighting should be the same as distance 'X' shown in Fig. 11.5.1(B).

Builder's lines or theodolite sightings would also be simultaneously strung or set up on the right angle profiles (see Fig. 11.5.1(Bi) and (Bii).

The alignment of columns on the first, second and subsequent floors are achieved by fixing the positions of the corner columns starter boxes first (if cast *in situ* concrete) by means of the simple plumb-bob, theodolite, or automatic optical plumbing instrument using the verticality techniques, as shown in Figs 11.5.1(C), 11.5.2 and 11.5.3(A).

Verticality

Where cast in-situ concrete columns, beams and floors are concerned at first and subsequent floor levels the important points to be established for column construction would be the concrete starters, using starter boxes, whose exact positions are located by plumbing up vertically from the ground floor to any floor above using recognised verticality techniques (plumb-bob, theodolite, automatic optical plumbing instrument or laser beam instrument). Once the concrete starters have set hard (after 24 to 48 hours) the reinforcement is fixed and the column boxes are assembled around the starters, and are plumbed using plumb-bob or the techniques shown in Figs. 11.5.1, 11.5.2 and 11.5.3. Note the preparation for concrete columns at each floor level in Fig. 11.5.1 (Ci), (Cii), (Ciii) and (Civ).

In erecting steel columns, beams, etc., the verticality control is easiest achieved using one or more of the four techniques previously described.

11.6 Setting-out roads

Where motorways and other major roads are proposed the follwing drawings are prepared by the designer and the civil engineer to enable the contractor to set out the road directions, widths and levels accurately:

1. Plan layouts.
2. Longitudinal sections.
3. Cross sections.

Where minor roads or access roads are contemplated it is argued that while the above drawings would be beneficial they are considered unnecessary because all of the information which is necessary could be shown on a plan layout of the proposed road, such as:

1. The start and finish of the road which may be denoted by site grid reference coordinates – Northings and Eastings.
2. Intersection points (IP) – where the road deflects or changes direction.
3. The tangent lengths of each curved section; or it may be more convenient to state the radius and radius angles of each circular curved road.
4. The radius of each curve.
5. The RL of the centreline of the road at suitable linear intervals.

Note: Obviously the drainage, thickness and construction and camber of the road would also be necessary.

Setting out process

Having established the site grid, as described in section 11.3, the major centreline points of the road would next be determined on site in the following order, and as outlined in Fig. 11.6.1 (A):

1. Access position 'a'.
2. Position 'b', using coordinates 210.5 m N and 201.0 m E.
3. Intersection points 1 (IP1) and 2 (IP2), using the coordinates 105.3 m N: 348.6 m E, and 121.5 m N: 250.8 m E respectively.
4. Set out other centreline pegs of the proposed road at suitable distances, say from 5 m to 20 m. Later sight rail pegs are offset at precise distances from those centreline pegs.
5. Calculate the data for the centreline pegs on the circular curve (see Fig. 11.6.3), and then set out the data, using a theodolite and tape, as shown in Fig. 11.6.4.
6. Secure pegs to mark out the road formation width (see Fig. 11.6.2).
7. Set out the sight rail pegs every 5–20 metres centres at the tangent points (TP) and starts and finish points. The sight rail pegs should be established at precise distances from the centreline pegs and protected against accidental disturbance with timber triangular frames. These sight rail pegs are later used to relocate centreline pegs once the excavator has dug to road formation levels.
8. Fix sight rail boards at correct heights (RL) to allow a predetermined traveller length to be used. see traveller length calculations in fig. 11.6.1. Also see Fig. 11.6.5 for calculated readings to be used in conjunction with a surveyor's level.

248

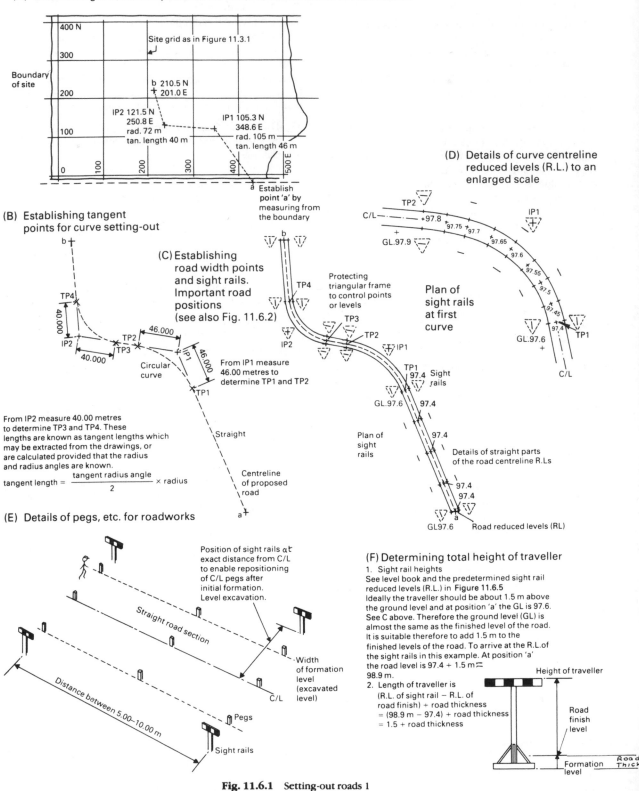

(A) Establishing intersection points of curves (IPs) and direction of road on-site

400 N

300

Site grid as in Figure 11.3.1

Boundary of site

200

b 210.5 N
201.0 E

IP2 121.5 N
250.8 E
rad. 72 m
tan. length 40 m

IP1 105.3 N
348.6 E
rad. 105 m
tan. length 46 m

100

0

100 200 300 400 500 E

a Establish point 'a' by measuring from the boundary

(D) Details of curve centreline reduced levels (R.L.) to an enlarged scale

TP2

C/L +97.8
+97.75 +97.7
GL.97.9
+97.65
+97.6
+97.55
+97.5
+97.45
+97.4
GL.97.6
TP1

IP1

C/L

(B) Establishing tangent points for curve setting-out

b

TP4

40.000

IP2

40.000

TP2
TP3

46.000

IP1

46.000

From IP1 measure 46.00 metres to determine TP1 and TP2

Circular curve

TP1

(C) Establishing road width points and sight rails. Important road positions (see also Fig. 11.6.2)

b

TP4

Protecting triangular frame to control points or levels

TP3

IP2

TP2

IP1

TP1
97.4 Sight rails

GL.97.6 97.4

Plan of sight rails

97.4

97.4

97.4
97.4

GL97.6

Plan of sight rails at first curve

Details of straight parts of the road centreline R.Ls

a Road reduced levels (RL)

From IP2 measure 40.00 metres to determine TP3 and TP4. These lengths are known as tangent lengths which may be extracted from the drawings, or are calculated provided that the radius and radius angles are known.

$$\text{tangent length} = \frac{\text{tangent radius angle}}{2} \times \text{radius}$$

Straight

Centreline of proposed road

a

(E) Details of pegs, etc. for roadworks

Straight road section

Distance between 5.00–10.00 m

Position of sight rails at exact distance from C/L to enable repositioning of C/L pegs after initial formation. Level excavation.

Width of formation level (excavated level)

C/L

Pegs

Sight rails

(F) Determining total height of traveller

1. Sight rail heights

See level book and the predetermined sight rail reduced levels (R.L.) in **Figure 11.6.5**
Ideally the traveller should be about 1.5 m above the ground level and at position 'a' the GL is 97.6. See C above. Therefore the ground level (GL) is almost the same as the finished level of the road. It is suitable therefore to add 1.5 m to the finished levels of the road. To arrive at the R.L.of the sight rails in this example. At position 'a' the road level is 97.4 + 1.5 m = 98.9 m.

2. Length of traveller is
(R.L. of sight rail − R.L. of road finish) + road thickness
= (98.9 m − 97.4) + road thickness
= 1.5 + road thickness

Height of traveller

Road finish level

Formation level

Road Thick

Fig. 11.6.1   Setting-out roads 1

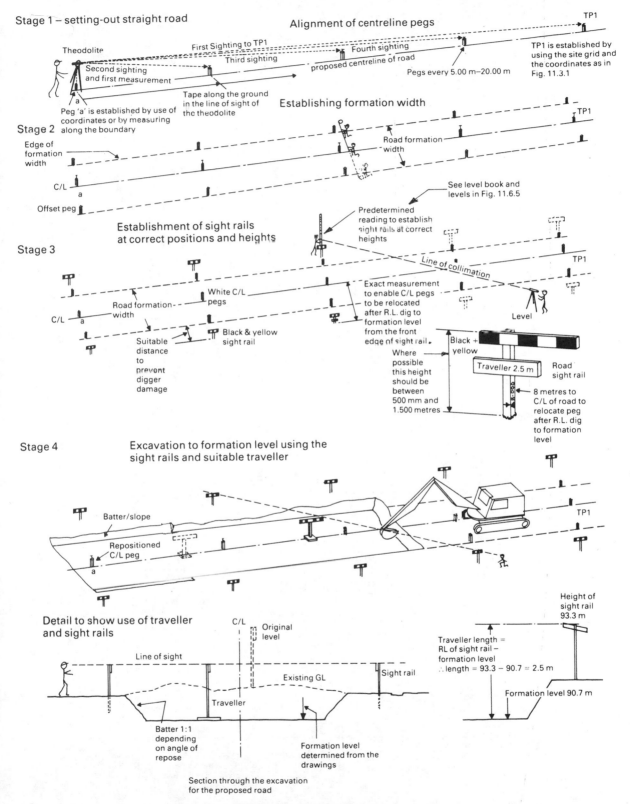

Fig. 11.6.2   Setting-out roads 2

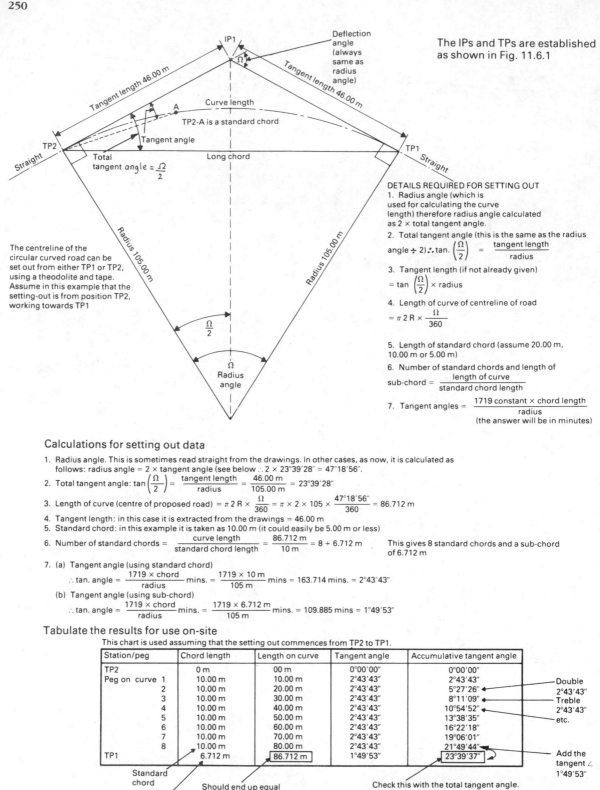

The IPs and TPs are established as shown in Fig. 11.6.1

DETAILS REQUIRED FOR SETTING OUT

1. Radius angle (which is used for calculating the curve length) therefore radius angle calculated as 2 × total tangent angle.

2. Total tangent angle (this is the same as the radius angle ÷ 2) ∴ tan. $\left(\dfrac{\Omega}{2}\right) = \dfrac{\text{tangent length}}{\text{radius}}$

3. Tangent length (if not already given)

= tan $\left(\dfrac{\Omega}{2}\right)$ × radius

4. Length of curve of centreline of road

= $\pi\,2\,R \times \dfrac{\Omega}{360}$

5. Length of standard chord (assume 20.00 m, 10.00 m or 5.00 m)

6. Number of standard chords and length of sub-chord = $\dfrac{\text{length of curve}}{\text{standard chord length}}$

7. Tangent angles = $\dfrac{1719\ \text{constant} \times \text{chord length}}{\text{radius}}$
(the answer will be in minutes)

The centreline of the circular curved road can be set out from either TP1 or TP2, using a theodolite and tape. Assume in this example that the setting-out is from position TP2, working towards TP1

Calculations for setting out data

1. Radius angle. This is sometimes read straight from the drawings. In other cases, as now, it is calculated as follows: radius angle = 2 × tangent angle (see below ∴ 2 × 23°39'28" = 47°18'56".

2. Total tangent angle: tan $\left(\dfrac{\Omega}{2}\right) = \dfrac{\text{tangent length}}{\text{radius}} = \dfrac{46.00\ \text{m}}{105.00\ \text{m}} = 23°39'28"$

3. Length of curve (centre of proposed road) = $\pi\,2\,R \times \dfrac{\Omega}{360} = \pi \times 2 \times 105 \times \dfrac{47°18'56"}{360} = 86.712\ \text{m}$

4. Tangent length: in this case it is extracted from the drawings = 46.00 m

5. Standard chord: in this example it is taken as 10.00 m (it could easily be 5.00 m or less)

6. Number of standard chords = $\dfrac{\text{curve length}}{\text{standard chord length}} = \dfrac{86.712\ \text{m}}{10\ \text{m}} = 8 + 6.712\ \text{m}$ This gives 8 standard chords and a sub-chord of 6.712 m

7. (a) Tangent angle (using standard chord)

   ∴ tan. angle = $\dfrac{1719 \times \text{chord}}{\text{radius}}$ mins. = $\dfrac{1719 \times 10\ \text{m}}{105\ \text{m}}$ mins = 163.714 mins. = 2°43'43"

   (b) Tangent angle (using sub-chord)

   ∴ tan. angle = $\dfrac{1719 \times \text{chord}}{\text{radius}}$ mins. = $\dfrac{1719 \times 6.712\ \text{m}}{105\ \text{m}}$ mins. = 109.885 mins = 1°49'53"

Tabulate the results for use on-site

This chart is used assuming that the setting out commences from TP2 to TP1.

Station/peg	Chord length	Length on curve	Tangent angle	Accumulative tangent angle
TP2	0 m	00 m	0°00'00"	0°00'00"
Peg on curve 1	10.00 m	10.00 m	2°43'43"	2°43'43"
2	10.00 m	20.00 m	2°43'43"	5°27'26"
3	10.00 m	30.00 m	2°43'43"	8°11'09"
4	10.00 m	40.00 m	2°43'43"	10°54'52"
5	10.00 m	50.00 m	2°43'43"	13°38'35"
6	10.00 m	60.00 m	2°43'43"	16°22'18"
7	10.00 m	70.00 m	2°43'43"	19°06'01"
8	10.00 m	80.00 m	2°43'43"	21°49'44"
TP1	6.712 m	86.712 m	1°49'53"	23°39'37"

— Double 2°43'43"
— Treble 2°43'43"
— etc.

Add the tangent ∠ 1°49'53"

Standard chord
Sub-chord

Should end up equal to the curve length

Check this with the total tangent angle. ∴ total tan. angle = 23°39'28". Therefore: there is a small difference of 9" which is reasonably accurate.

Fig. 11.6.3   Setting-out roads 3 – circular curve calculations

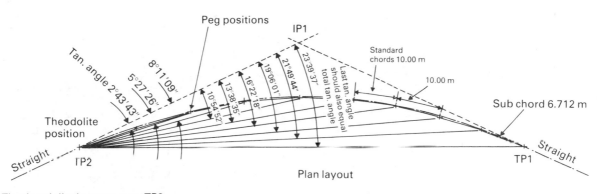

Peg positions

IP1

Standard chords 10.00 m

Tan. angle 2°43'43"

8°11'09"

5°27'26"

10.00 m

Sub chord 6.712 m

Theodolite position

13°38'35"

16°22'18"

19°06'01"

21°49'44"

23°39'37"

10°54'52"

Last tan. angle should also equal total tan. angle

Straight

TP2

Straight

TP1

Plan layout

The theodolite is set up over TP2
and is zero'd and orientated on to
IP1 and the lower instrument plate is
locked. By unlocking the top plate the
first tangent angle is secured of 2°43'43".
In line of sight of the telescope an assistant
then measures off the first chord which in this
case is a standard chord of 10.00 m. A peg (with nail) is inserted into ground.
Note: In most roadworks continuous chainage
is used, but not in this case.
The next tangent angle, 5°27'26" is fixed and
a further chord of 10.00 m is measured in the line of
sight so that between each peg on the curve there
is a measurement of 10.00 m. except at the last chord
(sub chord) which is 6.712 m.
See Fig. 11.6.3 for tabulated results

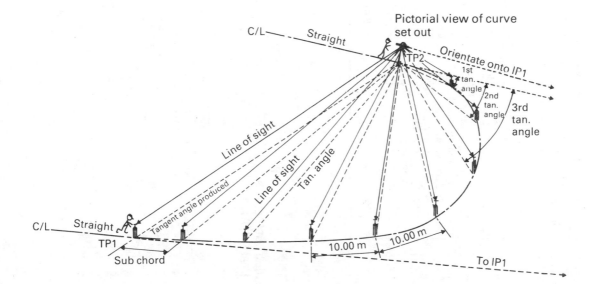

Fig. 11.6.4 Setting-out roads 4 – practical setting out of curved road

(A) Calculation of reduced levels

In Fig 11.6.1(C) and (D) the finished road levels are shown for one 'straight' and one circular curve.
In Figure 11.6.1(F) the height of a traveller has been determined and also the reduced levels (RLs) of the sight rails for the 'straight'.
The reduced levels (RLs) of the sight rails are calculated thus:

Position	Road reduced levels + sight rail height above road level		Reduced level of = sight rail
'a' 1st pair sight rails	97.4 m	+ 1.5 m	= 98.9 m
2nd pair sight rails	97.4 m	+ 1.5 m	= 98.9 m
3rd pair sight rails	97.4 m	+ 1.5 m	= 98.9 m
4th pair sight rails	97.4 m	+ 1.5 m	= 98.9 m
TP1. pair sight rails	97.4 m	+ 1.5 m	= 98.9 m
1st pair on curve	97.5 m	+ 1.5 m	= 99.0 m
2nd pair on curve	97.6 m	+ 1.5 m	= 99.1 m
3rd pair on curve	97.7 m	+ 1.5 m	= 99.2 m
TP2 pair on curve	97.8 m	+ 1.5 m	= 99.3 m

(B) Setting out for the horizontal board height on the profile pegs

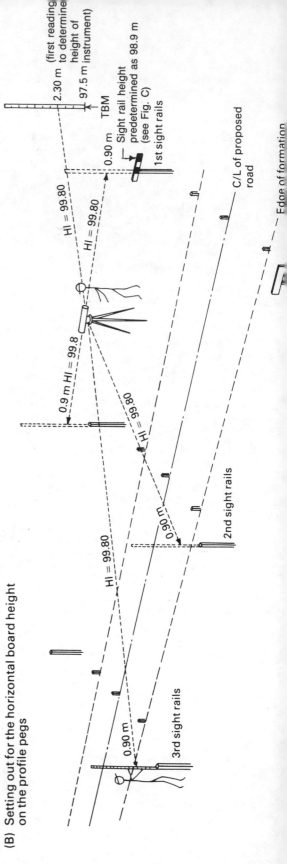

(C) Note: That the predetermined reduced
levels are inserted in the RL column
of the level book, and then by calculating
the height of instrument (HI), using the TBM + BS (97.5 m + 2.3 m = 99.8 m).
These RLs are deducted from the HI to determine the readings
necessary at each sight rail peg position (see IS and FS columns below).

These readings are expected on the sight rail pegs

BS (back sight)	IS (intermediate sight)	FS (fore sight)	HI (height of instrument)	RL (reduced level)	Distance	Remarks
2.30			99.80	97.5		TBM 97.5 m
	0.90			98.9		1st sight rails
	0.90			98.9		2nd sight rails
	0.90			98.9		3rd sight rails
	0.90			98.9		4th sight rails
	0.90			98.9		TP1 sight rails
	0.80			99.0		1st pair on curve
	0.70			99.1		2nd pair on curve
	0.60			99.2		3rd pair on curve
		0.50		99.3		TP2 sight rails
TOTALS = 6.60		0.50		901.1		

Answer from HI−RLs

Deduct RL from HI

The mathematical check on this height of instrument (HI) of booking is as follows:
Check: add up the RL column (except the first RL) + total IS + total F.S = HI × number of uses of each.
∴ 901.1 + 6.60 + 0.50 = 99.80 × 9

so, 908.20 = 908.20
Answer checks out. This means the mathematics of the level book are correct.

Fig. 11.6.5 Setting-out roads 5

Chapter 12
Supervisor's responsibilities when working in or near occupied buildings

12.1 Agreement with building user concerning work procedures, hours and methods prior to start of work

One should be sure before work commences that each party realises whether the builder is working as an independent contractor or purely with handyman status. If one works as an independent contractor, in law, one would be responsible for trespass, nuisance or negligence regarding site procedure, and would therefore have a duty of care to the occupier or building owner. Where a contract is signed, usually on a Joint Contract Tribunal Standard Form of Building Contract, or similar Standard Form, independent contractor status is assumed and the contractor would be responsible for carrying out the work diligently and in a workmanlike manner. This means that he/she would know what work is to be done, and the client/building owner/occupier would not state how it should be done; unlike a handyman, he is told what to do and, generally, how it should be done, and would not normally be liable in an action of tort in a court of law.

Work procedures

An agreement should be made between the contractor and the building owner/occupier as to which parts of a structure should be started on and completed first. The phased work and completions are of importance to many building owners, particularly if the existing building is a hotel, bank, shop, or factory. The least amount of disturbance and inconvenience to the building owner and his/her clients or customers is naturally essential.

Where the work is of a dismantling or demolition nature a firm agreement should be struck as to ownership of dismantled materials, and whether there are to be debits or credits to the contract sum for any materials saved.

The levels shown on plans should be checked and agreed by the site manager with the architect's representative, the clerk of works, before work gets under way. Any significant differences between the levels shown on the drawings and those encountered on the site should immediately be notified to the architect for instruction to be issued.

Inspections of existing buildings before commencing work

Existing damages to the building owner's property, jointly owned property, adjoining property or adjacent property should be agreed with the owners or their representative surveyors, and a schedule should be drawn up and signed by the parties (copies retained by each) before commencing building operations. If necessary, photographic records should be made in addition to the application of dated cement or glass tell-tales, which are affixed across any existing cracks in the brickwork as a check on a building's movements caused by building operations. These help to prevent unnecessary frivolous claims for damages being made by individuals or public authorities. The normal claims relate to damages to:

- paving slabs and driveways.
- blocked or cracked drains.
- cracking walls caused by vibrations or underpinning of foundations.
- broken fences.
- damaged flower/shrubs beds.
- water/gas/electricity/GPO mains damages.
- contamination of water supply.
- Mechanical damage to walls by vehicles and plant.
- Tree damages

Protection of adjoining properties is of paramount importance to prevent weather and other damages caused by removal of part or all of a client's building which abuts on to the adjoining property. Redirecting of gutters, or the provision of polythene/bituminous felt on exposed separating walls (partywalls) as a temporary weather protection is a normal requirement.

Allocation of contractor storage areas, and proof of boundaries

Careful note should be made of land or building space allocated by the building owner/occupier or architect for use by the contractor and subcontractors, etc., for access, hutting, storage and working facilities.

Proof of agreements between adjoining property owners and the building owner should be examined regarding boundary ownership. Any doubts about determination of boundaries and ownerships must be settled before the contractor carries out work on or adjacent to any boundary. The rights of each owner to carry out work on shared boundaries, party or separating walls, or even shared rights of way (alleys, passages or corridors) should be determined by the client's representative, but the contractor may in certain circumstances have this responsibility.

A simple personal approach to the adjoining property owners may lead to free access being agreed or liability for any damages being accepted. Alternatively, there could be a financial settlement before work commences. In any event there are conditions whereby notices must be served on adjoining property owners where work is contemplated which affects their boundaries particularly in London. Boundary determination is also a safeguard to prevent encroachment by the operatives and delivery drivers on to adjoining properties, or other contractors' adjoining contracts, which may lead to an action of trespass.

Stopping up highways would require a licence from the Highways Authority of the Local Authority under the Highways Act.

Public relations

Someone should represent the contractor in this capacity but, in general, the responsibility for public relations (PR) is delegated to the site supervisor, and links should be established with adjoining property owners – particularly those within the immediate vicinity of the site, so that they may feel they have free access for airing their complaints, if necessary. While having the courtesy to introduce himself/herself the contractor instils confidence into the neighbours and, additionally, neighbours should be encouraged to visit the site to see progress – which shows consideration, and generally enlists their cooperation and makes them more tolerant.

Hours of work

Operatives generally on average work 39 hours per week, but where work begins to fall behind schedule a site supervisor may feel it necessary to get permission to work overtime. This permission is obtainable from the client/ architect on many jobs, and where unions dominate the joint local overtime committee of the NJCBI would have to be approached if the contractor is a member of the BEC.

The client may have the condition written into the contract documents that permission for working over-time will only be given under special conditions – on certain days and between suitable hours.

It must be appreciated that on certain contracts to have noisy operations would affect the client's position with his/her customers. Take, for instance, alteration work to an existing bank: it would best be executed at nights, weekends or bank holidays, but this would be unsatisfactory due to the length of time the building work would take. Work, by necessity, would have to be done during normal bank hours. Therefore, conditions tend to be laid down as to the working hours, but particularly, with regards to the types of operations which are not allowed during, say, business time in the bank or shop.

One typical stipulation is that percussion tools should not be used between the hours of 9.30 and 3.30 p.m. (bank opening time); therefore, a contractor may find it convenient to find several alternative systems of work. Other conditions may require that special precautions are taken to minimise the problems of dust – extraction fans, damping down of floors, dust sheets, screens, etc. – and care should be exercised when handling toxic and other dangerous substances.

Other points

Where the work is undertaken inside an existing factory or business premises an agreement may be reached with the client for the contractor's employees to share the welfare facilities already in existence under the Factories Act 1961. This would minimise the cost of providing separate facilities by the contractor. The necessary registration and forms would be completed by the factory owner/client, and copies would be handed to the contractor's representative (site manager) as proof that facilities were provided.

Car parking by operatives and visitors to site could be objected to by the occupier of the work. It is therefore necessary to agree to the minimising of obstructions by preventing unauthorised parking by the contractor's team.

Advertising on hoardings, etc., must have the approval of the client. This is normally stipulated in the preliminaries section of the bills of quantities for the contractor to observe.

12.2 Safety of occupants

There are other safety problems in addition to those which normally exist on ordinary construction sites when building work is undertaken to an existing structure, particularly when the structure is in occupation such as a dwelling, offices, factory, or shop. The problems relate to the occupants who generally are unaware of the serious dangers which exist on construction sites or where construction work is undertaken: the work taking the form of extensions, alterations, improvements, refurbishments or demolition work.

It is normal procedure that conditions are laid down as to how the contractor should conduct the work while the client retains possession and continues using the structure while building work goes on, or possession is transferred for the duration of the contract. The contractor needs to know where he/she stands in relationship to possession so that the extent of liability, in law, can be ascertained – particularly with regards to the Occupiers Liability Act 1957.

Notices and rules should be agreed with the client or his/her representative to assist in preventing disputes and problems arising between the contractor's operatives and the user's employees. The notices usually take the form of warnings to the client's employees to minimise injuries and theft as follows:

1. Visitors to site should report to site office.
2. This building works area is dangerous – keep out.

3. Follow the direction finger notices for safe route.
4. Keep to the planked walkway and passage.
5. Emergency alarm button or fire bell.
6. Restricted headroom – be careful.

The client's employees and contractor's operatives should be given instructions on the areas which are out of bounds, or clearly visible notices should be displayed strategically around the works or close to hazardous areas.

Regarding safety precautions for the client's employees, the following could be provided.

1. Fences or barriers to define the work areas and dangerous obstacles.
2. Protective screens to prevent falling material injuring those outside the work areas – they also stop dust and excessive noises from causing an unnecessary nuisance, and unsightly work areas are hidden and reduce distractions to both the client's and contractor's employees.
3. Handrails and walkways to define safe routes for occupants which prevent them from straying on to dangerous work areas or near to hazardous processes and materials.
4. Dust sheets over existing fixtures and fittings.
5. Lighting/floodlighting to highlight protected ways or obstacles if necessary.
6. Doors or gates to retain operatives and the client's employees alike. They also define security areas or escape routes in cases of fire or other dangerous occurrences.
7. Gantries to protect those who pass under or close to work points.
8. Boarding or screens to windows to prevent materials falling through the glass where dismantling or dangerous work is undertaken close to communication routes or passages.
9. Protection around holes in the ground or floors.
10. Safe overhead cables fixed at suitable heights (do not allow unnecessary trailing wires/cables at ground or floor level).
11. Flying or other shoring when in doubt about structural stability of a building while undertaking structural alterations or removals.
12. Safe storage areas for equipment and materials to prevent occupants from injuring themselves.
13. Fire extinguishers or fire fighting equipment where necessary.
14. Clean and clear thoroughfares free from obstructions. The provision of a skip for rubbish – which is emptied regularly to prevent a hazard to health, and which limits the chances of vandals starting a fire.
15. Extraction fans to draw away from certain processes toxic or inflammable fumes and unpleasant smells from surrounding or adjoining buildings.
16. Safe ramps for access and egress where building operations have caused irregular steps in the pavement or floors – to prevent injury tripping and subsequent claims for damages from individuals other than the contractor's employees.
17. Bollards, road warning cones, protective covers, hoods, screens, safety nets and notices, such as: Access closed; No entry; Building workers only; Factory workers only; Access closed to builder's traffic; Overhead cables; Maximum speeds; Maximum headroom, etc.

It must be remembered that there is the Occupiers Liability Act 1957 which places the onus for safety of premises and land on those in occuption. This is to ensure that visitors are allowed to enter property (not trespassers) expecting safe condions to prevail, and if there is a dangerous condition and an injury occurs, generally, successful claim for damages would result – the occupier having a duty of care to visitors at all times.

Site supervisors/managers have a major responsibility to their employers to ensure that the general safety policy of the firm is constantly observed.

12.3 Security of adjoining properties

Although each firm or property owner generally obtains insurance cover on their property (land and buildings) as a safeguard against loss due to damages caused wilfully or by fire, flood and tempest, etc., they should advisably have made provisions for additional cover for the contents of their buildings, and for exceptionally valuable items. It is nonetheless important for a contractor to obtain an 'all-risks' insurance cover when undertaking work for a client against the possibility of structural and other damages, and against his/her negligence particularly with regards to the security of the works, the client's attached rooms and buildings and adjoining properties.

Trespass, interference with goods, theft, pilfering and malicious damages of a client's and his/her neighbour's property should be safeguarded against, and the following are but a few of the precautions to be taken on-site by the site supervisor.

1. Surround the works with a security fence or hoarding to retain one's own operatives within the construction site, and to keep client's employees and adjoining property owners off the site after working hours.
2. If access to the client's existing property is essential to the building operations from time to time during normal working hours, some means of securing the property should be available at the end of the work day (board up or provide a temporary access door which is lockable and durable against attempted unauthorised entry).
3. Security doors should be fixed for access where it may be necessary to periodically work from the adjoining property owners' land (with their approval).
4. The safe-keeping of keys for all access doors should be the responsibility of one person only – the site supervisor. Sometimes the client/adjoining property owners expect the keys to be lodged with them at the end of each work day.

5. If new, alteration or demolition works expose vulnerable parts of adjoining properties measures should be taken by the contractor's site representative to wire or board up low level windows or easily accessible routes. Warning notices can also be fixed to discourage trespass.

6. Stack materials sensibly to prevent them being used as access points to adjoining properties (away from boundaries).

7. Provide, where necessary, burglar alarms (buzzers, bells and flashing lights systems). Additionally, floodlight potential trespass points.

8. Scaffolding may be barbed wired at the lower ledgers and transomes to prevent them being used by trespassers to gain access to adjoining properties.

9. Ladders should be removed at the end of each day (and chained together) because they provide ease of access to the client's and other property owners' premises if left in position.

10. There should be separate access points for the builder's operatives and the client's employees, which also helps to prevent collusion between the less trustworthy on both sides.

11. Provide or agree with the client the form of after work security, e.g. notify police, or provide a night-watchman, visiting security firm or guard dogs with handlers. Under the Guard Dog Act 1975, guard dogs are not to be allowed to roam around freely within the confines of the works. Dogs should be under the supervision of a handler, or if absolutely necessary, should be tied up to prevent them mauling a trespasser – particularly children.

12. Close supervision of some operatives/employees is normally necessary to prevent pilfering/stealing from the adjoining property owner. Also builders materials and equipment if left carelessly around the works provides temptation for the client's employees to steal.

13. The client should be advised to remove valuable and easily transportable items away from the close proximity of the construction work to ease the temptation of theft by operatives.

14. Permission to enter on to the occupied client's areas by operatives, etc., should first be sought from the site supervisor, and if operatives are found wandering/trespassing they should be subjected to disciplinary action.

15. Unauthorised entry on to or through the contractor's work area by the client's employees should be similarly treated by the client.

16. Subcontractors should also follow the same procedures because the main contractor is responsible, generally, for their action.

Finally, if items of material or equipment are borrowed by either party to the contract from the other there should be a booking out system and returns system to prevent the items going astray without someone being liable.

12.4 Environment problems – noise, dust, smoke, etc., and vibration, and their minimisation

In an attempt to protect the general public, to a reasonable degree, from the problems of noise, dust, etc., Parliament introduced the Control of Pollution Act 1974. Contractors must therefore be watchful in preventing the contamination of the atmosphere, and should help to maintain it for the health and enjoyment of the public, and aggravation of the surrounding property owners would be reduced, thereby minimising claims for damages or prosecutions by the police or the Attorney General.

There are other Acts of Parliament which attempt to stamp out excessive pollution of the atmosphere and the environment in general which are:

1. Public Health Acts 1936 and 1961.
2. Clean Air Acts 1956 and 1968.
3. Health and Safety at Work, etc., Act 1974.
4. Alkali, etc., Works Regulations Act 1906 (and orders 1966 and 1971).
5. Factories Act 1961.
6. Highways Act 1980.
7. Office, Shops and Railway Premises Act 1963 and 1982.

The Code of Practice for Noise Control on Construction and Demolition Site is another document which should be studied by those in charge of work on-site.

Above all else, when working on existing structures which are in occupation, the control of noise, dust, etc., is essential for the occupants' comfort, and where there is a clause to this effect in the contract documents, care helps to maintain the status quo between the building owner (client), occupier, and the owners of the adjoining properties and beyond.

Noise

This unwanted sound may be controlled by the contractor as follows:

1. Choosing processes, techniques and machinery which limit noise pollution.
2. Muffling certain noisy machinery (pneumatic drills, etc.) which may be essential and irreplaceable for the work in hand (provide sound insulating jackets, or isolate the machine by surrounding it with a timber screen).
3. Providing hoarding or screens around the site work area or process, which will then restrict the airborne passage of noise to adjoining or surrounding properties.
4. Limiting noisy processes to periods of the day which are to the approval of the occupier/neighbours. The time chosen would be to give the least amount of annoyance (closing time of a bank; the same time as noisy processes are undertaken in a factory).
5. Turning off machinery when they are not required for immediate use.

6. Providing regular greasing, oiling or other mainten-ance to machines which can minimise reverberation and wear of working parts leading to smooth running.

7. Padding certain machine bases which rest on hard concrete surface which effectively reduces noise and vibration.

8. Preventing the use of radios by operatives, or by limiting the radios' output.

9. Advising operatives, when necessary, to use suitable levels of language, especially close to hospitals, nunneries.

10. Providing workshops or special work areas for noisy processes at places (on or off site) which will cause the least amount of disturbance and away from the close proximity of the occupiers.

11. Minimising the use of external telephone warning bells, or toning them down if they must be used on an extensive site.

12. Having certain work assembled off site before being brought and incorporated into the work.

Agreements can always be made with the client as to when various processes should be done to minimise the disturbance of the occupiers and adjoining owners.

Dust

This substance can be divided basically into two, i.e. (*a*) fine particles (dust); (*b*) coarse particles (grit).

The coarse particles settle out of the air quickly but can lead to serious problems in machines and to persons' eyes. The fine particles settle eventually but are irritants to the eyes, nose and throat; in addition, they contaminate surrounding exteriors and interiors of property and are therefore a nuisance, incurring substantial expense to remove. The amount of care one takes to minimise dust would depend on the risk of how much damage could be caused if careless, and the aceptable limits are laid down in law.

The following are a few precautions needed to minimise problems caused by dust.

1. Use a system of work which eliminates the production of dust.

2. Use dust covers (polythene) to protect fixtures, fittings and furniture, and even any goods which are in the process of being manufactured in the occupier's factory/property.

3. Provide soundly constructed and crack-sealed hoardings or screens to prevent the passage of dust from the work area to other areas.

4. Fit automatic door closers to all doors leading to or from work areas to reduce the transfer of dust.

5. Provide dust extractors or collectors to dust producing machines (sanding machines, etc.) if possible, where serious levels of dust are anticipated.

6. Transfer rubbish to collection points (preferably builders' skips) and remove off-site when full.

7. Clean up the work areas at regular intervals and dampen down before sweeping to prevent dust from rising.

8. Stripping out or demolishing structural elements close to other structures should be done with care and with the interests of the occupier and adjoining property owner/owners in mind. A safe and, as far as possible, dust-free system should be used – by col-lapsing the element in stages, perhaps by hand and not machine. Notification of the demolition of a building should be made to the local authority 28 days before the work is carried out – proof of notification by the client and approval should be inspected by the contractor otherwise he/she would be deemed as liable as the client.

Special Note. Asbestos dust is an additional hazard and extra care is necessary under the Asbestos Regulations 1969. (Also observe Asbestos (Licens-ing) Regulations 1983 and Asbestos (Prohibition) Regulations 1985.

Smoke

The minimum of rubbish should be burnt on-site. If dark smoke is produced the police could prosecute, and if the general public complain to the Attorney General it could lead to a fine and bad publicity.

Creating smoke in smokeless zones under the Clean Air Acts is an offence and careless, therefore site managers should make enquiries before allowing the burning of rubbish on-site. More important, if the burning of rubbish causes ash, soot, dust or fumes the existing occupier and adjoining property owners may sue for nuisance. These types of claims can be minimised by the disposal of rubbish on public or other tips without the need to create fires.

Instructions to operatives, or notices could be dis-played on-site, stating that fires or the burning of rubbish are prohibited, this not only maintains a suitable atmos-phere, but adds to the safety aspects because the risk of fire is minimised.

Fumes, vapours, gases, obnoxious smells

There are various safety procedures to be followed when using certain materials or substances which could be toxic or potentially explosive. The various regulations covering these should be checked before using such substances under the Health and Safety at Work, etc., Act 1974.

Extraction fans, piped exhaust systems or air pumps should be provided when using potentially hazardous substances to remove fumes and vapours from inside a building to prevent a nuisance to occupiers, etc., and to safeguard the works and surrounding properties from fires. Safety checks must be made by a qualified person at suitable intervals where dangerous substances (petroleum spirits, oxyacetylene bottles and gas containers) are being used or stored to prevent the build up of vapours, etc., which are potentially very dangerous to the workers, the building and the adjoining occupier.

Vibrations

Vibrations can be a source of discomfort for occupants of a building, and, worse still, can cause the fracture of gas and other service mains, cracking of buildings, and ultimately expensive remedial work and claims for damages by adjoining property owners and public undertakings.

To reduce vibrations on-site the following measures are recommended:

1. Turn off the engines of plant and machinery when they are not immediately required.
2. Provide insulation pads under certain machines which rest on concrete bases or hard stands.
3. Use processes or techniques which eliminate vibrations.
4. Heavy equipment or materials which have to be dismantled should be lowered to the ground/floor and not in an uncontrolled manner.
5. Where necessary use rotary drills instead of percussion drills.

The prevention of claims for damages from individuals is very important and site supervisors have the responsibility to ensure that such claims are minimised – remembering that it is the aim of all good contractors to pursue a sound public relations policy and to create a good image.

Insurance

The Joint Contract Tribunal Form of Building Contract, and other standard forms of contract, lays down that the contractor has an obligation to insure the works with an approved insurance company. Care must be exercised in obtaining the correct cover (all risks), particularly where work is to take place or conditions exist which are of an unusual nature.

12.5 Trespass, nuisance, negligence and reduction of claims

It is of necessity that site supervisors should have a satisfactory basic knowledge of the law of torts (civil wrongs) and how it effects them and their employer's positions on-site.

Too many problems have arisen in the past which have been both inconvenient and costly to the employer, thereby determining that the law cannot be ignored. Ignorance has led to the dismissal of the supervisor, but it has also meant that firms insurance premiums have had to be increased by insurance companies due to the fact that successful litigation has been made against the building companies by injured outside parties (plaintiffs).

On taking possession of a site which is to be developed the contractor generally becomes the occupier until the work is completed, and would be mainly responsible regarding trespass, nuisance and negligence, if any, which affects the land and the activities thereon.

The site supervisor must, therefore, understand the civil wrongs which could affect his/her position and the contractors on-site under the following three headings:

1. Trespass.
2. Nuisance.
3. Negligence.

Trespass

There are three forms of trespass, i.e.: (a) trespass to persons; (b) trespass to goods; (c) trespass to land.

(a) Trespass to persons
Under this kind of action there are three headings: (i) assault; (ii) battery; (iii) false imprisonment.

Assault: This is an attempt by one person to apply force or threaten to apply force to another person unlawfully. Words alone are not proof of an action of assault, but where a punch is thrown – even though it misses – it constitutes an assault; and a threatening speech with intent, such as, 'Get off this site or I will let my guard dog bite you'. Site supervisors should choose his/her words carefully at all times but especially where there is a disagreement with the operatives or others.

Battery: This is the applying of actual force by one person to another person. Examples being: pushing, holding an arm, punching or throwing an object at a person – even though the object does not cause physical damage; the mere hitting of someone with a folded newspaper is an act of assault.

It is of importance to know that if one is being assaulted or battered only the minimum of force should be applied in defence. If only one punch was thrown by an assailant it would generally be unlawful for the assaulted person to resist by defending himself/herself by kicking and repeatedly kicking the other person to the ground, causing serious injuries.

False imprisonment: If one attempts to restrain another person against his will, without proof of damage, it is actionable by the person being restrained. If, for example, an operative was suspected of stealing from his employer and he was led to a room and someone stood guard without evidence of guilt, that person could sue for false imprisonment. Where there is evidence of guilt, e.g. caught red-handed stealing, or there is forceful trespass, the suspect may be detained by reasonable force until the police arrive.

Another example of false imprisonment could be where a site manager finds a trespasser but there is no evidence of forceful trespass, and he threatens to hit him or hurt him if he moves from the spot on which he was discovered while someone calls the police. It would be more appropriate to ask the trespasser to accompany him/her to the office until investigations are made.

(b) Trespass to goods
If there is forceful intentional interference with the possessions of others it is classed as trespass to goods. The mere touching of the goods is not usually sufficient to be unlawful.

Site managers should ensure that their subordinates have no part in the interference with adjoining property owners' goods. Similarly, where work is undertaken to an existing occupier's building, operatives are warned not to be too inquisitive regarding things/goods that do not directly concern them.

Notices are best used as warnings to occupiers/clients, visitors and intending trespassers to make them refrain from stealing, etc, such as: Notice – Keep Off; Notice – Do Not Touch; Notice – Dangerous Substances; Notice – Not to be Opened; Warning – Thieves will be Prosecuted.

(c) Trespass to land

Trespass to land may occur on the land, under the surface of the land, or even in the air space above the land – although aeroplanes are allowed to fly overhead at a reasonable height provided they cause no damage.

There are three forms of trespass to land, which are: entry on to; remaining on; or placing or throwing of objects upon; the land of another.

When a contractor commences work it is necessary to ensure that the exits and entrances of adjoining properties are not obstructed. Site supervisors should refrain from allowing the operatives to walk on or work from adjoining properties. It is always prudent to get permission to enter on to other's properties, and one should convey why entry is necessary and for how long. Aggravations caused by trespass could affect the ease in which a contract can be executed. The more usual ways the contractor or his representatives would be trespassing – unless permission or an agreement can be reached with adjoining property owners – are given as follows:

1. Digging trenches or holes which encroach without permission on to another's land.
2. Allowing material to fall over the boundary of the site on to the adjoining site without removal.
3. Erecting scaffolding which overhangs the air space of another person's land.
4. Erecting a tower crane whose jib periodically swings over another person's air space.
5. Fixing a gutter, a beam or a notice which projects across the boundary of a site/building, or even allowing the foundations to project over the boundary of the site on to another person's property: each example being considered under one of the methods of trespass to land: trespass to the surface of the land, under the land or air space above the land.

Trespass of the public on to the contractor's site

The site supervisor must be vigilant to the fact that there is always the possibility of outsiders trespassing either intentionally or accidentally on to the construction site. Suitable notices should be displayed which serve as warnings to the general public and the employees – particularly where sections of the works or adjoining property are strictly out of bounds or are dangerous. Such notices are:

- Trespassers will be prosecuted.
- Warning – Keep out, dangerous.

- Notice – Keep off.
- Warning – Fragile roof.
- Warning – Not to be used, dangerous.
- Warning – Dangerous structure.

If the site manager ensures that such notices are displayed, when necessary, any accidents which do occur would not generally be actionable in negligence. There is of course the problem of children who, under the law, have a special status. Children tend to be allured to construction sites so contractors and their representatives on-site must be ultra cautious by bearing the following points in mind.

1. Keep boundary fences in good state of repair, and where there is evidence of a trespass route being used it should be sealed off as far as is practical. It would be no defence in a court of law where the site supervisor knew that children used a particular route to trespass on to the site, and he/she took no action to seal off that route, and a child trespasser injured itself (the courts however would take the age of the child into consideration when making an award – a fourteen-year-old being more responsible for its actions than an eight-year-old).
2. Highly dangerous situations on-site should be eliminated, e.g. provide barriers around holes; excavators and other dangerous machinery should be disengaged and made safe; ladders should be removed from the sides of structures or the rungs should be planked up, etc. This kind of action helps to prevent children from being injured in the event of them trespassing, and would help to reduce claims from the firm if accidents did occur.
3. Children should be discouraged at all times from coming on to the site (Physically warn them away each time they are seen on, or intending to come on to, the site.)

Note: A contractor does not necessarily have a duty of care to trespassers – except children, but he/she must not deliberately leave parts of the site in a dangerous state to teach trespassers a lesson or to trap them.

Nuisance

There are two types of nuisance: (a) public nuisance (not a tort); (b) private nuisance.

(a) Public nuisance

This is a crime, and therefore one who commits this type of nuisance can be prosecuted by the police, or the Attorney General (public prosecutor) may bring an action on behalf of the public. It is generally classed as an act which affects the comfort and lives of persons living in a locality. Example of public nuisances caused by a contractor are:

1. Obstructing a public highway (road, verge, pavement, footpath or bridleway) by digging holes or depositing materials which causes unnecessary inconvenience, or by depositing builders' skips without a licence.

2. By emitting excessive smoke or fumes, etc., into the atmosphere.
3. Causing flooding to occur on the highway.
4. Causing excessive vibrations or noise.

These points are covered under the Control of Pollution Act 1974, Clean Air Acts 1959 and 1968, Highways Act 1980, Public Health Acts 1936 and 1961.

The site supervisor can also be classed as a public relations officer for the firm for whom he/she works, and a measure of success in this capacity could be ascertained by the number of prosecutions, over the years, which have been brought against the employer regarding the sites under his/her control.

(b) Private nuisance

Where there is an unlawful interference by one person with the enjoyment and use of another person's land, it may be termed a private nuisance and is actionable only if there are damages.

It is more usual for a contractor to be sued for damages under private nuisance than for a contractor to sue members of the public. Bearing this in mind a contractor, through his/her site supervisor, should prevent nuisances occurring if good public relations are to be maintained and claims are to be minimised from adjoining property owners. This is best achieved by ensuring that:

1. The emission of smoke, fumes, dust is limited.
2. There is a prevention of obstructions to the entrances of other person's drives and shops by careless parking of lorries, depositing of materials, digging of holes/trenches or flooding.
3. Vibrations and noises are minimised.

It must be emphasised that the aforementioned are the more obvious nuisances, and that damages may be claimed through a court of law by individuals, or an injunction may be sought to restrain further nuisance. Also, the act of nuisance generally must be continuous or of a recurring nature (it must happen more than once for the courts to believe the action by a plaintiff is not trivial).

Negligence

Negligence is the omission of duty or the want of proper care by one person for the interests of others as the law may require.

If there appears to be negligence on the part of one person leading to injury or damage to another or his/her property, the injured party, if to be successful in suing in negligence, must prove three points: that

1. The defendant had a duty of care to the plaintiff.
2. There was a breach of that duty.
3. The plaintiff suffered damages as a result.

It may generally be said that a duty of care exists in the following situations, and where there is a breach of that duty and damages result an individual may sue:

(a) Architects owe a duty of care in conducting their affairs on behalf of the client, and therefore must be professionally competent at all times in discharging their duties.
(b) The contractor (generally referred to as independent contractor) has a duty of care to employees that the site and equipment on the site are reasonably safe.
(c) The contractor has a duty of care to the general public where the erection of the following are contemplated: hoardings and gantries; barriers to excavations adjoining the site; and scaffolding parallel to the highway – particularly where children and even blind persons are concerned.
(d) There is a duty of care to site visitors (not trespassers) that there are safe routes and situations provided around the site, and where dangers do exist these are high-lighted by notices.
(e) Faulty construction which ultimately causes injury or damages is also a negligent act and prospective occupiers are owed a duty of care. If through his/her own negligent act a plaintiff contributed to his/her own injuries the courts would reduce any awards due to 'contributory negligence'.

There are many other situations in negligence where, if damages result, injured parties may sue.

The Occupiers Liability Act 1957 covers some of the points previously outlined.

Index